The Common Good and Ecological Integrity

Proponents of the concept of ecological integrity argue that it is a necessary component of global governance on which the sustainable future of the planet and its inhabitants depends. This book presents the latest research and current thinking on the role of ecological integrity in support of life on Earth and the importance of governance for the common good, or the benefit of all.

The book considers whether present forms of governance support the common good, or whether they are endangering its very foundations. It explores the connection between consumerism and capitalism, the destruction of natural resources and with it, the elimination of many of the ecosystem services that support life in general, and human life in particular. Chapters focus on the defence of human rights, and in particular the rights to key resources such as food, water and general health/wellbeing, as well as energy and security.

Topics covered include climate change, biodiversity, migration and conflict resolution, with approaches from various perspectives such as politics, ethics, sociology and law. Overall the book provides a stimulating insight into the multifaceted debates surrounding ecological integrity, global governance and sustainability.

Laura Westra is Professor Emerita (Philosophy), University of Windsor, Canada, and Sessional Instructor at the Faculty of Law, University of Milano (Bicocca), Italy.

Janice Gray is Senior Lecturer in the Faculty of Law and an Affiliate of the Connected Waters Initiative Research Centre at the University of New South Wales, Australia. She is also a Visiting Scholar at the Centre for Socio-Legal Studies, University of Oxford, and Queen Mary University of London, UK.

Antonio D'Aloia is Professor of Constitutional Law at the Faculty of Law, University of Parma, Italy.

The Common Good and Ecological Integrity

Human rights and the support of life

Edited by Laura Westra,
Janice Gray and Antonio D'Aloia

 Routledge
Taylor & Francis Group

LONDON AND NEW YORK

 earthscan
from Routledge

First published 2016
by Routledge

2 Park Square, Milton Park, Abingdon, Oxfordshire OX14 4RN
711 Third Avenue, New York, NY 10017

Routledge is an imprint of the Taylor & Francis Group, an informa business

First issued in paperback 2018

British Library Cataloguing-in-Publication Data
A catalogue record for this book is available from the British Library

Library of Congress Cataloging in Publication Data
Names: Westra, Laura, editor. | Gray, Janice, 1956– editor. |
D'Aloia, Antonio, 1965– editor.
Title: The common good and ecological integrity : human rights and the support of life / edited by Laura Westra, Janice Gray, and Antonio D'Aloia.
Description: New York, NY : Routledge, 2016. | Includes bibliographical references and index.
Identifiers: LCCN 2015048762 | ISBN 9781138668225 (hbk) |
ISBN 9781315618746 (ebk)
Subjects: LCSH: Ecological integrity—International cooperation. |
Common good—International cooperation. | Environmental protection—
International cooperation. | Environmental policy—International cooperation. |
Human security—International cooperation.
Classification: LCC QH541.15.E245 C66 2016 | DDC 333.72—dc23
LC record available at http://lccn.loc.gov/2015048762

ISBN: 978-1-138-66822-5 (hbk)
ISBN: 978-1-138-36403-5 (pbk)

Typeset in Goudy
by Keystroke, Station Road, Codsall, Wolverhampton

Contents

Contributors

Vladimír Bencko, MD, PhD
Country of affiliation: Czech Republic
Professor (Hygiene and Epidemiology)
Charles University in Prague, First Faculty of Medicine
vladimir.bencko@lf1.cuni.cz

Susana Borràs, PhD (law)
Country of affiliation: Spain
Professor of Public International Law and International Relations
Research Center for Environmental Law Studies of Tarragona, Rovira i Virgili
University (Tarragona-Spain)
Susana.borras@urv.cat

Klaus Bosselmann, PhD
Countries of affiliation: New Zealand, Germany
Professor of Law; Director, New Zealand Centre for Environmental Law
University of Auckland
www.law.auckland.ac.nz/people/k-bosselmann

Donald A. Brown, JD, MA (liberal studies, philosophy and art)
Country of affiliation: USA
Scholar in Residence and Professor of Sustainability Ethics and Law
Widener University Commonwealth Law School
Ethicsandclimate.org
Dabrown57@gmail.com

Mery Ciacci, PhD (law)
Country of affiliation: Italy
PhD (Law)
European University Institute (Florence) and University of Siena
Mery.ciacci@eui.eu

Sheila D. Collins, PhD
Country of affiliation: USA
Professor Emerita (Political Science)
William Paterson University
sheila.collins65@verizon.net

Pavel Cudlín, PhD (biology)
Country of affiliation: Czech Republic
Associated Professor (Ecology)
Global Change Research Centre, Academy of Sciences of the Czech Republic
www.czechglobe.cz
cudlin.p@czechglobe.cz

Eva Cudlínová, PhD (economics)
Country of affiliation: Czech Republic
Associate Professor
Head of the Department of Regional Management, Faculty of Economics,
University of South Bohemia in Ceske Budejovice
evacu@centrum.cz
evacu@ef.jcu.cz

Antonio D'Aloia
Country of affiliation: Italy
Full Professor (Constitutional Law)
Department of Law, University of Parma (Italy)
antonio.daloia@unipr.it

Joseph W. Dellapenna, BBA, JD, LLM (international and comparative law),
LLM (environmental law)
Countries of affiliation: USA, Italy
Professor of Law
Villanova University
dellapen@law.villanova.edu

Rose A. Dyson, EdD
Country of affiliation: Canada
Media Education Consultant
President of Canadians Concerned about Violence in Entertainment
www.C-CAVE.com
rdyson@oise.utoronto.ca or rose.dyson@alumni.utoronto.ca

Franz-Theo Gottwald
Country of affiliation: Germany
Professor (Environmental Ethics)
Executive Director of Schweisfurth Foundation

www.schweisfurth-stiftung.de
info@schweisfurth.de

Janice Gray BA, LLB, (Grad) Dip Ed, (Grad) Dip Leg Prac, MA
Countries of affiliation: Australia, UK
Senior Lecturer, Faculty of Law, UNSW
Visiting Scholar, University of Oxford and Queen Mary University of London
j.gray@unsw.edu.au

Tomáš Hák, PhD (applied ecology)
Country of affiliation: Czech Republic
Head of Department
Charles University Environment Center, Czech Republic
www.czp.cuni.cz
tomas.hak@czp.cuni.cz

Miloslav Lapka, PhD (philosophy)
Country of affiliation: Czech Republic
Vice-Dean of Faculty of Economics, University of South Bohemia in Ceske
Budejovice
www.ef.jcu.cz
miroslav.lapka@ef.jcu.cz

Kathleen Mahoney, QC, FRSC, LLM, JD
Country of affiliation: Canada
Professor at University of Calgary
kmahoney@ucalgary.ca

Massimiliano Montini
Country of affiliation: Italy
Associate Professor, European Union Law and Sustainable Development Law
Co-director of R4S Regulation for Sustainability Research Group
University of Siena
www.r4s.unisi.it
massimiliano.montini@gmail.com

Francesca Mussi, MA
Country of affiliation: Italy
PhD candidate (International Law)
University of Milano Bicocca
f.mussi2@campus.unimib.it

Petra Nováková, MSc (geography)
Country of affiliation: Czech Republic
PhD student

Faculty of Humanities, Charles University in Prague, Czech Republic
novakova.petra.cz@gmail.com

John Quinn, MD, MPH
Countries of affiliation: Czech Republic, United States
PhD student, Global Health Researcher
Prague Center for Global Health
Institute of Hygiene and Epidemiology
First Faculty of Medicine
Charles University in Prague
http://pcgh.lf1.cuni.cz/
john.quinn@lf1.cuni.cz

Colin L. Soskolne, PhD (epidemiology)
Countries of affiliation: Canada, Australia
Professor Emeritus and Adjunct Professor
Universities of Alberta and Canberra
www.colinsoskolne.com
colin.soskolne@ualberta.ca

Ngozi Stewart, LLB, LLM, BL, Dip. IEL, PhD
Country of affiliation: Nigeria
Senior Lecturer, Faculty of Law, University of Benin
www.earthnicepc.com
ngozistewart@yahoo.com

Sabrina Urbinati, PhD (law)
Country of affiliation: Italy
Post Doctoral Researcher in Public International Law
University of Milano-Bicocca, Department of Jurisprudence, School of Law
www.dsgni.unimib.it/?personale-type=sabrina-urbinati
http://unimib.academia.edu/SabrinaUrbinati
sabrina.urbinati@unimib.it

Valentina Vadi, PhD (law), MJur, MRes, JD, MPolSc
Countries of affiliation: United Kingdom, Italy
Professor of International Economic Law
Lancaster University
v.vadi@lancaster.ac.uk

Jan Vávra, PhD
Country of affiliation: Czech Republic
Assistant Professor
Faculty of Economics, University of South Bohemia in Ceske Budejovice
jvavra@ef.jcu.cz

Anne Venton, BA, MA
Country of affiliation: Canada
Former member, Immigration and Refugee Board
Anne.venton@sympatico.ca

Peter Venton, BA (economics), MA (economics)
Country of affiliation: Canada
President JPV Associates (Economics and Public Policy)
Retired, former Senior Economist, Ministry of Finance in the
Government of Ontario
Peter.venton@bell.net

Francesca Volpe, PhD
Country of affiliation: Italy
Research Fellow, Sustainable Development Law
University of Siena
www.r4s.unisi.it
francescavolpe@gmail.com

Laura Westra, PhD, PhD (law)
Countries of affiliation: Canada, Italy
Professor Emerita (Philosophy), University of Windsor
Sessional Instructor, Faculty of Law
www.ecointegrity.net
GEIG: www.globalecointegrity.net

Tomáš Zelený
Country of affiliation: Czech Republic
Institute of Hygiene and Epidemiology, First Faculty of Medicine, Charles
University in Prague
Zeleny.t@gmail.com

Introduction

The common good and the May 2015 papal encyclical

Laura Westra and Janice Gray

> Patriarch Bartholomew has spoken in particular of the need for each of us to repent of the ways we have harmed the planet ... "For human beings ... to destroy the biological diversity of God's creation; for human beings to degrade the integrity of the earth by causing changes in its climate, by stripping the earth of its natural forests or destroying its wetlands; for human beings to contaminate the earth's waters, its land, its air, and its life – these are sins". For "to commit a crime against the natural world is a sin against ourselves and a sin against God".[1]

Perhaps we cannot speak of "sins", but we can certainly speak of crimes, ecocrimes to be precise (Westra 2004). These are the very crimes which the Global Ecological Integrity Group has studied, researched and discussed over the last twenty-four years. After considering various aspects of eco-integrity from the standpoint of different disciplines, and later considering the interface between it and international legal regimes for the protection of the environment and of human rights, we have now reached the point when we attempt to define the shape and form of green global governance.

However, despite the emergence of many recent scholarly works linking ecology and human rights, it was not until 2015, that a religious and world leader, Pope Francis, came forth with powerful and far-reaching words about this connection. In his *Laudato Si'* encyclical, from which the quotation above is taken (as are further quotations below), he reached similar conclusions to those reached by our group, although starting from a very different point of view. Few if any religious leaders and heads of state before Pope Francis have dared to condemn openly and clearly capitalism, globalization and consumerism, as well as the clear consequences that follow upon ongoing overuse and abuse of the Earth.[2]

The encyclical's main merit is the link it traces between all the "sins" it enumerates, that is, all the environmental crimes on one hand, and the life and dignity of humankind, particularly the most vulnerable, but also all life, on the other. We know of no other leader who has searched for "the deepest causes of this planetary attack" (para. 15) and found answers embedded in the anthropocentric assault on "ecosystem services" (para. 34); services that we, and all life need for survival. Several scholars have interrogated the link between eco-system degradation and life and the dignity of humankind by tying it to the inequality between the rich and poor of the world. Francis goes further by openly decrying "the premature death of the poor" (para. 48) and acknowledging that the "true ecological approach always becomes a social approach" (para. 49). In fact, Francis even affirms that the climate is "a common good" (para. 23).

This volume's focal point is precisely the connection between various environmental issues and human rights, collectively, the common good. That includes the right to food and water in Part I; climate change and the right to health in Part II; and several aspects of environmental governance in Part III. Thus the work of the twenty-fourth meeting of the Global Ecological Integrity Group emphasizes once again some of the major themes which emerge strongly from the Papal Encyclical, with the focus not only on international legal instruments, but on desirable forms of global governance.

Part I discusses the right to food and water, and Franz-Theo Gottwald's chapter leads the way, offering a detailed discussion of bio-integrity. Not many members of the Global Ecological Integrity Group devote their work to such a thorough and deep conceptual analysis of our founding notion, integrity, than does Gottwald. He cites Juan Enriquez-Cabot's definition of "bioeconomy" as the "economic field which uses novel biological knowledge for commercial and industrial purposes", thus shedding light on the problem that is basic to the impact generated by the worldwide embracing of "disintegrity". Gottwald points out that, if everything including life "has a monetary value, the integrity of life is compromised". To him, the economic approach comprises all aspects of agriculture, including the use of genetically modified organisms (GMOs) and other technologies, so that the very dignity of human life is at stake. Hence radical change is required to establish novel forms of governance, based on respect for the common good.

Eva Cudlínová et al. (Chapter 2) address the question, often neglected, of the importance of the soil, which they define as a "key to food security", because soil represents "a kind of interface between agriculture and nature". Agriculture cannot be practised successfully without a fertile soil, but today erosion is an ongoing threat, primarily because of industrial agriculture. In addition, the presence of globalization entails that harmful practices increase and are multiplied everywhere in the world. The authors propose changing agricultural practices and policies, and ensuring that food prices reflect the basic need for soil protection.

In discussing the management of water, Tomáš Hák, Petra Nováková and Pavel Cudlín (Chapter 3) explore the application and potential benefits of the water footprint concept (Allan 1994); the footprint being a metric to measure the amount of water directly and indirectly consumed to produce goods and services all along the supply chain. By relying on the experiences of the Czech Republic and observations about the soft roots that the concept has put down in United Nations (UN) agencies (and evidenced in some UN reports), Hák et al. come to the conclusion that the challenge ahead is to shift water footprint thinking from academia to public administration in order better to serve ecological outcomes.

In Chapter 4 Valentina Vadi addresses the issue of food as more than nourishment: she argues it expresses "deeply held cultural practices", and therefore it is part of the intangible cultural heritage (ICH). To Vadi, globalization, and especially trade law do not respect the cultural importance of food as many of the disputes regarding hormones, GMOs and the like, oppose both safety regulations and cultural concerns. The World Trade Organization (WTO) simply views such concerns as "forms of protectionism", as they are in opposition to so-called "free trade".

Mery Ciacci's Chapter 5 exposes the multiple harms arising from the widespread use of GMOs, ranging from the use of carcinogenic substances such as glyphosate, to the increasing number of suicides of Indian farmers, who are unable to survive and follow their time-honoured traditional practices. As we saw in the previous chapter, the WTO's policies work against countries' food sovereignty but to date, only some Latin American countries have attempted to ban or limit the use of GMOs by constitutional provisions.

Meanwhile, Ngozi Stewart (Chapter 6) argues that the introduction of genetically modified crops represents a potential adaptive measure to address the negative effects of climate change on food production. She acknowledges that the introduction of genetically modified crops, while supporting food security, may also have deleterious effects on the environment but argues that robust legal frameworks are able to manage those negative outcomes. Her chapter relies on deontological rather than utilitarian reasoning because she sees deontological reasoning as promoting a duty of care (and values-based) principles. She concludes that where genetically modified crops are used, the precautionary principle will have a very active role to play.

Joseph Dellapenna returns us to the subject of water in Chapter 7. In particular, he seeks to make the connection between water and energy. He argues that if water availability declines, economic activity "grows or remains stable" and populations increase, then more energy will be needed to "abstract, store, treat, move and use water". In order to pursue these interconnections more fully, he examines how well the two competing systems of water law in the United States (those of riparianism and prior appropriation) are equipped to respond to increased pressures on water, especially those arising in the energy context. Setting these two systems up as a contest between common property and private property he comes down in favour of a modified version of riparianism (operating in some of the United States) and known as regulated riparianism. He concludes that it "scores well for introducing flexibility and for protecting public values compared to appropriative rights, important dimensions of adaption to global climate disruption" although he also acknowledges weaknesses in the system's capacity to respond to some potential problems.

Part II starts with Donald Brown's chapter on climate change (Chapter 8). Brown argues that, given the ongoing consequences of climate change, it is unfortunate that the ethical issues which emerge in that context are ignored by governments. Thus it is imperative that not only state policy-makers, but also non-governmental organizations (NGOs), and the related scientific organizations, who are charged with making "policy recommendations on environmental issues" commit to approaching such issues from a "deep applied ethics perspective".

Kathleen Mahoney (Chapter 9) also addresses the question of climate change as she discusses the legal, as well as the moral responsibilities involved. She names these responsibilities as those "to future generations of humanity"; "to different populations around the world" and "to the natural world and its natural state". She observes that despite the fact that there is a plethora of legal instruments which could help mitigate the effects of climate change, few of those instruments have thus far been employed by agencies and actors. Mahoney remarks that it is not only Indigenous leaders who have the moral obligations to be environmental stewards, but that the general imperatives of distributive justice regarding the poor must be implemented. She notes that the poor is a group which has not, comparatively speaking, contributed as greatly to climate change as have wealthier nations. In particular, she comments on the Inuit in the arctic regions noting that they should not be ignored, as they are also disproportionally affected by climate change.

In Chapter 10, Colin Soskolne considers another aspect of responsibility when he explores the role of the professional epidemiologists who are charged by society with bringing to the public reliable evidence about both environmental and health issues needed to inform health policies. Their evidence is also helpful to the development of just jurisprudential outcomes. The "voluntary professional societies" of epidemiology are viewed as "Davids" attempting to stand up to the "industrial juggernaut", or the "Goliaths". The latter's power manifests itself through the manufacture of doubt and uncertainty, in order to advance their own interests rather than provide valid science for the protection of the public interest.

Sabrina Urbinati's Chapter 11 discusses the unusual situation created by Security Council (SC) resolution 2177, "Peace and Security in Africa", which states that "the Ebola outbreak in Africa constitutes a threat to international peace and security". The SC had addressed the threat of infectious disease before Ebola, but its increasing power and the full extent of its role need to be examined. The WHO's activities included surveillance, case management, laboratory services, infection control and even assistance with safe burial practices, among others. The UN, that is, both the Secretary General and the General Assembly itself worked to appoint a Deputy Ebola Coordinator, as the Ebola virus was viewed as a threat to the affected states' "peace and security". Previously, the SC had approached the HIV/AIDS problem in a similar way, hence in the future, it could also recommend several "measures involving the use of force", both binding and non-binding, thus addressing public health and humanitarian needs in a novel way that might be invoked for the harms imposed by climate change as well.

Chapter 12, by John Quinn, Tomáš Zelený and Vladimír Bencko, deals with human security in the context of conflict and disaster and takes Ukraine as its case study. The authors argue that Ukraine is in a state of transition having severed many institutional ties with Russia and presently being on the path towards a new, modern, stable, sovereign state. The authors unpack how violent conflict in Ukraine has impacted on the physical and mental health of Ukraine's people and how it has also impacted on the physical environment (through the use of explosive remnants and water contamination associated with warfare, for example). Additionally, they observe the economic impacts, including unemployment, of the conflict on Ukraine.

Part III addresses various aspects of governance and the common good. In Chapter 13 Sheila Collins acknowledges the role played by militarism in connection with climate change, and its role in the enormous influx of refugees and asylum seekers arriving in Europe. Collins argues that "climate change and war . . . are symbiotically related". Climate change and violent conflict proceed hand-in-hand. The largest producer and purveyor of armaments, the US, is also the largest user of fossil fuels in the world, and, as its products make their way to all participants and areas of conflict, it bears a great deal of responsibility. The results of the US's uncritical profit-seeking manifest themselves not only in the presence of bombs and arms in the hands of many groups of belligerents, but also tend to foster the droughts and famines in Africa and elsewhere, which also produce local violence and conflict. Collins argues convincingly that the interface between those aspects of climate change and armed conflicts must be researched and brought to light.

Anne Venton's Chapter 14 discusses the loss of democracy present in the Canadian Harper government (Steven Harper was the former Prime Minister of Canada, before the 2015 elections which elected Justin Trudeau to replace him). His policies not only affected the life of all Canadians, but they also were an impediment to the acceptance of the many refugees who had hoped to find asylum in Canada, in order to escape from the conflicts in their native countries.

In Chapter 15 Francesca Mussi addresses the grave problems arising from the "protection gap" present between the asylum seekers covered by the Convention on the Status of Refugees (1951) and the Internally Displaced Persons. The worse situation, aside from those escaping armed conflicts, is that of climate migrants, whose homelands become increasingly unliveable, or are destroyed by extreme weather events. These refugees have no protection in law. Some efforts are being made on their behalf, but they have not yet produced any real progress. In 2009, the General Assembly submitted a comprehensive report on "Climate Change and its Possible Security Implications" to the UN Secretary General. Aside from the UN's efforts, the Nansen Conference on Climate Change and Displacement in the 21st Century, covered by the Government of Norway, appears to be the most significant move to address this issue thus far.

Meanwhile Peter Venton's chapter, which offers an essay on democratic capitalism, takes a broader approach to issues of the common good and ecological integrity (Chapter 16). In it he argues for the role of democratic capitalism in serving the tenets of the Earth Charter, particularly the elements of the common good and ecological integrity. He offers a reflection on the integration of political, economic and moral systems and posits some interesting views on issues such as the availability of time which people have to devote to labour. He concludes by suggesting that now is the moment to leave behind oligarchic capitalism (the present model which embraces "consumerism, entertainment, spectacle and the acquisition of wealth") and move towards democratic capitalism.

Antonio D'Aloia (Chapter 17) analyses the difficulties of including the concept of future generations in legal theory, starting from the numerous legal sources which address our obligation to future generations. Nevertheless, D'Aloia acknowledges that there are no binding legal instruments, and several aspects of the question militate against the easy insertion of the rights of future generations into existing legal instruments. D'Aloia proposes that constitutionalism, as it is oriented to the future, may provide the best way to use international law to establish domestic obligations.

Massimiliano Montini and Francesca Volpe (Chapter 18) argue for the need for a "new regulatory approach" to promote and support ecological sustainability. The authors view both "green washing" and the presence of "sustainability development consumerism" as the major obstacles to the sustainability that Pope Francis's encyclical advocates, and the world needs. Missing both a common understanding of "sustainable development", and a serious—in depth—understanding of sustainability we must start by eliminating the goal of unlimited growth as part of a desirable future, because viable sustainability can only be ecologically based. Hence a new regulatory approach is needed, one that includes a long-term perspective and a thorough re-assessment of current international legal environmental instruments.

In Chapter 19 Susana Borràs addresses the very real issue of forced migration due to climate change particularly that caused by rising sea levels. She highlights the irony that developed nations are commonly more responsible for increased levels of greenhouse gas emissions but developing nations, who are less well resourced, are more likely to feel the impacts of those increased levels in terms of lifestyles. She examines migration as an adaptive strategy and interrogates whether the refugee frame, international human rights law and stateless persons status are helpful ways with which to deal with these climate change related problems observing that the role of human dignity will be important to protection of climate migrants.

Rose Dyson (Chapter 20) notes that, given the extent to which we now live in a global village, it is increasingly clear that what is needed is global governance. The public media's present role is one of distracting the public from the real issues we face, such as attacks on public health, climate change and the like, all of which arise from the accepted current belief in capitalism, its rules and its goals. Given the consequence of that acceptance, radical change appears necessary, such as that advocated by Pope Francis's encyclical.

In the concluding chapter, Klaus Bosselmann raises the question: who owns the Earth? The question suggests another urgent issue: how best can the commons be governed? He observes that the interests of single states do not always align well with their simultaneous responsibility for the Earth and all its inhabitants. He remarks, "countries do not intentionally destroy the earth, of course, but they allow it to happen". Indeed, he argues, states must become environmental trustees, so that the absolutes of their sovereign interest are no longer the final word on their policies. He sees the necessary limits on state interests being provided by a global commons governance approach, supported in turn by engaged civic activism which will need to "operate at all levels . . . locally, nationally and globally".

Finally, as will be seen from the above, this collection covers diverse ground from food, water and health case studies to more conceptual approaches interrogating foundational principles but common to all the chapters is the importance of ecological integrity; a principle which the Papal Encyclical has embraced albeit in different language. We commend the book to you and hope that you will be challenged by engaging with its contents.

Notes

1 Pope Francis, "Encyclical Letter *Laudato Si'* of the Holy Father Francis on Care for Our Common Home", 24 May 2015, available at http://w2.vatican.va/content/francesco/en/encyclicals/documents/papa-francesco_20150524_enciclica-laudato-si.html, para. 8, quoting Patriarch Bartholomew's Address in Santa Barbara, California, on 8 November 1997. Subsequent quotations in the text giving paragraph numbers refer to the same encyclical letter.
2 We note that the Dalai Lama is an exception. He blamed capitalism for many of the world's problems while delivering his "A Human Approach to World Peace" lecture for the Presidency University in India in 2015. See the report by C. K. Chumley (2015), where the Dalai Lama is quoted as stating, "We must have a human approach . . . As far as socioeconomic theory, I am Marxist." He continued, "In capitalist countries, there is an increasing gap between the rich and poor . . . In Marxism, there is emphasis on equal distribution . . . [and] many Marxist leaders are now capitalists in their thinking." He has also been reported by A. C. Brooks (2014) as commenting "while free enterprise could be a blessing, it was not guaranteed to be so. Markets are instrumental, not intrinsic, for human flourishing. As with any tool, wielding capitalism for good requires deep moral awareness. Only activities motivated by a concern for others' well-being, he declared, could be truly "constructive", leading Brooks to conclude that "Washington needs to be more like the Dalai Lama. Without abandoning principles, we need practical policies based on moral empathy."

References

Allan, J. A. (1994) "Overall Perspectives on Countries and Regions". In P. Rogers and P. Lydon (eds), *Water in the Arab World: Perspectives and Prognoses*. Harvard University Press, Cambridge, MA, pp. 65–100.

Brooks, A. C. (2014) "Capitalism and the Dalai Lama". *New York Times*, 17 April, available at www.nytimes.com/2014/04/18/opinion/capitalism-and-the-dalai-lama.html?_r=0 (accessed 18 November 2015).

Chumley, C. K. (2015) "Dalai Lama Slams Capitalism for World's Ills: 'I Am Marxist'". *Washington Times*, 16 January, available at www.washingtontimes.com/news/2015/jan/16/dalai-lama-slams-capitalism-for-worlds-ills-i-am-m (accessed 18 November 2015).

Westra, L. (2004) *Ecoviolence and the Law*. Brill, Leyden, the Netherlands.

Part I

The common good and the right to food and water

1 Bioeconomy

A challenge to integrity?

Franz-Theo Gottwald

Bioeconomy: the economization of life

The green economy has reached top levels of political attention: Even during the G7 Summit in June 2015, world leaders were debating resource efficiency and energy supply security. In the long term, politics and industry aim at substituting a fossil oil-based economy with a bio-based one.

The term bioeconomy was introduced as early as 1997 by the geneticists Juan Enriquez-Cabot and Rodrigo Martinez during a meeting of the American Association for the Advancement of Science. In a summary of their contribution about the economic potentials of genomics, Juan Enriquez-Cabot defined bioeconomy as an "economic field which uses novel biological knowledge for commercial and industrial purposes" (Enríquez-Cabot and Martínez 1998: 925–926). This definition reveals that bioeconomy does not constitute the ecological alignment of economics, but the economical alignment of ecology – or in other words – the economization of all living entities (Gottwald and Krätzer 2014: 12).

The German Federal Ministry for Education and Research defines bioeconomy as follows:

> The concept of bioeconomy covers agricultural economy and all manufacturing sectors and associated service areas that develop, produce, process, handle, or utilize any form of biological resources, such as plants, animals, and microorganisms. This spans numerous sectors, such as agriculture, forestry, horticulture, fisheries and aquaculture, plant and animal breeding, the food and beverage industries, as well as the wood, paper, leather, textile, chemical and pharmaceutical industries, and aspects of the energy sector. Bio-based innovations also provide growth impetus for other traditional sectors, such as in the commodity and food trade, the IT sector, machinery and plant engineering, the automotive industry, environmental technology, construction, and many service industries.
>
> (Federal Ministry for Education and Research 2010: 2)

This is one example from a governmental perspective which is in line with other national programmes worldwide. Its goal is to use research and innovation to facilitate a structural transition from an oil-based to a bio-based industry, which will also offer much-lauded opportunities for growth and employment. At the same time, research and innovation will be the bases for taking on more international (or better: industrial) responsibility for global nutrition, the supply of commodities and energy from biomass, as well as for climate and environmental protection. This research strategy sets five priorities to continue a politically designed path towards a knowledge-based, internationally competitive bioeconomy: global food security, sustainable agricultural production, healthy and safe food, industrial use of renewable resources and biomass-based energy sources (ibid.).

Bio-based economy, bioeconomy, green economy – on first perusal, these terms sound harmless and environmentally friendly. Only a closer scrutiny reveals the intention of exploiting the biosphere commercially, while employing genetically modified organisms to implement far-reaching changes in the fields of nutrition, chemistry and pharmaceuticals. This enforced conformity in many areas aims at the total economization of all living organisms: alterations of the genetic structure of plants and animals as well as genetically engineered medications and therapeutics are means to these ends. Synthetic biology which creates artificial organisms in the lab takes this one step further. This is a very unsettling development, given that the consequences of our present genetic engineering practices are totally unknown and unknowable. In addition, there is no way of predicting how artificial life will interact with the environment.

Integrity of life: a bioethical perspective on bioeconomy

Integrity is commonly understood as being honest and morally upright. It describes the congruence of a person's moral views and his or her actual behaviour. The concept of integrity also includes intactness of a person's mental and physical state.

Following natural law, all living beings and organisms have inherent values such as dignity and integrity. Activities for self-preservation, sustenance and intactness are connected to the concept of life as well as intrinsic values. In ethics, integrity means honesty, truthfulness or accuracy of one's actions. Individuals show ethical integrity if their actions, beliefs and principles all derive from a core group of values.

Integrity focuses on correct behaviour, not on correct principles: a person's integrity is measured by his or her actual behaviour. Integrity-based ethics may refer to rules that prohibit "bad" or illegal behaviour. However, the main attention is focused on achieving proper actions or behaviour rather than behaviour to be avoided.

However, not only humans have integrity, but also animals. The concept of animal integrity defines integrity as "the wholeness and completeness of the animal and the species-specific balance of the creature, as well as the animal's capacity to maintain itself independently in an environment suitable to the species" (Rutgers and Heeger 1999).

Industrial production, western consumption patterns, overpopulation and other effects caused by humans impair life on earth for all living beings.

For decades economic growth has been the measure for wealth. However, this one-dimensional focus on economic growth severely affects ecological integrity and ethical principles such as precaution, responsibility and generational justice. The challenge of the future will not be greening industries, but shifting attention of industry and governance from growth fixations to sustainability and ecological integrity (Bosselmann 2010: 92).

Besides integrity of humans and animals, ecological integrity is a term mostly used in nature conservation. Integrity in this context means not only intactness. Ecological integrity describes an ecosystem that is "whole, intact, sound, unimpaired, and well-functioning" (Westra et al. 2000). Integrity hence is a synonym for intactness and integration. There are different concepts of Ecological Integrity and they encompass various dimensions:

- state of an ecosystem;
- long-term functions and processes/intact natural ecological processes;
- freedom from human activity stresses;
- capacity of self-sustaining and self-renewal;
- biodiversity;
- ecosystem health;
- sustainable development;
- self-regulation and self-maintenance;
- goods and services of an ecosystem; and
- protection against ecological risks and future anthropogenic demands.

From a cultural perspective, integrity and dignity are properties of human beings describing their distinct value and status and demonstrating their extraordinary responsibility for all life on earth. Pope Francis states in his May 2015 encyclical:

> We need to see that what is at stake is our own dignity. Leaving an inhabitable planet to future generations is, first and foremost, up to us. The issue is one which dramatically affects us, for it has to do with the ultimate meaning of our earthly sojourn.
>
> (Pope Francis 2015: 68)

The encyclical claims that "together with our obligation to use the earth's goods responsibly, we are called to recognize that other living beings have a value of their own in God's eyes" (ibid.: 160).

Integrity and dignity are fundamental principles broadly agreed on in modern societies. While recognition and protection of human dignity is included in many constitutions worldwide, the constitution of Switzerland even protects the dignity of creation with regard to genetic engineering (art. 120 on non-human gene technology).

Unlike utilitarianism or other ethical concepts, deontology agrees on general duties and rights. Whereas teleological argumentations appraise a certain behaviour or action by assessing their intention and benefit, deontology assumes certain behaviour as morally good or bad by itself, regardless of the consequences of the behaviour. The British philosopher John Locke, for instance, claims that humans must allow each other the right to exist, the right to liberty and property. This deontological perspective also builds the basis for the human rights declaration. With respect to ecological integrity, deontological bioethics assumes this inherent integrity in all life forms right up to ecosystems.

From a behavioural point of view, all actions or behaviour harmful to integrity (human, animal, plant, or eco-system) must be avoided. If integrity is understood as the core value of each living entity, it cannot be subject to utilitarian horse-trading. Otherwise, life has an economic price that is negotiable.

Forms of biotechnology

However, using living organisms, their tissues, by-products or components for industrial purposes is the basic principle of biotechnology. Societal acceptance of biotechnology depends on its utilization and the benefits it achieves. This also can be determined by considering the five main fields of application.

Red biotechnologies refer to all uses connected to medicine. It ranges from the production of drugs, vaccines, antibiotics and diagnostic tools to genetic manipulation of cells for developing novel therapies. Due to the potential life-saving qualities of red biotechnologies, there is – at least in Germany – a broad acceptance for these technologies in research and practical application. This excludes the use of biotechnologies in reproductive medicine, which is frequently subject to social and political debates (see Renn 2005).

White or industrial biotechnology involves the use of biotechnological methods (primarily based on enzymes) for industrial processes; that is, the development of more efficient and environmentally friendly products, energy forms and processes, the use of (genetically modified) microorganisms, and the production of chemicals and biofuels.

Grey biotechnology refers to environmental protection (i.e. preservation of biodiversity as well as elimination of pollutants and contaminants). Methods used in grey biotechnology range from cloning and gene storage techniques to genetically modified organisms and plants. Blue biotechnology is based on the utilization of marine life forms in order to develop processes and products used in industrial contexts.

Green biotechnology is used in agriculture. It includes the development of genetically modified plants in order to enhance yield, drought tolerance or the resistance against certain pathogens, vermin or pesticides. It is the technology facing the most distinct social controversy, especially in Europe.

Creating life: synthetic biology

All these applications of biotechnology have in common that they utilize life in order to create monetary values. Particularly alarming are the efforts made in synthetic biology. Synthetic biology is a relatively young field of research, evolving so rapidly that no widely accepted definitions exist.

"Synthetic biology is an emerging area of research that can broadly be described as the design and construction of novel artificial biological pathways, organisms or devices, or the redesign of existing natural biological systems" (Royal Society 2007: 1). Application of synthetic biology reaches from analysing and manipulating existing genes to constructing new genes and gene arrangements. In other words, scientists are now able to develop sequences of synthetic DNA, which are unknown in nature – creation ex nihilo. Synthetic biology covers various applications. It is applied in biomedicine, the synthesis of biopharmaceuticals, the development of efficient biotransformation in chemical industry, in inventing smart materials/biomaterials for the environment and energy supply. Although the risks of this technology are potentially high (e.g. bioterrorism), synthetic biology has hardly been noticed by the public.

Experience has shown proof that most innovations, which are technically achievable, have come into practical use. Moral concerns or risk assessment play a minor role, as examples from present research prove:

- Genetically modified goats are provided with spider genes. These GM goats no longer give milk but produce stable silk threads in their udders. The invention is not necessary in order to save lives, improve human or animal lives – it is just a clever idea. Does a smart idea justify such a deep invasion into the integrity of an animal?
- April 2015: According to a new scientific publication, Chinese researchers have genetically engineered human embryos for experimental purposes. They were using new methods, so-called nucleases or DNA scissors (CRISPR/Cas). The results ring alarm bells: several side effects were observed within the human genome, indicating severe health risks. Scientists warn that these new methods might be introduced too fast and also demand a broader ethical debate.
- May 2015: French researchers produce organic asphalt made from algae. The algae are heated up to 200 degrees and can replace components, which are commonly made from fossil oil. This algae asphalt is a good example of a sustainable alternative – nevertheless, practicability and durability have to be proven.

Modern biotechnologies obviously harm the integrity of life and no longer refer to its dignity. By ignoring ethical values and the right to integrity of humans, animals, plants and other life forms and their components, universal, culturally grounded and statutory concepts of dignity and integrity are violated.

From life to DNA-software-driven systems

"The short answer is: life is a DNA software-driven system, at least on this planet, as far as we know", answers Craig Venter, an American biotechnologist, to the question of what life is (Venter 2015). This quotation reveals one of the main ethical problems of biotechnologies and bioeconomical approaches: they degrade life. From a bioeconomical point of view, life is nothing mysterious and nature has no value itself. Life is conceived simply as a biological system, which can be manipulated by switching genes on and off, by developing new life forms and by adding favoured traits or eliminating others. Biotechnologies tend to simplify life and its components. A living entity is more than the sum of its genes; its life and its integrity has a value itself. In bioeconomy, life is reinterpreted into biomass. The dominant bioeconomic paradigm finds its expression in a shift away from nature, away from creation, away from mankind as guardian protecting and caring for the world. The scientific rational paradigm declares life as a biological system, which must be effectively integrated into the economic value chain. This conception, where economy is beyond all doubt, is the basis of controversial technologies and applications such as fracking and GM crops as well as political and legal decisions (i.e. granting patents for life or trade agreements like CETA, TTIP and others).

Biotech and integrity: some unsolvable dilemmas for bioethics

However, dignity and integrity do not only prohibit certain actions, research or applications; the concepts also can constitute a moral duty in order to take action and apply a technology.

Examples from intensive medicine, euthanasia, and prenatal diagnostics show that dignity and its perception can be individually different. For instance, for some humans serious handicaps can be bearable, while others would suffer a loss of dignity. For some incurably ill people, euthanasia might be the last chance to preserve their dignity while for others assisted suicide is no option at all.

Not only at the end of life but also at its inception questions might arise concerning which technologies are morally imperative and which are morally intolerable. If prenatal medical tests indicate severe mental and/or physical deficiencies of an embryo – who can decide whether this life ought to live? Same with premature infants: if survival is possible only with future irreversible impairment – are medical interventions reasonable in that case? One can argue that dignity and integrity are inseparably connected to an intact mental state. Following this argument, medical treatment of infants who will never be able to live a self-determined life is morally not imperative. However, from another perspective, every living being has an unrestricted right to life. Even if a person is in a coma and cannot understand the concept of dignity and integrity, the individual's dignity is inviolate.

Dilemmas between the use of biotechnologies and ethical principles also concern agrarian production. In many parts of the world, GM plants are widely used. In 2013, cultivation of GM crops worldwide rose to a total of 174 million hectares (GMO Compass 2014). The producers promise higher yields, resistance to pesticides, fungal and insect pests or other traits such as better drought-tolerance. However, risks for the ecosystems, for biodiversity, soils and animal and/or human health still remain unclear. There is a major dearth of independent research and risk assessment. This is particularly alarming given that co-existence between organic and conventional agriculture is hardly possible. From a consumer perspective, freedom of choice is not guaranteed when it comes to GM food since there are no distinct labelling standards: animal products derived from animals fed with GM feed do not need to be labelled. Technical progress rapidly advances. Most consumers do not understand gene technology and other biotechnologies. They depend on politics and a legal framework, which protects them from harmful technologies, but confront humankind also with ethical questions or problems affecting human dignity and integrity as well as those of other life forms.

Pro integrity of life – what civil society can and should do

The promotion of the bioeconomic strategy by industries, science and the political administration is ultimately a re-interpretation of the concept of sustainability for the benefit of technological solutions to major corporations. Bioeconomy is described as sustainable, progressive, innovative and indispensable for the good of all humanity. Critics have been silenced. This happens – increasingly in the media – by defaming sceptics as reactionary, ideologically blinded or overprotective.

Civic engagement is annoying for the whole biotech industry. The former CEO of BASF, Jürgen Hambrecht, expressed his views bluntly:

> Europe must not miss the opportunities of green biotechnology. We must not be guided by irrational fears. There is not a single scientific proof that plant biotechnology is harmful to the environment or human beings. On the contrary, we will open up many doors to better health and quality of life! . . . Europe must remain a driver of innovation! . . . But above all, I wish that politics explicitly acknowledges new technologies. A mere moderation of public opinion is not enough.
>
> (Hambrecht 2010: 27)

However, scepticism is indicated when considering the hard facts: Risk research? Technology assessment? In practice none. It should be legitimate in a democratic society to demand accountability, the opportunity to recall new technologies and forestall any of their consequences. Finally, it is society that ultimately has to bear the costs.

Policy fosters bioeconomic strategies and grants billions of tax revenues to industrial stakeholders. To give an example, which stands for many others in the industrialized world, the German Federal Government is very generous when it comes to the implementation of bioeconomy. The High-Tech Strategy for Germany was supported with 27 billion euros between 2010 and 2013. In addition, 2.4 billion euros were invested by the BMBF for the concept of BioEconomy 2030. The close connection of state institutions and industry and the missing democratic legitimacy of these payments is problematic considering this kind of governmental funding practice. For instance, there is a public-private partnership between the EU and an industry group, provided with nearly four billion euros from 2014 to 2020, including one billion from the EU. The Federal Government supports companies in the implementation of a high-tech strategy, while other alternative forms, whose sustainability and benefits to the public have been proven, are simply disregarded. The Federal Organic Farming Scheme (BÖLN), for instance, which is part of the sustainability strategy of the Federal Government and the Federal Ministry of Food and Agriculture (BMEL) is chronically underfinanced. The BÖLN was set up to conduct research in organic agriculture and promote sustainable forms of agricultural land use. Target groups of the various projects, training and information measures of BÖLN are not only producers, processors and trade, but also consumers, teachers, the media and other multipliers. This is precisely why this programme is also socially and ecologically relevant. However in recent years, funding for the BÖLN has been cut further: while 35 million euros a year flowed into the programme in 2003, the annual budget today amounts to only 16 million euros (Gottwald and Krätzer 2014: 145–146). Due to this preference for bioeconomy, there is a dearth of funds for research into organic forms of agriculture as well as alternative forms of decentralized economic and manufacturing activities.

Civil society therefore must organize itself or join NGOs, which:

- fight for transparency and democracy and against corruption;
- plead for independent risk assessment; and
- explain chances and risks of new technologies and process information.

The persistent and multifarious protest against patents on life or against TTIP shows that civil society is able to converge into a strong countermovement in order to ensure proper consumer protection.

Another way of getting involved is inventing or promoting alternatives. There are many ways of truly sustainable, low-cost and easily imitable solutions within the energy and food sector.

Viable alternatives to the economization of life

There are many ways to develop a truly sustainable future of agriculture, land use and energy supply. In contrast to bioeconomy policies a policy is possible in line with the guiding principles of increasing sustainability through efficiency, consistency and sufficiency, which furthers research and economic activities in keeping with the principles of precaution, responsibility, intergenerational fairness as well as biodiversity. Diverse political steering mechanisms are in place and range from regulatory laws, to tax laws all the way to planning laws, which could be used for fostering alternatives. Failure to implement alternatives to bioeconomy would constitute dire political negligence. Politics and industries, especially major corporations and global players are mired in an outdated comprehension of economic wealth, growth and power constellations. Hence small and innovative businesses, which cooperate, network and break fresh ground in order to realize an alternative economic growth, might play a key role in this process. This applies particularly to agriculture, which is challenged by climate change, scarce resources, decreasing soil fertility and availability as well as a dramatic decline in biodiversity.

There is a strong need for sustainable agricultural systems, which are based on principles like Common Goods, cooperation and co-existence, ecologically and socially sound methods, and on respecting the integrity of life. These systems are locally adapted and committed to a closed loop economy. Jules Pretty states four key principles for a sustainable agriculture:

1 integrate biological and ecological processes such as nutrient cycling, nitrogen fixation, soil regeneration, allelopathy, competition, predation and parasitism into food production processes,
2 minimize the use of those non-renewable inputs that cause harm to the environment or to the health of farmers and consumers,
3 make productive use of the knowledge and skills of farmers, thus improving their self-reliance and substituting human capital for costly external inputs, and
4 make productive use of people's collective capacities to work together to solve common agricultural and natural resource problems, such as for pest, watershed, irrigation, forest and credit management.

(Pretty 2008: 451)

Sustainability of agricultural systems, however, does not mean "ruling out any technologies or practices on ideological grounds. If a technology works to improve productivity for farmers and does not cause undue harm to the environment, then it is likely to have some sustainability benefits" (ibid.: 451).

Likewise, the Food and Agriculture Organization of the United Nations (FAO) assumes that chances and benefits of small-scale rural communities for ensuring food security and contributing to social welfare and rural development are far underestimated. It describes the Smallholders' Approach to Agroecology as a concept that:

- Is site-specific and its performance is not due the techniques per se, but rather the ecological processes that underlie sustainability. It avoids dependence on external inputs, emphasizing use of agro-diversity and beneficial synergies
- Is a culturally acceptable approach, as it builds upon traditional and indigenous knowledge, in improving agro-biodiversity and local natural resources while increasing food availability and improving nutrition
- Is socially beneficial, as its diffusion requires constant farmers' participation and community building

- Is a promoter of processes of governance as it is built on greater participation in decision-making, social empowerment, inclusiveness and locally adequate measures and approaches
- Is ecologically sound, as it does not attempt to modify the flows of energy and nutrients of existing systems, but rather tries to optimize their performance through adaptation
- Is economically beneficial, as it increases the real value of capital input, while constituting a major source of income and jobs for farmers and families, helping to reduce poverty.

(Wolfenson 2013: 21)

Sustainable creation of value derives from civil society. It is a grassroots movement, which must be strengthened, maintained and supported by politics and societies.

References

Bosselmann, K. (2010) "Earth Democracy: Institutionalizing Sustainability and Ecological Integrity". In J. R. Engel, L. Westra and K. Bosselmann (eds), *Democracy, Ecological Integrity and International Law*, Cambridge Scholars Publishing, Newcastle, pp. 91–114.

Enríquez-Cabot, J., Martínez, R. (1998) "Genomics and the World's Economy". *Science Magazine* 281, 925–926.

Federal Ministry of Education and Research (2010) "National Research Strategy Bioeconomy 2030". Available at www.bmbf.de/pub/Natinal_Research_Strategy_BioEconomy_2030.pdf (accessed 20 November 2014).

GMO Compass (2014) "Genetically Modified Plants: Global Cultivation on 174 Million Hectares". Available at www.gmo-compass.org/eng/agri_biotechnology/gmo_planting/257.global_gm_planting_2013.html (accessed 22 October 2015).

Gottwald, F.-T., Krätzer, A. (2014) *Irrweg Bioökonomie. Kritik an einem totalitären Ansatz*, Edition Unseld, Suhrkamp, Berlin.

Hambrecht, J. (2010) "Ohne Risiko kein Fortschritt: Interview mit Frank Stäudner". *Wirtschaft & Wissenschaft* 1: 27.

Pope Francis (2015) "Encyclical Letter *Laudato Si'* of the Holy Father Francis on Care for our Common Home". Available at http://w2.vatican.va/content/francesco/en/encyclicals/documents/papa-francesco_20150524_enciclica-laudato-si.html (accessed 15 October 2015).

Pretty, Jules (2008) "Agricultural Sustainability: Concepts, Principles and Evidence". *Philosophical Transactions of the Royal Society B* 363: 447–465. Available at http://rstb.royalsocietypublishing.org/content/363/1491/447 (accessed 23 October 2015).

Renn, Ortwin (2005) "Technikakzeptanz: Lehren und Rückschlüsse der Akzeptanzforschung für die Bewältigung des technischen Wandels". *TATuP: Zeitschrift des ITAS zur Technikfolgenabschätzung* 3(14): 29–38.

Royal Society (2007) "Synthetic Biology: Call for Views". Available at https://royalsociety.org/~/media/Royal_Society_Content/policy/projects/synthetic-biology/CallForViews.pdf (accessed 21 October 2015).

Rutgers, B., Heeger R. (1999) "Inherent Worth and Respect for Animal Integrity". In M. Dol, M. Fentener van Vlissingen, S. Kasanmoentalib, T. Visser and H. Zwart (eds), *Recognizing the Intrinsic Value of Animals*, Van Gorcum, Assen, pp. 41–51.

Venter, Craig (2015) "J. Craig Venter on DNA and Life's Mysteries". *Wall Street Journal*, 9 February, available at www.wsj.com/articles/j-craig-venter-speaks-about-dna-and-the-mysteries-of-life-1423540853 (accessed 21 October 2015).

Westra, Laura, et al. (2000) "Ecological Integrity and the Aims of the Global Integrity Project". In D. Pimentel et al. (eds), *Ecological Integrity: Integrating Environment, Conservation, and Health*, Island Press, Washington, DC, pp. 23–26.

Wolfenson, Karla D. Maass (2013) *Coping with the Food and Agriculture Challenge: Smallholders' Agenda: Preparations and Outcomes of the 2012 United Nations Conference on Sustainable Development (Rio+20)*, FAO, Rome.

2 Soil as a key to food security

Social perception of soil erosion in the Czech Republic (a case study)

Eva Cudlínová, Jan Vávra and Miloslav Lapka

> Motto: It is not enough to describe the soil as a country's greatest source of wealth; it is more than that; it is a country's life.
>
> (Kelley 1990)

Introduction

This chapter presents a sociological case study of the perception of soil erosion by various social groups (urban and rural dwellers and farmers) carried out in South Moravia, a region with intensive agriculture and a high level of soil erosion in the Czech Republic. This chapter is embedded in the broader context of the role of soil for human beings in general and what kind of risks for our civilization soil erosion means in terms of both losses of amounts and fertility of soil.

Soil could be considered as a kind of interface between agriculture and nature. It is a non-renewable resource continually transformed by human activity. The comparison of the time taken for soil creation with the speed of its degradation is fascinating. It takes many years to create a layer of soil, but it can be destroyed in almost no time at all. The human abilities to grow food crops and graze animals, to produce fibre and forests are lost with the loss of soil. "We are losing 30 soccer fields of soil every minute, mostly due to intensive farming", said Volkert Engelsman, an activist with the International Federation of Organic Agriculture (Arsenault 2014).

Soil has many functions, all of them important for human life:

> Healthy soils are the basis for healthy food production. Soils are the foundation for vegetation which is cultivated or managed for feed, fiber, fuel and medicinal products. Soils support our planet's biodiversity and host a quarter of the total surface of the planet. Soils help to mitigate and adapt to climate change by playing a key role in the carbon cycle. Soils store and filter water, improving our resilience to floods and droughts.
>
> (FAO 2015a)

Soil and food

The most visible and sensitive is soil's role in food production. Humans world-wide obtain more than 99.7 per cent of their food (calories) from the land and less than 0.3 per cent from the oceans and aquatic ecosystems (Pimentel and Burgess 2013).

The high value of soil as a source of food production is embedded in the fact that hunger as a worldwide problem is still very actual in the present time. Even if the 2013 Global Hunger Index (GHI) core has fallen by 34 per cent compared to the 1990 GHI score, yet world hunger still remains "serious", with 19 countries suffering from levels of hunger that are either "alarming" or "extremely alarming". South Asia has the highest regional GHI score, followed by Africa south of the Sahara, with Burundi, Eritrea and Comoros having the highest levels of hunger (International Food Policy Research Institute 2013).

More people die each year from hunger and malnutrition than from AIDS, tuberculosis and malaria combined (FAO 2015b) and the World Bank estimates that cereal production needs to increase by 50 per cent and meat production by 85 per cent between 2000 and 2030 to meet the demand of the population (Cabinet Office 2008). Reduction of poverty and hunger are also in the significant first place among the eight Millennium Development Goals (MDG 2015).

We know that approximately a third of all food produced is discarded, and whenever food is thrown out it is as if it were stolen from the table of the poor. Still, attention needs to be paid to imbalances in population density, on both national and global levels, since a rise in consumption would lead to complex regional situations, as a result of the interplay between problems linked to environmental pollution, transport, waste treatment, loss of resources and quality of life (Pope Francis 2015).

The global food market makes the problem of hunger more intense. The global interconnection of the world through the world market that has reflected itself in the global financial crisis of 2007 was combined with a food crisis as well. In 2007, a rapid increase in oil prices increased fertilizer and other food production costs. The food prices spike of 2008 was a warning of what was to come. Staple food prices rocketed – wheat rose by 130 per cent, sorghum by 87 per cent and rice by 74 per cent – and caused riots in 36 countries. In early 2009, food crises persisted in 32 countries from the 36 affected in 2008 (FAO 2009b).

Soil erosion

Soil covers most of the land surface of the earth in a thin layer, ranging from a few centimetres to several metres. The total land area of the world exceeds 13 billion hectares, but less than half can be used for agriculture, including grazing. A much smaller fraction – about 1.4 billion hectares – is presently suitable for growing crops. Only 3.2 billion ha of potentially arable land is currently cropped and 41 per cent is considered moderately or highly productive (Scherr 1999).

This valuable resource is unfortunately still decreasing due to the soil erosion caused by water and wind, chemical pollution, etc. According to Oldeman (1997), water erosion is the predominant form of soil degradation (55.7%) followed by wind erosion (27.8%), chemical (12.3%) and physical damage. "Thirty three per cent of our global soil resources are under degradation and human pressures on soils are reaching critical limits, reducing and sometimes eliminating essential soil functions", said the UN General Assembly of FAO in Rome, when it declared 5 December World Soil Day (FAO 2015c; UN News Centre 2015).

Soil erosion is a natural process but it is extremely intensified by human activity, especially agriculture. Soil erosion occurs whenever humans remove vegetative cover, though world agricultural production accounts for about three-quarters of the soil erosion worldwide (Lal and Stewart 1990; FAO 2002; Pimentel 2006).

Soil erosion caused by humans is not a new phenomenon, it is as old as human economic activity. The loss of productive soil was many times the main reason for the vanishing of old civilizations. We can mention the collapse of the civilization in Guatemala around 900 AD or the Mayan civilization's decline around 1700, which were probably caused by a similar reason. Unfortunately the same processes of soil degradation which destroyed civilizations in the past are still present today (Kelley 1990).

Currently, about 80 per cent of the world's agricultural land suffers moderate to severe erosion, while 10 per cent experiences slight erosion (Pimentel 1993; Lal 1994; Speth 1994). Worldwide, erosion on cropland averages about 30 t/ha/yr and ranges from 0.5 to 400 t/ha/yr (Pimentel et al. 1995). As a result of soil erosion during the last 40 years, about 30 per cent of the world's arable land has become unproductive and much of that has been abandoned for use in agriculture (Kendall and Pimentel 1994; WRI 1994). Each year an estimated 10 million ha of croplands are abandoned worldwide due to lack of productivity caused by soil erosion (Faeth and Crosson 1994). Worldwide, soil erosion losses are highest in the agro-ecosystems of Asia, Africa, and South America, averaging 30–40 t/ha/yr of soil loss (Taddese 2001).

It is certainly an irony that the production of food, which we are most dependent on as human beings and which is connected with the soil, is one of the main culprits of the degradation of soils.

The new features of soil erosion today are its global character and the speeding up of the process. During the last fifty years the speed of soil erosion increased due to the technological progress and rapid growth of the population and its dietary needs. The main driving forces were the post-war "second agricultural revolution" in developed countries, and the "green revolution" in developing nations in the mid-1960s (FAO 2009a). These two forces have transformed agricultural practices and raised crop yields dramatically, but this process was accompanied by soil degradation, mostly caused by water and wind erosion.

Besides all these factors, according to Eurobarometer 2014, the perception of soil degradation as an environmental problem is relatively low in the EU. Most of the respondents were worried about air pollution (56 per cent) and water pollution (50 per cent). In comparison, relatively few Europeans worry about noise pollution (15 per cent), or land and soil degradation (13 per cent) (European Commission 2014).

Different results are reported in terms of different types of information. Respondents in the Eurobarometer survey were given a list of 14 environmental issues, and asked to pick the main five about which they particularly lack information. Almost three out of ten people (29 per cent) mentioned soil degradation, which makes it the second most often mentioned problem that citizens lack information about (health impacts of everyday chemicals in daily use being the first).

Soil erosion as a socio-economic problem

Although land degradation is a physical process, its underlying causes are firmly rooted in the socio-economic, political and cultural environment in which land users operate (Stocking and Murnaghan 2001). Despite this recognition, the socio-economic aspects of the problem have been neglected in most technical studies of the erosion problem in agriculture, which have instead emphasized geological and geographic aspects of the problem (e.g. Crole-Rees et al. 1990; Van Wesemael et al. 2006; Martín-Rosales et al. 2007; Romero Díaz et al. 2010).

There are many publications regarding soil conservation and degradation from the biological or geological point of view. Cerdan et al. (2010) compiled a large database of erosion rates under various land use types in Europe. Many EU projects and modelling methods such as EUROSEM and PASERA also exist, which focus on the risk and resilience of soil under erosion.

The situation in soil erosion research and its focus is changing gradually and the studies of the views or perceptions of stakeholders have been increasing. There are already some studies focused on the social perception and farmers' perceptions of soil degradation in Africa and the Mediterranean area (Karltun et al. 2013; Nabahungu and Visser 2013; Sop and Oldeland 2013; Pereira et al. 2014; Vila Subirós et al. 2014).

Most of these studies focus on different perspectives of farmers' attitudes (e.g. willingness to change conventional agriculture practices, perception of erosion due to flooding, etc.; Marques et al. 2014; Bielders et al. 2003). Farmers' attitudes and ideas are important for many reasons. They give us important information needed when new forms of agricultural policy are planned to be implemented. Other socio-economic research of soil erosion deals not only with farmers but with the whole rural community and its perception of landscape changes within a region (Kelly et al. 2015). Groups of respondents could be even broader than the rural community; researchers and policy makers can be included as well, as for example in the study entitled "Runoff and Soil Erosion in Arable Britain: Changes in Perception and Policy since 1945" (Evans 2010).

There are also several studies and overviews on soil management knowledge from a historical perspective (McNeill and Winiwarter 2004; McNeill and Winiwarter 2006, 2010; Winiwarter 2006; Winiwarter and Blum 2006; Brevik and Hartemink 2010). These studies show that long before the introduction of scientific research, subsistence farmers had a good practical knowledge of how to manage soils sustainably. The study of Brevik and Hartemink (2010) presents examples of the perception of soil erosion in historical times in the context of a precise understanding of soil and the birth and development of soil science.

At present the majority of the EU population has an indirect connection with soil. It is through agricultural policy, food prices and economic incentives. If human behaviour regarding agricultural practice is to be changed to decrease soil erosion, it is necessary to change the perception and understanding of the problem.

Soil erosion in the Czech Republic: case study

Soil erosion in many areas of Western Europe can be seen as a result of agricultural modernization with typical intensive large-scale farming. Soil erosion in the Czech Republic is also a consequence of large-scale farming, but the size of fields was even increased by the political process of communist collectivization of agriculture during the 1950s. Within less than 10 years, almost all private farmers were forced to give up their land and the small-scale field structure changed to large fields managed by the state or cooperative farms (Hájek 2008).

The size structure of agricultural enterprises in the Czech Republic differs significantly from the structure of enterprises in the other EU member countries. Businesses with more than 50 ha of agricultural land occupy 92.2 per cent of the total acreage of farmed agricultural land (Ministry of Agriculture of the Czech Republic 2015). The average size of farm exceeds 200 ha, which is the second highest value in the EU, after Slovakia. Additionally, more than 70 per cent of the overall area of agricultural land is rented by the farmers – they do not work on their own land (European Commission 2014; Ministry of Agriculture of the Czech Republic 2015).

According to the database of the Research Institute for Soil and Water Conservation in the Czech Republic (2008), individual types of degradation differ between the areas affected, the most notable being water erosion which threatens more than 45 per cent of agricultural areas. This is followed by wind erosion (11 per cent), extreme soils – clay soils (4.5 per cent), and soils affected by dryness (1.5 per cent) (Šarapatka et al. 2010).

Description of our study area

Our research area historically belongs to the so-called wine region and wine is still an important product in this region, but nowadays the area is dominated by corn and wheat, which are totally unsuitable crops for this local rolling landscape. The area of our interest is situated in the Morava River basin northwest of the town of Hodonín and belongs to the Kyjovské platforms. It is a slightly to strongly sloping terrain with an average altitude of 250 m above sea level. The climate of the region is warm and dry, with mild winters. The bedrock of the area is shaped

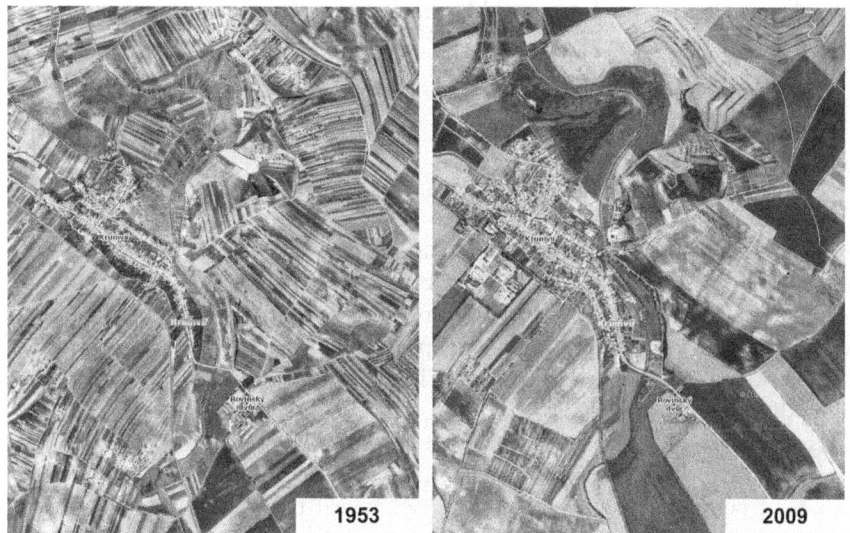

Figure 2.1 Example of agricultural field patterns before and after collectivization (example from Moravia, our studied area)

Sources: http://kontaminace.cenia.cz; historical orthophotomap © CENIA 2010 and GEODIS BRNO, spol. s r.o. 2010; background aerial photographs provided by Military Geographic and Hydrometeorology Office (VGHMÚř) Dobruška © Ministry of Defence and Armed Forces of the Czech Republic (MO ČR) 2009; actual orthophotomap © Czech Office for Surveying, Mapping and Cadastre (ČÚZK).

Figure 2.2 Description of our study area

Source: Own research and processing.

by a Quaternary cover of loess. This area is one of the most endangered by water erosion in the whole Czech Republic.

Methods

The questionnaire method was used to address three groups of people within the surveyed area of South Moravia. It was the same area where physical measurements of soil erosion were carried out by a research team from the Faculty of Agriculture, University of South Bohemia. The groups sampled consist of the urban group, the rural non-farming group and rural farmers. The data collection by questionnaire field work was done by the Augur Consulting Company in 2013 and 2014.

From a methodological point of view, our study focuses on the social constructions of the real problem (i.e. the reflection of soil degradation as it exists in the minds of people). This approach does not reveal the picture of the physical reality, just its social perception.

The original questionnaire focused on three areas of questions: awareness of soil erosion and its perception as a problem; causes and actors of soil erosion; and actions to be done to improve the situation. Due to limited space, this chapter focuses only on the first area of the questions, the latter two will be analysed in further publications.

The data were statistically analysed in the IBM SPSS Statistics 19 program, using descriptive methods and chi-square (χ^2) to analyse categorical data.

Sample description

Three groups of respondents were surveyed in the questionnaire study: the urban population, the rural non-farming population and rural farmers. These groups differ regarding their place of living and their personal experience with land management. We use the term proximity for labelling these subgroups.

The urban population consists of a representative sample of the citizens living in Brno-city district ($n = 117$). The rural population is represented by a sample from the municipalities living in Brno-country district ($n = 101$). To differentiate between these groups and farmers, the farmers were not included in the rural sample, which makes it representative for the rural non-farming population. Farmers as representatives of actors who directly work with land and who could potentially be specific in terms of the perception of soil erosion were sampled in the municipalities of Brno-country district ($n = 42$). The group of farmers consisted of private farmers, managers and employees of farm holdings and winemakers.

The total number of all respondents was 260. The urban and rural groups were regionally representative regarding their gender, age and education. 50 per cent of the urban group were males, while 52 per cent of the rural group were. The average age in the urban group was 45 years and in the rural group was 47 years. The educational level of the urban sample is as follows: 11 per cent low (grammar or no school), 64 per cent middle (including vocational) and 25 per cent university (any degree). The rural sample's educational ratio is 13 per cent low, 69 per cent middle and 18 per cent high. The farmers' group was, naturally, different due to the specifics of the farmers' population. The average age was 47 years, 91 per cent of the sample were men and their education is significantly higher in general: 7 per cent low, 48 per cent middle and 45 per cent high.

Hypotheses

Following especially the results of the Eurobarometer (European Commission 2014), we assume that the knowledge of soil erosion and perception of the importance of the problem will be relatively low, and the amount of information will be perceived as insufficient. We hypothesize that there is a strong link between the awareness of the problem and the perception of its importance.

We also assume that the awareness of soil erosion is not limited to the information from the media, but that respondents' everyday experience linked to their place of living is important. We expect the three groups of respondents (urban, rural, farmers) to report different levels of awareness of soil erosion and its evaluation due to their proximity to the problem. We expect that farmers will be aware of soil erosion and perceive it as an important problem. The urban population is expected to stand in a relatively opposite position to the farmers and the rural population somewhere between these two groups, reflecting the physical distance of the groups from the problem.

Results

The responses to the three questions are presented in this section. All respondents were asked about:

- their awareness of soil erosion ("Are you aware that the soil in South Moravia is being washed away during rain and therefore loses its fertility?");
- their perception of soil erosion as a problem ("Do you agree that the soil erosion in South Moravia is a big problem?"); and
- their assessment of information about soil erosion ("How much information is there about the soil erosion in South Moravia?").

Respondents were given several possible answers (close-ended questions).

We present the results of the answers to these three questions and analyse the effects of socio-demographics (including the differences between the subgroups) on the results. Finally, we analyse the relationship between the three questions. For the purpose of the analysis of categorical data, the ages were divided into five groups: 18–30, 31–45, 46–60, 61–75 and over 75 years.

In the sample as a whole ($n = 260$), 73 per cent of respondents were aware of soil erosion; only 27 per cent were not. There were no statistically significant effects of gender, age and education on the awareness of respondents.

However, there were significant differences caused by the proximity of respondents to the problem of soil erosion ($\chi^2(2) = 14.610$, $p = 0.001$). The number in brackets stands for the degrees of freedom, p represents statistical significance. Answers of urban dwellers and farmers significantly differed from the average results gained from the whole sample of respondents; rural respondents did not. Members of the urban sample were less often aware of soil erosion (less than 2/3), while almost all farmers (more than 90 per cent), were aware of this phenomenon (see Table 2.1 for the results).

The question on perception of soil erosion as a big problem resulted in three answers: 57 per cent of all respondents agreed that soil erosion is a big problem, 11 per cent disagreed and 33 per cent did not feel competent to answer. As in the previous question, characteristics such as gender, age and education had no effect on the perception of soil erosion as a problem, while the proximity from the soil erosion phenomenon due to place of living and contact with soil were significant factors ($\chi^2(4) = 10.651$, $p = 0.031$). Urban dwellers agreed significantly less that soil erosion is a big problem, while farmers did agree significantly more (see Table 2.1 for the results). In contrast to farmers, members of the urban sample also did not tend to assess the problem more often.

Regarding the amount of information, 67 per cent of all of the addressed sample of respondents ($n = 249$) answered that there is too little information about soil erosion, 31 per cent thought that the amount of information is adequate and only 2 per cent supposed that there is too much information. The answers were again not significantly affected by gender, age or education, as in the two previous cases. Also as in the previous questions, the only significant factors in terms of the type of answers were the place of living and contact with the problems of agricultural soil ($\chi^2(4) = 16.466$, $p = 0.002$). Farmers significantly more often assessed the amount of information as adequate and less often as too little. Rural respondents tended to answer in the opposite way: more often too little information, less often adequate information (see Table 2.1 for results).

We analysed the relationship between the three answers for the sample as a whole. It is not surprising that there is a significant relationship between

Table 2.1 Respondents grouped according to their proximity to soil erosion

		Urban	Rural non-farming	Farmers
Awareness of soil erosion	Yes	63%	76%	93%
	No	37%	24%	7%
Soil erosion as a big problem	Yes	47%	60%	76%
	No	14%	9%	5%
	Hard to assess	39%	31%	20%
Amount of information about soil erosion	Little	70%	74%	41%
	Adequate	28%	24%	57%
	Too much	2%	2%	2%

Source: Own research and processing.

Note: The sum can exceed 100 per cent due to rounding.

the awareness of soil erosion and its perception as a problem ($\chi^2(2)$ = 62.689, p = 0.000). From those who are aware of soil erosion, 70 per cent agreed that soil erosion is a big problem, but only 4 per cent did not agree and 26 per cent could not assess the problem. The situation is different among the respondents who are not aware of soil erosion: only 20 per cent agreed that it is a big problem, 29 per cent disagreed and 51 per cent could not assess it. There is no statistically significant relationship between the awareness of soil erosion and perception of the amount of information. Similarly, there is no significant relationship between the perception of soil erosion as a problem and the perceived level of information. In this case, those who feel that there is a little information about soil erosion tended to less often disagree that it is a big problem and more often found it hard to assess. Interestingly, those who said that the level of information is adequate tended to say soil erosion is not a big problem more often and less often did they answer that they could not assess it. However, none of these results are statistically significant at the value of 5 per cent.

Our hypothesis that the awareness of soil erosion and its perception as a problem would be relatively low is not supported by our data. Almost three quarters of the whole sample are aware of soil erosion and more than half of the respondents perceive it as a big problem. Our findings cannot be directly compared with the results of the Eurobarometer (European Commission 2014) due to different wording, but we can say that our sample could be more concerned about soil erosion due to the regional aspect of its public awareness. (We can expect the respondents from larger cities and less agricultural regions to be even less aware of soil erosion). The number of people who find it hard to assess the problem is quite high in our study (two thirds of respondents said there is little information), which is in line with the generally perceived lack of information as reported in the Eurobarometer. The relationship between the amount of information and perception of the problem is interesting, though it was not statistically significant in our study. Those who tended to perceive a lack of information do not tend to underestimate the problem, but they less often disagreed that it is a problem and more often did not assess it. This could be interpreted as a principle of precaution, but more research needs to be done on this aspect to make such a generalization. The relationship between the awareness and perception of the severity of the problem was expected and is quite logical.

We find it interesting that there is no effect of the socio-demographic characteristics of the sample (gender, age, education), but very significant differences caused by the proximity to the problem (place of living/occupation).

Conclusions

In the concrete case of South Moravia, we tried to describe the perception of soil erosion in the minds of people living in the area suffering from soil degradation. Our results revealed that distance from areas of soil degradation is an important factor influencing perception of this problem. The hypothesis that people living in urban areas perceive the problem of soil degradation as less important than people living in rural areas and farmers has been confirmed. As regards the role of available information about the problem, the urban population more frequently than the other two groups mentioned that they need more information. This corresponds with their position of alienation from agricultural production and less contact with agricultural land.

In our opinion, more information in the media will not change the attitudes of the urban population to soil and its degradation. Food prices play one of the most important roles, in contrast to the media and political documents speaking about the necessity of soil protection (FAO, CAP, etc.).

The practical source of information which comes to the urban population is via food prices, and this does not reflect soil quality and its degradation. Low prices of food produced by industrial agriculture provide distorted information about the real situation of soil and its quality.

Based on our results, it seems that changes in food prices could more adequately be related to the costs of soil degradation, and could also change the perception of soil degradation as a problem by the urban population.

Acknowledgements

We would like to thank Greg Moore very much for help with the review of the English of the chapter. This work reflects the work of research project NAZV QJ1230066 "Land Degradation and its Impact on Complex Soil Properties, Including the Proposal of Corrective Measures to Restore the Agro-ecological Functions Of Soil", 2012–2016, supported by the Ministry of Agriculture.

References

Arsenault, C. (2014) "Only 60 Years of Farming Left if Soil Degradation Continues". Available at www.reuters.com/article/2014/12/05/us-food-soil-farming-idUSKCN0 JJ1R920141205 (accessed 11 September 2015).

Bielders, C. L., Ramelot, C. and Persoons, E. (2003) "Farmer Perception of Runoff and Erosion and Extent of Flooding in the Silt-Loam Belt of the Belgian Walloon Region". Environmental Science and Policy 6(1): 85–93.

Brevik, E. C. and Hartemink, A. E. (2010) "Early Soil Knowledge and the Birth and Development of Soil Science". Catena 83(1): 23–33.

Cabinet Office (2008) Food Matters: Towards a Strategy for the 21st Century, Crown, London

Cerdan, O., Govers, G., Le Bissonnais, Y., Van Oost, K., Poesen, J., Saby, N., Gobin, A., Vacca, A., Quinton, J., Auerswald, K., Klik, A., Kwaad, F. J. P. M., Raclot, D., Ionita, I., Rejman, J., Rousseva, S., Muxart, T., Roxo, M. J. and Dostal, T. (2010) "Rates and Spatial Variations of Soil Erosion in Europe: A Study Based on Erosion Plot Data". Geomorphology 122(1–2): 167–177.

Crole-Rees, A., Baril, P. and Schaub, D. (1990) "Cartographie des risques d'erosion: une approche multidisciplinaire" ["Soil Erosion Risk Mapping: A Multidisciplinary Approach"]. Soil Technology 3(4): 351–366.

European Commission (2014) *Special Eurobarometer 416: Attitudes of European Citizens Towards the Environment Report*, EU Directorate-General for Environment and Directorate-General for Communication, Brussels.

Evans, R. (2010) "Runoff and Soil Erosion in Arable Britain: Changes in Perception and Policy since 1945". *Environmental Science and Policy* 13(2): 141–149.

Faeth, P. and Crosson, P. (1994), "Building the Case for Sustainable Agriculture". *Environment* 36(1): 16–20.

FAO (2002) "Restoring the Land". Available at www.fao.org/inpho/vlibrary/u8480e/U8480E0d.htm (accessed 8 October 2015).

FAO (2009a) "How to Feed the World in 2050". www.fao.org/fileadmin/templates/wsfs/docs/expert_paper/How_to_Feed_the_World_in_2050.pdf (accessed 11 September 2015).

FAO (2009b) *Crop Prospects and Food Situation*, no. 1(2009), Food and Agriculture Organization of the United Nations, Rome.

FAO (2015a) "Soil is a Non-renewable Resource". Available at www.fao.org/soils-2015/news/news-detail/en/c/275770 (accessed 11 September 2015).

FAO (2015b) "The State of Food Insecurity in the World 2015". Available at www.fao.org/hunger/en (accessed 11 September 2015).

FAO (2015c) "World Soil Day 2015: Soils a Solid Ground for Life". Available at www.fao.org/globalsoilpartnership/world-soil-day/en (accessed 11 September 2015).

Hájek, P. (2008) *Jde pevně kupředu naše zem: Krajina českých zemí v období socialismu 1948–1989*, Malá Skála, Prague.

International Food Policy Research Institute (2013) "Global Hunger Index Calls for Greater Resilience-Building Efforts to Boost Food and Nutrition Security". Available at www.concern.net/sites/default/files/media/resource/2013_ghi_press_release_final_2.pdf (accessed 11 September 2015).

Karltun, E., Lemenih, M. and Tolera, M. (2013) "Comparing Farmers' Perception of Soil Fertility Change with Soil Properties and Crop Performance in Beseku, Ethiopia". *Land Degradation and Development* 24(3): 228–235.

Kelley, H. W. (1990) *Keeping the Land Alive: Soil Erosion – Its Causes and Cures*, Soils Bulletin 50, Food and Agriculture Organization of the United Nations, Rome.

Kelly, C., Ferrara, A., Wilson, G. A., Ripullone, F., Nolè, A., Harmer, N. and Salvati, L. (2015) "Community Resilience and Land Degradation in Forest and Shrubland Socio-ecological Systems: Evidence from Gorgoglione, Basilicata, Italy". *Land Use Policy* 46 (July): 11–20.

Kendall, H. W. and Pimentel, D. (1994) "Constraints on the Expansion of the Global Food Supply". *Ambio* 23(3): 198–205.

Lal, R. (1994) "Water Management in Various Crop Production Systems Related to Soil Tillage". *Soil and Tillage Research* 30(2–4): 169–185.

Lal, R. and Stewart B. A. (1990) *Soil Degradation*, Springer-Verlag, New York.

Marques M. J., Bienes, R., Cuadrado, J., Ruiz-Colmenero, M., Barbero-Sierra, C. and Velasco, A. (2014) "Analysing Perceptions Attitudes and Responses of Winegrowers about Sustainable Land Management in Central Spain". *Land Degradation and Development* 26(5): 458–467.

Martín-Rosales, W., Pulido-Bosch, A., Vallejos, A., Gisbert, J., Andreu, J. M. and Sanchez-Martos, F. (2007) "Hydrological Implications of Desertification in Southeastern Spain". *Hydrological Sciences Journal* 52(6): 1146–1161.

McNeill, J. R. and Winiwarter, V. (2004) "Breaking the Sod: Humankind, History, and Soil". *Science* 324(5677): 1627–1629.

McNeill, J. R. and Winiwarter, V. (eds) (2006) *Soils and Societies. Perspectives from Environmental History*, White Horse Press, Isle of Harris.

McNeill, J. R. and Winiwarter, V. (eds) (2010) *Soils and Societies: Perspectives from Environmental History*, White Horse Press, Cambridge, available at www.environmentandsociety.org/node/3489 (accessed 13 October 2015).

MDG (2015) "The Millennium Development Goals Report 2015". Available at www.un.org/millenniumgoals/news.shtml (accessed 11 September 2015).

Ministry of Agriculture of the Czech Republic (2015) "Zemědělství / Agriculture". Available at http://eagri.cz/public/web/mze/zemedelstvi (accessed 11 September 2015).

Nabahungu, N. L. and Visser, S. M. (2013) "Farmers' Knowledge and Perception of Agricultural Wetland in Rwanda". *Land Degradation and Development* 24(4): 363–374.

Oldeman, L. R. (1997) "Soil Degradation: A Threat to Food Security?". At International Conference on Time Ecology: Time for Soil Culture – Temporal Perspectives on Sustainable Use of Soil, 6–9 April, Tutzing, Germany.

Pereira, P., Mierauskas, P. and Novara, A. (2014) "Stakeholders' Perceptions about Fire Impacts on Lithuanian Protected Areas". *Land Degradation and Development* early view (doi: 10.1002/ldr.2290).

Pimentel, D. (1993) *World Soil Erosion and Conservation*, Cambridge University Press, Cambridge.

Pimentel, D. (2006) "Soil Erosion: A Food and Environmental Threat". *Environment, Development and Sustainability* 8(1): 119–137.

Pimentel, D. and Burgess, M. (2013) "Soil Erosion Threatens Food Production". *Agriculture* 4(3): 443–463.

Pimentel, D., Harvey, C., Resosudarmo, P., Sinclair, K., Kurz, D., McNair, M., Crist, S., Sphpritz, L., Fitton, L., Saffouri, R. and Blair, R. (1995) "Environmental and Economic Costs of Soil Erosion and Conservation Benefits". *Science* 267(5201): 1117–1123.

Pope Francis (2015) "Encyclical Letter *Laudato Si'* of the Holy Father Francis on Care for our Common Home". Available at http://w2.vatican.va/content/francesco/en/encyclicals/documents/papa-francesco_20150524_enciclica-laudato-si.html (accessed at 15 October 2015).

Romero Diaz, A., Belmonte-Serrato, F. and Ruiz-Sinoga, J. D. (2010) "The Geomorphic Impact of Afforestation on Soil Erosion in Southeast Spain". *Land Degradation and Development* 21(2): 188–195.

Šarapatka, B., Bednář, M. and Novák, P. (2010) "Analysis of Soil Degradation in the Czech Republic: GIS Approach". *Soil and Water Research* 5(3): 108–112.

Scherr, S. J. (1999) *Soil Degradation: A Threat to Developing-Country Food Security by 2020?*, Food, Agriculture, and the Environment Discussion Paper 27, International Food Policy Research Institute, Washington, DC.

Sop, T. K. and Oldeland, J. (2013) "Local Perceptions of Woody Vegetation Dynamics in the Context of a 'Greening Sahel': A Case Study from Burkina Faso". *Land Degradation and Development* 24(6): 511–527.

Speth, J. G. (1994) *Towards an Effective and Operational International Convention on Desertification*, Intergovernmental Negotiating Committee on the International Convention on Desertification, United Nations, New York.

Stocking, M. A. and Murnaghan, N. (2001) *Handbook for the Field Assessment of Land Degradation*, Earthscan, London.

Taddese, G. (2001) "Land Degradation: A Challenge to Ethiopia". *Environment Management* 27(6): 815–824.

UN News Centre (2015) "Spotlighting Humanity's 'Silent Ally', UN Launches 2015 International Year of Soils". Available at www.un.org/apps/news/story.asp?NewsID= 49520#.VfLFRxHtlBd (accessed 11 September 2015).

Van Wesemael, B., Rambaud, X., Poesen, J., Muligan, M., Cammeraat, E. and Stevens, A. (2006) "Spatial Patterns of Land Degradation and Their Impacts on the Water Balance of Rain-Fed Tree-Crops: A Case Study in South East Spain". *Geoderma* 133(1): 43–56.

Vila Subirós, J., Rodríguez-Carreras, R., Varga, D., Ribas, A., Úbeda, X., Asperó, F., Llausàs, A. and Outeiro, L. (2014) "Stakeholder Perceptions of Landscape Changes in the Mediterranean Mountains of the Northeastern Iberian Peninsula". *Land Degradation and Development* early view (doi: 10.1002/ldr.2337)

Winiwarter, V. (2006) "Soil Scientists in Ancient Rome". In B. Warkentin (ed.), *Footprints in the Soil: People and Ideas in Soil History*, Elsevier, Amsterdam, pp. 3–16.

Winiwarter, V. and Blum, W. E. (2006) "Souls and Soils: A Survey of Worldviews". In B. Warkentin (ed.), *Footprints in the Soil: People and Ideas in Soil History*, Elsevier, Amsterdam, pp. 107–122.

WRI (1994) *World Resources Institute*, Oxford University Press, New York.

3 The Czech water footprint in the European context

Tomáš Hák, Petra Nováková and Pavel Cudlín

Introduction

Water: natural resource and ecosystem

Water is a precondition for human, animal and plant life as well as an indispensable resource for the economy. Only very few organisms can exist almost without water (e.g. the kangaroo rat never drinks water in its lifetime and obtains water internally by oxidizing food fat). Besides its resource use, water also plays a fundamental role in the climate regulation cycle and provides other ecosystem services (MA 2005). Therefore, protection of water resources and water ecosystems is one of the cornerstones of environmental protection worldwide. There are already many signs that humanity is globally consuming more resources than the planet is able to produce and regenerate (Ewing et al. 2010; Galli et al. 2012). Today's global situation may be characterized as a water crisis: "But the crisis is not about having too little water to satisfy our needs. It is a crisis of managing water so badly that billions of people – and the environment – suffer badly" (WWC 2000).

Water from European and Czech perspective

Additionally, in Europe there has been and continues to be an uneven distribution of water resources and an uneven distribution of water-related natural hazards (for example water scarcity, drought and floods). The main aim of European and national water policies is to ensure that throughout Europe, a sufficient quantity of suitable quality water is available for people's needs and for the environment. However, the most recent pan-European environmental assessment says that Europe is far from meeting water policy objectives and enjoying healthy aquatic ecosystems (EEA 2015). Only about half of the surface water bodies reaches "good" ecological status (a target set by the EU's Water Framework Directive); Europeans abstract on average around 13 per cent of all renewable and accessible freshwater from natural water bodies, including groundwater. The water exploitation index (WEI) (a ratio of total freshwater abstraction to the total renewable resources) shows Cyprus (45 per cent) and Bulgaria (38 per cent) to have the highest WEI scores in Europe; also high values were reported from Italy, Spain, the former Yugoslav Republic of Macedonia and Malta. More detailed data show real severity of water scarcity in sub-national regions: while Spain's national WEI is approximately 34 per cent, the southern river basins of Andalusia and Segura have extremely high WEIs of 164 per cent and 127 per cent, respectively (EEA 2009). WEI above 20 per cent implies that a water resource is under stress and values above 40 per cent indicate severe water stress and clearly unsustainable use of the water resource.

Water is so far not a critical resource in the Czech Republic (CR). It is a landlocked country in Central Europe (area of 78,866 km² and 10.2 million inhabitants) situated at the watershed of three seas – the North, Baltic and Black Seas – which separates the country territory into three main catchment areas: the Elbe, the Oder and the Danube. The country's specifics are that no significant watercourse bears water to the country's territory from the neighbouring countries: the consequence of this is the absolute dependence of the Czech Republic's water resources on atmospheric rainfall. It is therefore essential to systematically develop and protect these resources. Besides water ecosystems protection it is also necessary to implement sustainable management policies and governance for water resources (MoA 2013).

Water monitoring and reporting

Water sources, abstraction, use, efficiency, wastewater . . . these are important elements of the water supply chain and the water management system. Related data and indicators on household, company, city, region, river basin, nation and/ or planet levels help provide a clearer picture of these issues. At a global level, the UN-Water serves as the United Nations inter-agency coordination mechanism for all freshwater related issues (UN Water undated). It is supported by the UN Food and Agriculture Organization's (FAO) Global Information System on Water and Agriculture, AQUASTAT (www.fao.org/nr/water/aquastat), designed to improve the comparability of water information at regional and global levels.

In Europe, the key water policies aiming at ensuring good quality water of sufficient quantity are the Directive for "Community action in the field of water policy" (European Union Water Framework Directive) and the "Blueprint to safeguard Europe's water resources" (EC 2000, 2012). Supporting data and indicators are produced by the Water Information System for Europe (WISE), a partnership of four European institutions: Directorate-General Environment, European Environment Agency (hosts the Water Data Centre), Joint Research Centre (conducts water resources modelling) and Eurostat (collects and disseminates water statistics). Significant progress in national water reporting has been supported by the System of Integrated Environmental and Economic Accounts – SEEA (UN 2014). SEEA contributes to better understanding the interactions between economy and environment by putting detailed description of stocks and changes in stocks of environmental assets into satellite accounts. A water satellite account complements the economic metrics with statistics in physical terms.

Accounting goes far beyond standard statistics and considers also flows such as evaporation from lakes and artificial reservoirs and flows between water bodies. Pilot projects were carried out in 14 EU countries jointly by Eurostat, OECD and the UN Economic Commission for Europe (Eurostat 2002). A structured framework allows the water data to be linked to socioeconomic data as well as calculating aggregated indicators as Gross Water Input, Net Domestic Water Use and Final Water Use. Firms provide obligatory data on water abstraction, discharges etc. to statistical and environmental agencies; however, in particular large corporations having a substantial share of responsibility for water on the planet may choose other suitable water reporting instruments such as the Global Reporting Initiative, Global Water Tool (by the World Business Council for Sustainable Development), UN Global Compact, Corporate Social Responsibility (e.g. WaterAid in the UK), CDP Water Disclosure, water accounting, etc.

Water footprint concept

In 1994, Allan introduced the concept of virtual water to understand how arid countries in the Middle East can deal with the water shortages and feed their people (Allan 1994). Building on this concept, Hoekstra developed the water footprint (WF) in 2005 as a metric to measure the amount of water consumed to produce goods and services along the full supply chain (Hoekstra and Hung 2005). WF is an indicator of freshwater use that looks not only at the direct water use of a consumer or producer, but also at the indirect water use (i.e. the freshwater consumption and pollution "behind" products being consumed or produced); it is equal to the sum of the water footprints of all products consumed by the consumer or of all (non-water) inputs used by the producer. WF can be expressed for an agricultural commodity, a product, a service, a specific process step, or for an entire supply chain. Spatially, WF may be analysed within a delineated area such as a municipality, province, nation, catchment or river basin (Hoekstra et al. 2011). National WF accounts extend restricted statistics on water withdrawals within their own territory by including data on rainwater use and volumes of water use for waste assimilation and by adding data on water use in other countries for producing imported products, as well as data on water use within the country for making export products (Hoekstra and Mekonnen 2011). It has three components – blue, green and grey water (consumption of surface and groundwater water resources, rainwater insofar as it does not become run-off, and freshwater required to assimilate pollutants; Hoekstra et al. 2009). WF calculations follow a common methodology by Water Footprint Network (Hoekstra et al. 2011). In order to secure full standardization (underlying data, terminology, reporting formats etc.), the International Organization for Standardization developed standard ISO 14046: Environmental management–Water footprint–Principles, requirements and guidelines (ISO 2014). It is anticipated that ISO 14046 will support water disclosure by conducting and reporting WF as a stand-alone assessment, or as a part of more comprehensive environmental assessment.

Water footprint applications

In spite of political demand, methodologies developed and data mostly available (in industrial countries), most nations, regions and companies have still based their water policies and plans on conventional data and indicators on water consumption, water intensity and water pollution. However, "hidden" water use behind products or services is instrumental in understanding the real water consumption and the global character of fresh water sources. WF provides deep insight into the water use in an entire supply chain and by linking producers, suppliers and retailers explicitly highlights the responsibility of individual actors. Numerous examples of the WF use at various scales, by various actors for various purposes are in place. Enterprises and governments, but also cities, various organizations as schools have used WF in their strategies, policies, and in reporting on their own environmental performance. Governmental agencies, civic society organizations, academia etc. have used WF as a relevant measure in state-of-the-environment or sustainability analyses and reports.

International organizations

Supranational, multilateral, and other organizations of both governmental and non-governmental character are interested in stimulating resource efficiency and sustainable use of natural resources. They set global goals, develop and evaluate policies etc. (Witmer and Cleij 2012).

The United Nations (UN) promotes efforts to fulfil international commitments made on water-related issues by 2015 by declaring the International Decade "Water for Life". WF has received increasing attention in the UN's World Water Development Reports (see WWDR 2012, 2014) and it is anticipated that the WF indicator will likely be published regularly between 2020 and 2030. "UN's Operational Strategy for Freshwater" envisages a further development of harmonized methodologies accounting for water use; for example, footprint techniques (UNEP 2012).

The FAO has developed a global water information system AQUASTAT and a decision support tool CROPWAT (used, inter alia, for WF calculation). FAO has consistently promoted WF (see FAO 2012); besides providing input data, it has also analysed complex water-related issues such as a "food wastage footprint" (FAO 2013).

The Organisation for Economic Co-operation and Development (OECD) has not used WF in its reports yet. In its opinion WF analyses cannot set optimal policy alternatives as they do not take into account the opportunity cost of water in production, environmental impacts of water use etc. (OECD 2010, 2012).

Governments engaged with water footprint accounting

For governments, the WF indicator is useful as a tool in developing new strategies of national/regional sustainable water allocations and protection of freshwater resources (Hoekstra et al. 2011). In Europe, Spain was the first country to adopt the WF in national policy making. In 2008, the Spanish Water Directorate General approved a regulation on WF analysis of various socio-economic sectors as a technical criterion for the development of the River Basin Management Plans (Aldaya et al. 2010). Meanwhile the German Statistical Office conducted a project on water consumption from the perspective of consumers. It calculated and officially published the WF of agriculture production including imports and exports (DStatis 2012). The Swedish Environmental Agency also incorporates the WF into management and reporting. Its website informs citizens on their excessive consumption of water resources: every Swede uses roughly 5,500 litres (WF per capita) a day compared with the global average of 4,000 litres (SEPA 2010). The Netherlands Environmental Assessment Agency has examined suitability of the WF for national policy-making and recommended that this very useful concept be still developed and tested (Witmer and Cleij 2012).

The EU is aware of its excessive use of ecological assets that are locally available (EC 2005). Therefore, the European Commission (EC) collects data and indicators on availability of natural resources, their location, efficiency of use, waste generation and recycling rates etc. The WF is included there as an optional indicator for global resource demand (Witmer and Cleij 2012). In 2010, the EC funded the "One Planet Economy Network: Europe" project to compile evidence and practical tools for assessing policy interventions to transform resource-efficient Europe by 2050. The project drew on ecological, carbon and water footprints to define a "footprint family" of indicators tracking human pressure on the planet (Galli et al. 2012).

Corporate water footprint accounting

Besides governments, some producers have also already included the WF in their annual reports. In particular food and beverage corporations became increasingly aware of their water dependence and the water-related risk facing their companies (Franke and Mathews 2013). Despite immense corporations' responsibility for water, only one Czech company has been identified as using the WF accounting: the cheese producer Orrero uses it for its key product – the Gran Moravia cheese. Gran Moravia is said to be the first cheese producer worldwide calculating the WF of the entire production chain (Orrero undated). The low WF penetration in the Czech business environment corresponds with KPMG's survey that 60 per cent of major businesses across the globe have not established a reporting policy including water scarcity and just a few of the world's largest 250 companies have employed the WF (KPMG 2012). Many Czech companies report various water statistics and indicators (mostly focused on water savings and pollution) (Business for Society 2014); only in the case of multinationals (such as Unilever, SABMiller, Nestle, Heineken and Dow Chemical) are the Czech branch facilities included in the WF accounting.

Interesting results are reported by SABMiller on a beer value chain in its major production facility in the Czech Republic, Plzeňský Prazdroj, comprising three breweries, two malting plants and 13 sales and distribution centres

(SABMiller and WWF 2009). Beer is a very important Czech export commodity – total beer exports accounted for 3.65 million hectolitres in 2014 (Czech Association of Breweries and Malting 2015). SABMiller selected the Prazdroj brewery for this assessment because the brewery significantly contributes to the company's European volume of beer produced (20%), its water resources are classified as being under stress and it is home to the iconic brand Pilsner Urquell. The WF (including all three WFs) was 39 billion litres (i.e. 46 litres of water per 1 litre of beer in 2008). The most significant component of the WF was crop cultivation: the local crops accounted for about 70 per cent and the imported crops accounted for 24 per cent of the overall WF. The net green water component formed by far the largest element, over 90 per cent of the final volume, since most of the crops were reliant on rainwater. Blue water accounted for only 6 per cent of the overall WF; it is mostly related to water consumed during the brewing and bottling process while grey water only accounted for 2 per cent and it was associated with the crop production phase (SABMiller and WWF 2009). The water intensity of the Czech beer production was then compared with beer produced in South Africa (since many regions there were classified as water scarce). The results are comparable in terms of the percentage split of the total WF across four main phases in the value chain (crop cultivation, crop processing, brewing and bottling, and waste disposal). Both WFs, however, differ significantly in totals: the WF of the South African beer was 191 litres per one litre of beer which was four times higher than the WF of the Czech beer. Different climatological profiles (evapo-transpiration), reliance on irrigated crops and the larger proportion of imported highly water-intensive agricultural imports to South African business were among the main reasons for this difference (Figure 3.1; SABMiller and WWF 2009).

Water footprint research in the Czech Republic

Despite a long tradition of water monitoring and the existence of water research institutes, the WF-related research (aimed both at methodological aspects of the WF and at the WF applications) is underdeveloped in the CR. It is centred mostly at universities. This chapter proceeds by offering a few examples of recent studies and projects.

Selected case studies

Agricultural case study

The first study we consider is a pioneering study calculating the WF of selected agricultural crops in the CR – wheat, maize, sugar beet and tomatoes. This study followed the common methodological framework (Hoekstra et al. 2009) and it used the CROPWAT 8.0 model for calculation of crop evapotranspiration. Results showed that wheat had the largest WF (867 m³/t), then maize with the WF of 565 m³/t, tomatoes with 178 m³/t and lastly sugar with 106 m³/t. Wheat

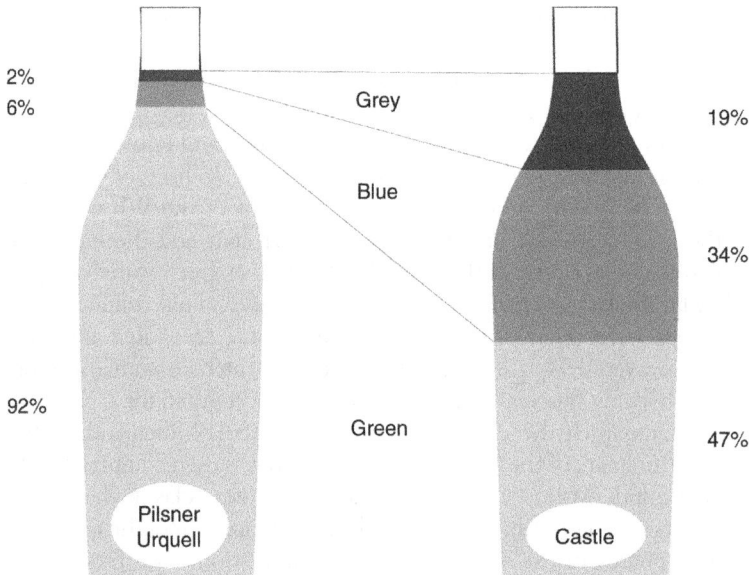

Figure 3.1 Split between net green, blue and grey water for SAB Ltd. in the Czech Republic
         (Pilsner Urquell brand) and in South Africa (Castle Lager brand), 2007

Source: data from SABMiller and WWF (2009).

and maize production showed good potential to reduce their grey water component (the grey WF of wheat in the CR more than doubled a global average;
Landová 2011).

Biofuels case study

Another study assessed biofuels in terms of their water demands. First, blue,
green and total WFs of agricultural crops used for the first generation biofuels
production in the CR were calculated. The particular crops were oilseed rape
(production of biodiesel) and sugar beet, Indian corn and winter wheat (bioethanol). Their total water footprints were ranked: sugar beet (102 m³/t), Indian corn
(387 m³/t), winter wheat (737 m³/t) and oilseed rape (1,492 m³/t). The WF indicator was then related to the appropriate low heating values – standard indicators
of energy concentration in the fuel (in GJ/t) – and the final indicator "WF per
low heating value units" was calculated. The results showed the same ranking
with again remarkable differences among the crops: Sugar beet (6 m³/GJ), Indian
corn (18 m³/GJ), winter wheat (43 m³/GJ) and oilseed rape (56 m³/GJ). One
of the conclusions was that production of biodiesel in the CR puts considerably higher demands on water resources than the production of bioethanol.
Nevertheless, oilseed rape is extensively cultivated and it is the main crop for
biofuels production. In 2012, Indian corn appeared for the first time among the

biofuels crops and mostly replaced winter wheat. Since the water/energy intensity of Indian corn is less than a half of winter wheat, this emerging trend is favourable (Žlábková 2013).

Consumption of agricultural commodities case study

A study in the field of regional and political geography investigated consumption of selected agricultural commodities in the CR and its impacts on water resources in producing countries. Food consumption causes a pressure on water resources in different countries through international trade. It is assumed that (i) there is a direct link between the demand for water-intensive products (notably agricultural ones) in importing countries and the water used for production of export goods in other countries, and (ii) water used for producing export goods may significantly affect the local water systems. The study applied a geographical approach focusing on national consumption of two imported commodities – avocado and tomato. Their WFs were adopted from global databases (Hoekstra et al. 2011) and adjusted to take into account the state of water resources in exporting countries (subtracting green WF, applying water stress characterization factors etc.; Ridoutt and Pfister 2010). The basic volumetric WF of tomatoes pointed to Slovakia (265 m³/t) and Poland (173 m³/t), but after revision it appeared that tomato production in these countries did not cause major impacts on water resources (Slovakia: 8 m³/t; Poland: 0.2 m³/t). The revised WFs revealed, on the other hand, the exporting countries with the highest pressure on water resources by production of tomatoes: Turkey (60 m³/t), Israel (59 m³/t) and Morocco (53 m³/t). Results were strongly affected by a water stress level and the use of blue and grey water in exporting countries. As an example, the Czech import of tomatoes from Spain and the Netherlands in 2011 was similar in quantity (25 thousand and 28 thousand tonnes, respectively). However, cultivation of that amount of exported tomatoes in Spain had 50 times higher potential to contribute to water scarcity than cultivation of about the same amount of tomatoes in the Netherlands (the revised WF of Spanish tomatoes exported to the CR was 861 thousand m³ while only 17 thousand m³ in the Netherlands). The results showed that the CR, as a country with a very low level of water stress, externalizes (due to consumption of imported crops) a considerable part of its impact on water resources to the countries with serious problems of water scarcity (Nováková 2014).

Additional case studies

There are also several ongoing research projects in the CR. One is a current project which focuses on the impact of Czech households' food wastage on water resources. This topic links two global problems: food wastage and an increasing pressure on water resources. Other research seeks to improve the grey WF methodology and assess the impacts of bio-gas stations expansion in the CR (about 60% are agricultural bio-gas stations processing maize, therefore production of

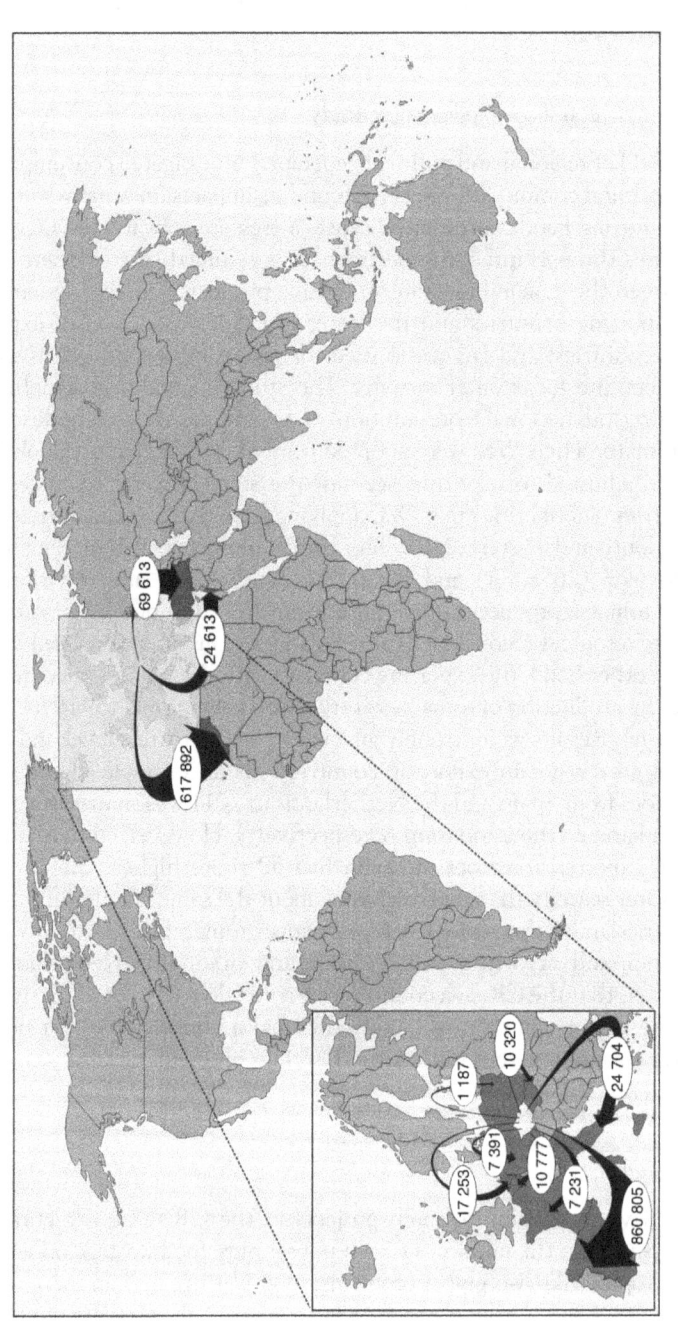

Figure 3.2 The revised external water footprint of the Czech consumption of tomatoes (m³), 2011

Source: Nováková (2014).

Note: The external water footprint of national consumption is defined as the volume of water resources used in other nations to produce goods and services consumed by the population in the nation considered.

bio-ethanol has shifted from processing sugar beet to maize). Today's grey WF studies focus on nitrogen and phosphorus only but the complexity of their cycles highly influences the actual impact of fertilizers. The study will, therefore, calculate the grey WF with the data on herbicides used for energy crops cultivation and thus provide a more comprehensive assessment of biogas production. Also, there is Czech participation in the EU-funded project AGRIWAT (COST 2011) testing various model approaches towards the WF, in particular the green component since it constitutes a significant share of the total WF yet uncertainty and regional differences are not well known. The WFs of all major Czech field crops will be, therefore, estimated by three different crop models at three climatically different sites. Finally, a new research proposal plans to test applicability of life cycle assessment for the WF analysis. Because of different traditions and purposes, the two methodologies have developed slightly different solutions for impact assessment; the proposed SimaPro software allows for several methods of spatially differentiated water scarcity impact assessment.

Conclusions and recommendations

This chapter has only touched on some known activities in which the water footprint concept occupies a central place. Space does not permit a more detailed coverage. The WF accounting is a new analytical and reporting tool in the CR; however, as demonstrated in this chapter, it has already put down soft roots. It is obvious that the main role in the WF concept introduction is currently within the scientific community. Universities and Academies of Science have been developing various aspects of the WF methodology, conducting pilot calculations for country-specific products and teaching WF in sustainability courses for example. The key challenge for the near future is to shift the "footprint thinking" on water management from academia to public administration and include it in policy making. Based on the still developing Czech experience – in terms of both the number of WF-related initiatives and the period of experimentation with the WF concept – some tentative conclusions may be drawn:

- *Assessment tool.* The WF provides an innovative tool useful for formulating national water plans. It allows governments to consider the global dimension of water demand patterns. Knowledge of the dependency on water resources elsewhere is relevant for a national government for many reasons – global responsibility, food security, economic benefits etc. The Czech Ministry of Agriculture has recently commissioned a three year research project aimed at adapting the WF methodology according to the new ISO norm, development of a database of local and national data needed for the WF calculations and demonstration of some pilot WF calculations in various economic sectors.
- *Corporate reporting.* For many companies, freshwater is a fundamental substance for their operations, while effluents may lead to pollution and degradation of local ecosystems. The WF offers a tool for expanding the non-financial reporting on the sustainability impacts of business on society.

Smaller companies may benefit from development of standards for environmental management systems (ISO, EMAS, etc.) and assistance of organizations promoting Corporate Social Responsibility schemes.

- *Statistical agenda.* Well-informed national policy needs timely and reliable data and indicators. The System of Integrated Environmental and Economic Accounting (SEEA) comprises water accounts that will become obligatory for national statistical offices. Until then, statisticians might check possibilities to adjust SEEA-Water to "national water footprint accounts" (by including data on green water use and water use for waste assimilation, by adding data on water use for producing imported products, for example).
- *Education.* The WF is an effective tool to improve water literacy. There is much international experience on the use of the WF concept in education and many materials on human footprints for both teachers and students are available. Czech schools may benefit from the inclusion of the WF in their educational programs as part of environmental or sustainability education and educational experts should develop relevant didactical tools.
- *Publicity.* The WF concept is an analogue of the older ecological footprint (EF) concept. EF has become one of the most cited alternatives to the gross domestic product indicators. Solid methodological work, development of huge amount of applications (for all levels, sectors, products, etc.) and the EF and WF employment in respected analyses (e.g. Living Planet Report by WWF) are behind that. The WF promoters in the CR may follow suit (so far, even a simple WF calculator in Czech is missing).

Acknowledgement

This work has been supported by the project of the Ministry of Education, Youth and Sports of the CR within the National Sustainability Program I (NPU I; grant number LO1415 and grant number CZ.1.07/2.3.00/20.0248) and by the project "Various Aspects of Anthropogenic Material Flows in the Czech Republic" (no. P402/12/2116) funded by the Czech Science Foundation.

References

Aldaya, M. M., Garrido, A., Llamas, R. M., Varela-Ortega, C., Novo, P. and Rodríguez-Casado, R. (2010) "Water Footprint and Virtual Water Trade in Spain". Available at www.waterfootprint.org/media/downloads/Aldaya-el-al-2010-WaterFootprintSpain.pdf (accessed 10 September 2015).

Allan, J. A. (1994) "Overall Perspectives on Countries and Regions". In P. Rogers and P. Lydon (eds), *Water in the Arab World: Perspectives and Prognoses*, Harvard University Press, Cambridge, MA, pp. 65–100.

Business for Society (2014) *Companies & Water: Sustainable Use of Water by Companies.* Business for Society, Prague.

COST (2011) "Assessment of European Agriculture Water Use and Trade under Climate Change". Available at www.cost.eu/COST_Actions/essem/ES1106 (accessed 28 September 2015).

Czech Association of Breweries and Malting (2015) "Report on the State of Czech Brewing and Malting for 2014" (in Czech). Available at www.ceske-pivo.cz/sites/default/files/dokumenty_tz/2015_zprava_hospodarske_vysledky_oboru_za_2014_final_20150410.pdf (accessed 21 September 2015).

DStatis (2012) *Water Footprint of Food Products in Germany, 2000–2010*. Federal Statistical Office of Germany, Wiesbaden.

EC (2000) "Directive 2000/60/EC of the European Parliament and of the Council of 23 October 2000 Establishing a Framework for Community Action in the Field of Water Policy". *Official Journal of the European Communities* 327: 1–72.

EC (2005) *Thematic Strategy on the Sustainable Use of Natural Resources*. COM (2005) 670 final, European Commission, Brussels.

EC (2012) *Blueprint to Safeguard Europe's Water Resources*. COM/2012/0673, European Commission, Brussels.

EEA (2009) *Water Resources across Europe – Confronting Water Scarcity and Drought*. EEA Report No 2/2009, European Environment Agency, Copenhagen, Denmark.

EEA (2012) *Water Resources in Europe in the Context of Vulnerability*. EEA Report No 11/2012, European Environment Agency, Copenhagen, Denmark.

EEA (2015) *The European Environment – State and Outlook 2015: Synthesis Report*. European Environment Agency, Copenhagen, Denmark.

Eurostat (2002) *Water Accounts – Results of Pilot Studies*. Luxembourg: Office for Official Publications of the European Communities.

Ewing, B., Moore, D., Goldfinger, S., Oursler, A., Reed, A. and Wackernagel, M. (2010) *The Ecological Footprint Atlas 2010*. Global Footprint Network, Oakland, CA.

FAO (2012) *Coping with Water Scarcity: An Action Framework for Agriculture and Food Security*. Food and Agriculture Organization of the United Nations, Rome.

FAO (2013) *Food Wastage Footprint. Impacts on Natural Resources*. Summary report, Food and Agriculture Organization of the United Nations, Rome.

Franke, N. and Mathews, R. (2013) "Grey Water Footprint Indicator of Water Pollution in the Protection of Organic vs Conventional Cotton in India". Available at www.waterfootprint.org/media/downloads/Grey_WF_Phase_II_Final_Report_Formatted_06.08.2013.pdf (accessed 18 September 2015).

Galli, A., Weinzettel, J., Cranston, G. and Ercin, E. (2012) "A Footprint Family Extended MRIO Model to Support Europe's Transition to a One Planet Economy". *Science of The Total Environment* 461–462: 813–818.

Hoekstra, A. Y. and Hung, P. Q. (2005) "Globalisation of Water Resources: International Virtual Water Flows in Relation to Crop Trade". *Global Environmental Change* 15(1): 45–56.

Hoekstra, A. Y and Mekonnen M. M. (2011) "Water Footprint of Humanity". *Proceedings of the National Academy of Sciences of the USA* 109(9): 3232–3237.

Hoekstra, A. Y., Chapagain, A. K., Aldaya, M. M. and Mekonnen, M. M. (2009) *Water Footprint Manual: State of the Art 2009*. Water Footprint Network, Enschede, the Netherlands.

Hoekstra, A. Y., Chapagain, A. K., Aldaya, M. M. and Mekonnen, M. M. (2011) *The Water Footprint Assessment Manual: Setting the Global Standard*. Earthscan, London.

ISO (2014) *Environmental Management – Water Footprint – Principles, Requirements and Guidelines*. ISO 14046:2014, Edition 1, International Organization for Standardization, Geneva, p. 33.

KPMG (2012) "Sustainable Insight, Water Scarcity: A Dive into Global Reporting Trends". Available at www.kpmg.com/ES/es/ActualidadyNovedades/ArticulosyPublicaciones/

Documents/sustainable-insight-water-scarcity-global-reporting-trends.pdf (accessed 20 September 2015).

Landová, L. (2011) "Water Footprint. Calculation of Water Footprint of Wheat, Maize, Sugar Beet and Tomatoes in the Czech Republic". Master's thesis, Charles University, Prague, Czech Republic (in Czech).

MA (2005) *Ecosystems and Human Well-Being: Wetlands and Water Synthesis.* Millennium Ecosystem Assessment, World Resources Institute, Washington, DC.

MoA (2013) *Report on Water Management: Blue Report,* Ministry of Agriculture, Prague (in Czech).

Nováková, P. (2014) "Consumption of Selected Agricultural Commodities in Czechia and its Impacts on Water Resources in Producing Countries". Master's thesis, Charles University in Prague, Czech Republic (in Czech).

OECD (2010) *An Economic Analysis of the Virtual Water Concept in Relation to the Agri-food Sector.* Organisation for Economic Co-operation and Development, Paris.

OECD (2012) *Environmental Outlook to 2050, the Consequences of Inaction.* Organisation for Economic Co-operation and Development, Paris.

Orrero (undated) "About Company: History". Available at www.orrero.cz/o-firme/historie (accessed 18 September 2015) (in Czech).

Ridoutt, B. G. and Pfister, S. (2010) "A Revised Approach to Water Foot Printing to Make Transparent Their Pacts of Consumption and Production on Global Freshwater Scarcity". *Global Environmental Change* 20: 113–120.

SABMiller and WWF (2009) "Water Footprinting, Identifying & Addressing Water Risks in the Value Chain". Available at www.sab.co.za/sablimited/action/media/download File?media_fileid=918 (accessed 21 September 2015).

SEPA (2010) "Water – Water Use". Available at www.miljomal.se/Environmental-Objectives-Portal/Undre-meny/Publications-and-presentations/Swedens/Swedish-consumption/Water—water-use (accessed 12 September 2015).

UN (undated). "The United Nations Inter-Agency Mechanism on All Related Issues, including Sanitation". Available at www.unwater.org (accessed 15 October 2015).

UN (2014) *System of Integrated Environmental and Economic Accounting 2012.* Central Framework, United Nations, New York.

UNEP (2012) *Healthy Waters for Sustainable Development.* Operational Strategy for Freshwater (2012–2016), United Nations Environment Programme, Nairobi.

Witmer, M. C. H. and Cleij, P. (2012) *Water Footprint: Useful for Sustainability Policies?* PBL publication no 500007001, PBL Netherlands Environmental Assessment Agency, The Hague.

WWC (2000) *World Water Vision Report, Making Water Everybody's Business.* Earthscan, London.

WWDR (2012) *Managing Water under Uncertainty and Risk.* United Nations World Water Development Report 4, UNESCO, Paris.

WWDR (2014) *Water and Energy.* United Nations World Water Development Report 2014, vol 1, UNESCO, Paris.

Žlábková, J. (2013) "Evaluation of Environmental Burden in the Czech Republic by Using Indicator Water Footprint". Master's thesis, Charles University in Prague, Czech Republic (in Czech).

4 Food wars

Food, intangible cultural heritage and international trade

Valentina Vadi

Who controls the food controls the people.

(Henry Kissinger)

Introduction

Food preparation and even certain types of food can express deeply held cultural practices and values and constitute a type of intangible cultural heritage (ICH). The United Nations Educational Scientific and Cultural Organization (UNESCO)[1] – the specialized agency of the United Nations which aims to contribute to peace and security by promoting international collaboration through education, science, and culture – has recently acknowledged the link between food and intangible heritage, listing a range of diverse diets in its list of intangible cultural heritage. Despite formal recognition at the international level, food preparation and associated cultural practices remain vulnerable to irreparable loss.

Economic globalization risks jeopardizing the protection of food as intangible cultural heritage and associated cultural practices. While economic globalization has spurred a more intense dialogue and interaction among nations – potentially promoting cultural diversity and even access to food – it can also jeopardize cultural traditions. The expansion of foreign direct investments in agriculture; the increasing food processing and marketing through multinational corporations; the development of biotechnology and genetic engineering, and a growing international trade in food facilitated by the reduction of trade barriers can affect local cultural practices including those associated with food preparation. Moreover, trade in cultural products can lead to cultural commodification, cultural homogenization and even cultural hegemony. Dominant cultures – also reflecting the global distribution of power – tend to predominate in the global markets (Friedman 1999: 8).

Food-related controversies have arisen during trade negotiations and have been brought before the World Trade Organization (WTO)[2] Dispute Settlement Mechanism (DSM) where states have claimed that regulatory measures affecting their economic interests are in breach of the relevant international trade law provisions. These disputes highlight the emergence of a clash of cultures between global economic governance and the protection of local cultural practices concerning food. Has an international economic culture emerged that emphasizes productivity and economic development at the expense of cultural wealth? Does the existing legal framework adequately protect cultural diversity vis-à-vis economic globalization? How have WTO dispute settlement bodies dealt with food wars? Should they take into account the cultural concerns of the affected communities? Or are cultural concerns merely to be conceived as a disguised form of protectionism? What steps can be taken to ensure mutual supportiveness between different legal regimes protecting food on the one hand, and trade on the other?

This chapter addresses these questions, illustrating the specific "clash of cultures" between economic globalization and domestic food policies and focusing on the interplay between traders' rights and the protection of "food heritage". The chapter proceeds as follows. First, it briefly examines the conceptualization of food as a form of intangible cultural heritage. Second, it discusses and critically assesses the main features, promises and pitfalls of the UNESCO Convention on the Safeguarding of Intangible Cultural Heritage. Third, it examines recent food wars. Fourth, the question as to whether trade "courts" are taking intangible cultural heritage into account will be addressed. Finally, some conclusions will be drawn. By effectively analysing food through the perspectives of cultural heritage law and international trade law this chapter productively brings two overlapping but frequently separate theoretical frameworks into conversation.

Food as intangible cultural heritage

Parodoxical as it may seem, something as perishable as food constitutes a living legacy of the past. Food "constitutes a basic element of the culture of a people or of a community" (Maffei 2012: 83). The transmission of food practices from one generation to the other contributes to constructing identity and memory of both individuals and communities. Food – its cultivation, preparation and communal consumption – can be a form of intangible cultural heritage (Di Giovine and Brulotte 2014).

The protection of intangible cultural heritage – meant as the wealth of cultural traditions and practices passed on from one generation to another – has long been neglected by international law. Early expressions of such protection were incorporated in minority protection treaty provisions and – albeit sparingly – surfaced in the jurisprudence of the Permanent Court of International Justice (Pentassuglia 2009). In the aftermath of World War II, aspects of intangible heritage have been governed and/or touched upon by a number of international law instruments. For example, human rights instruments require states "to take into account, as far as possible, cultural values attached to, *inter alia*, food and food consumption" (CESCR 2009: para. 16(e)). Cases adjudicated before the International Court of Justice (ICJ) have touched upon the cultural practices and subsistence harvest of local communities.[3]

Only in the past decades, however, have specific instruments and programmes been devoted to the protection of food as a form of intangible heritage. In 1989 UNESCO issued a Recommendation on the Safeguarding of Traditional Culture and Folklore,[4] listing policies that countries could implement to preserve their intangible cultural heritage. Yet, the recommendation was a "soft" international instrument and had little impact due to its "top-down" and "state oriented" approach (Kurin 2004: 68). Very few states took action in this regard. In 2001 the launch of the Masterpieces of the Oral and Intangible Heritage programme – which established three rounds of proclamations of given traditions as representative "Masterpieces" to raise awareness about intangible heritage – was very well received and paved the way for the elaboration of the CSICH (Aikawa-Faure 2009). Although no specific food preparations were included among the Masterpieces of Intangible Heritage, food figured prominently in the performance of some traditional cultural practices.[5]

The promises and pitfalls of the CSICH

The 2003 UNESCO Convention on the Safeguarding of the Intangible Cultural Heritage (CSICH)[6] constitutes the principal instrument governing intangible cultural heritage at the international level. It defines intangible cultural heritage as "the practices, representations, expressions, knowledge, skills – as well as the instruments, objects, artefacts and cultural spaces associated therewith – that communities, groups and, in some cases, individuals recognize as part of their cultural heritage".[7] "[T]ransmitted from generation to generation", intangible cultural heritage provides groups and communities with "a sense of identity and continuity, thus promoting respect for cultural diversity and human creativity".[8] The convention considers solely "such intangible cultural heritage as is compatible with existing international human rights instruments, as well as with the requirements of mutual respect among communities, groups and individuals, and of sustainable development".[9] Although the convention does not expressly refer to food, its definition of intangible heritage is broad enough to encompass it.

The CSICH requires state parties to draw inventories of their intangible cultural heritage and to collaborate with local communities on various appropriate means of "safeguarding" those traditions (Blake 2009). The UNESCO Committee established under the CSICH oversees two international lists:

1 the "Representative List of the Intangible Cultural Heritage of Humanity", which includes, *inter alia*, the items already designated as Masterpieces of Oral and Intangible Heritage by UNESCO and is comparable with the World Heritage List;[10] and
2 the "List of Intangible Cultural Heritage in Need of Urgent Safeguarding".[11]

Food practices figure in both lists. Among the items included in the Representative List are such food practices as the "Gastronomic Meal of the French" (France),[12] the "Traditional Mexican Cuisine" (Mexico),[13] the "Mediterranean Diet" (Spain, Greece, Italy and Morocco)[14] and "Washoku", traditional dietary cultures of the Japanese.[15] Food also figures prominently as an element of other protected items, showing the deep connection between food and civilizations. In this respect, the Paach Ceremony, a corn-veneration ritual celebrated in Guatemala, appears among the items included in the list of endangered intangible heritage.[16]

The Convention aims to remedy two structural imbalances within international law. First, the CSICH aims to remedy a gap in global cultural governance which has traditionally favoured the protection of tangible heritage such as monuments and sites over the protection of intangible heritage. Within UNESCO, the 1972 World Heritage Convention (WHC)[17] has focused on the conservation of monuments and sites. Only recently has the WHC expanded its purview to comprehend elements of intangible heritage, and it now protects mixed forms of cultural heritage – such as cultural landscapes – which include both tangible and intangible features (Vadi 2012). Yet, the World Heritage List remains imbalanced, including more cultural than natural sites (Brown 2005).

Second, the CSICH aims to counterbalance the regulation of cultural resources by international trade law. In fact, globalization and trade in cultural products have the potential of promoting cultural exchange, but can also jeopardize local and regional cultural practices. The diffusion of a global mass culture has raised the fundamental question of whether "valuable traditions, practices, and forms of knowledge rooted in diverse societies would survive the next generation" (Kurin 2004: 68). In this regard, the CSICH would counter the perceived commodification of culture (i.e. its reduction to a good or merchandise to be bartered or traded). Rather, the CSICH proposes an alternative view perceiving oral traditions and expressions, knowledge and practices concerning food as forms of intangible cultural heritage.

Despite its achievements, the CSICH risks both "substantive overreach" (Broude 2015: 3) and procedural underachievement. Substantively, the definition of intangible cultural heritage is too broad and descriptive (Scovazzi 2008) risking an unwelcome "politicization of culture" (Broude 2015: 3). Procedurally, the listing approach "convert[s] selected aspects of localized . . . heritage into . . . the heritage of humanity" (Kirshenblatt-Gimblett 2004: 57). Yet, inventories do not do justice to intangible cultural heritage meant as a living phenomenon. Rather, the listing approach risks "misconceiving culture as atomistic items" and freezing this living heritage (Lubina 2009: 50). Furthermore, while inventories and lists may recognize various traditions, they will hardly save such traditions (Kurin 2004: 74).

Problems of conflict and/or coordination between the CSICH and other international norms – whether customary or conventional – have shown additional procedural shortcomings of the CSICH. The interaction between the CSICH and other international law regimes relates to the general question as to whether international law is a fragmented system or not (ILC 2006) and raises the specific question as to whether the protection of intangible cultural heritage should be taken into account in the implementation of other international law regimes. In fact, while the CSICH can (and has) overlap(ped) with international trade law, it does not provide binding dispute settlement mechanisms. Not surprisingly, food-related disputes have been brought before the WTO DSM. The next section will explore some food wars where the protection of intangible cultural heritage clashes with the promotion of free trade.

Food wars

Food-related trade conflicts and disputes do not have a typical form and may relate to different areas of international trade law, ranging from international intellectual property law to agricultural law. For instance, although not all intellectual property constitutes intangible heritage, and vice versa (Gervais 2003–2004: 633), there is significant interaction between the CSICH and the Agreement on Trade Related Aspects of Intellectual Property Rights (TRIPS Agreement).[18] In this regard, the CSICH provides for a compatibility clause, stating that none of its provisions can be interpreted as affecting "the rights and obligations of States Parties deriving from any international instrument relating to intellectual property rights".[19] Yet, the CSICH lacks specific rules concerning ownership and control over intangible cultural heritage (Kearney 2009: 216). In turn, the TRIPS Agreement fails to recognize the non-economic interests associated with intangible assets.

The inadequacy of the TRIPS Agreement to safeguard intangible cultural heritage is particularly evident with regard to controversies and disputes raised by the patenting by multinational corporations of ethnic food traditionally consumed by local communities (Woods 2002–2003: 123). Without a sensible remodelling, intellectual property rights risk overprotecting individual economic interests, while ignoring the collective entitlements of the relevant cultural communities.

Another area of connection among food, cultural practices and intellectual property is that of geographical indications. Geographical indications (GIs) – such as Parmigiano Reggiano, Champagne, and Gouda – are signs used to identify products which come from these places and whose quality, reputation or other characteristic depends on these specific geographical locations. GIs can be a possible vector for protecting intangible cultural heritage (Gangjee 2015). By protecting regional food products which have acquired a strong reputation among consumers and have been produced using centuries-old manufacturing techniques, GIs can indirectly protect the intangible cultural heritage associated with the production of these goods. Three dimensions of culture are relevant to GIs:

1   the culture of producing a given type of food;
2   the culture of consuming certain foods; and
3   "the culture of identity in which a good is somehow representative of a group's cultural identity" (Broude 2015: 15).

However, there is a transatlantic divide over the protection of GIs. While the TRIPS Agreement provides the protection of GIs in order to avoid misleading the public and to prevent unfair competition,[20] and provides for some limited exceptions,[21] it does not provide a detailed regulation of such issues. While European states have protected certain foodstuffs originating from given geographical locations since the fifteenth century (Shimura 2010: 129), other countries consider them as obstacles to trade. For instance, when the European Commission considered a US request to drop a ban on import into Europe of American wines bearing the label *chateau*, a term used mainly on wines from Burgundy in eastern France, French wine producers contended that American competitors should not be allowed to sell *chateau*-type wine in the EU, as their production standards differ from the French ones. In France, wine labelled *chateau* is made entirely from grapes grown on a *terroir* – a specific area of land – supposedly giving it a unique character and flavour (Lauter 2012). According to French rules, only wine from grapes grown on the property and made into wine in facilities on the property can bear the *chateau* label (Cody 2012). According to French wine-makers, "[t]his is a guarantee of quality . . . a declaration to the buyer that [s]he is sharing in the heritage that gave rise to h[er] wine" (ibid.). Instead, American wines are made with a mixture of grapes purchased from different growers, as the American labelling system "traditionally highlights grape variety, rather than where the fruit was grown" (ibid.). Therefore, French producers claim that American producers should not be allowed to have *chateau* on the label.

The divergent approaches to geographical indications of the US and the EU have fostered intense conflicts at different venues. At the multilateral level, WTO members are debating the adoption of a multilateral register for wines and spirits. Some countries, including the EU, are pushing for a register with binding effects, while other countries, including the United States, are pushing for a non-binding system under which the WTO would simply be notified of the members' respective geographical indications. This divergence is also playing a central role in the negotiations of the Transatlantic Trade and Investment Partnership (TTIP) – the free trade agreement currently being negotiated between the European Union and the United States. While the European Union wants to stop American producers from being able to commercialize and label products with their protected names, the United States would favour free trade (Bonadio 2015). This lack of protection – the European negotiators argue – allows an unacceptable exploitation of European intangible heritage, and affects the economic interests of European producers. Conversely, the US negotiators contend that such names have become generic and cannot be monopolized by anyone. Moreover, EU-style legal protection would constitute a barrier to trade, allow monopolies and ultimately increase final prices for consumers. Finally, the EU system would be unfair because European immigrants have long produced such products, thus sharing the same intangible cultural heritage.

Not only does the intense debate over GIs have an evident economic component, but it is also characterized by cultural aspects as well. In fact, proponents of GIs conceive food as something more than a tradable commodity but as an artefact characterized by both visible features and intangible cultural qualities related to the traditional manufacturing processes and place of origin.

In other words, "as a forged painting and the original one may not differ at all materially, while still being quite different artworks, so a GI cannot be equated to its material constitution: some aspects of its making are key to its identity" (Borghini 2014: 1118).

The interaction between international economic law and the protection of food as intangible heritage includes other areas beyond the purview of intellectual property. For instance, the EU ban on the commercialization of seal products caused a cultural skirmish across the Atlantic.[22] As Europeans perceive the hunting of seals to be morally objectionable, the EU has banned the trade in seal products except those derived from hunts traditionally conducted by the Inuit and other indigenous communities for cultural and subsistence reasons.[23] Seals constitute the most important component of an Inuit diet (Searles 2002) and indigenous hunting practices constitute a form of intangible cultural heritage deemed essential to preserve indigenous way of life.

Canada and Norway brought claims against the EU before the WTO Dispute Settlement Body (DSB), contending that the EU seal regime was inconsistent with the European Union's obligations under the General Agreement on Tariffs and Trade 1994 (GATT 1994)[24] and under the Technical Barriers to Trade (TBT) Agreement.[25] Specifically, Canada and Norway argued, *inter alia*, that the indigenous communities condition (IC condition) violated the non-discrimination obligation under Article I:1 and III:4 of the GATT 1994. According to Canada and Norway, such conditions accord seal products from Canada and Norway treatment less favourable than that accorded to like seal products of domestic origin as well as those of other foreign origin, in particular from Greenland.[26] In fact, the majority of seals hunted in Canada and Norway would not qualify under the exception, while most, if not all, of Greenlandic seal products would satisfy the requirements under the IC exception.[27] The panel and the Appellate Body held, *inter alia*, that the exception provided for indigenous communities under the EU Seal Regime accorded more favourable treatment to seal products produced by indigenous communities than that accorded to like domestic and foreign products[28] in breach of Articles I:1 and III:4 of the GATT 1994. Little reference was made to the various instruments which protect indigenous cultural practices at the international law level,[29] including the 2007 United Nations Declaration on the Rights of Indigenous Peoples (UNDRIP).[30] Despite the reference to these instruments, however, the panel concluded that the design and application of the IC measure was not even-handed because the IC exception was available *de facto* to Greenland.[31]

Another sector in which cultural clash between free trade and food-related cultural practices takes place is that of agriculture. The WTO's Agreement on Agriculture[32] is based on the market liberalization model and efficiency criteria (Smith 2011: 159). It does "not allow farmers to maintain their current methods of production solely on cultural or environmental grounds, if those methods prevent the farmers from efficiently adjusting their production in line with market forces" (ibid.: 168). Rather, the WTO regards agriculture "as an economic sector like any other industrial sector" (ibid.: 159). In some countries, the liberalization of the market – opening the markets to highly subsidized agriculture – has meant that local farmers have to compete with heavily subsidized imports (Hauter 2007: 1071). Competition has driven down the price and forced these farmers out of business. This phenomenon has also involved products central to a country's culture (Smith 2009). For example, the influx of highly subsidized corn from the United States undermined the ability of Mexican farmers to grow corn, a crop that Maxican have cultivated for centuries (Smith 2011: 161).

Another area of cultural resistance, in which the clash between free trade and cultural attitudes is particularly evident, is that related to food safety. The WTO Sanitary and Phytosanitary Agreement[33] addresses the interest of member states in assuring that their citizens are being supplied with safe food encouraging member states to base sanitary and phytosanitary measures on internationally accepted scientific standards (Beghin 2014: 7). Problems have arisen with regard to the interpretation of scientific evidence. While the precautionary approach to risk management is a general principle of EU law, entailing that given products are prohibited until they are proven safe, on the other side of the Atlantic, it goes the other way around and products have to be proven unsafe to be banned. These different approaches to risk and food safety – based on different cultural understandings of food – have given rise to a number of disputes at the WTO, concerning hormones, genetically modified organisms (GMOs) and others. Trade experts tend to see safety regulations and cultural concerns as forms of protectionism and technical trade barriers rather than legitimate concerns.

Conclusions

Food is at the heart of civilizations, given its importance for human subsistence, well-being and human flourishing. Nutrition plays a cultural role in shaping a community's and a person's specific identity (Korthals 2008: 445) and, in certain cases, food preparation and consumption can constitute a form of intangible heritage. Since 2003, ICH has been ingrained in a dedicated international law instrument, the CSICH. Despite some achievements, the CSICH is characterized by substantive overreach and procedural underachievement. Not only does it fail to ensure adequate safeguarding of intangible heritage, as it constitutes a "*de facto* soft law instrument in formal hard law clothing" (Broude 2015: 3), but it also fails to provide a meaningful forum to address ICH-related trade disputes. Despite the recognition of food as an element of intangible cultural heritage, economic globalization risks jeopardizing cultural practices by altering food consumption patterns.

In general terms, food-related trade disputes are characterized by the need to balance two very different issues: the protection of cultural practices associated with food, and economic interests associated with trade. Many such controversies arise during trade negotiations or are brought before the WTO dispute settlement mechanism. The WTO is a legally binding and highly effective regime which demands states to promote and facilitate free trade. The WTO is based on a free-market paradigm and its rules are about trade (Smith 2011: 176). It is not interested in local communities, food, culture and farming techniques as such. These issues are considered as non-economic concerns and therefore remain at the margins of the regime (ibid.: 176). Not surprisingly, trade "courts" have paid very little attention to the cultural aspects of food-related disputes.

Notes

1 Constitution of the United Nations Educational, Scientific and Cultural Organization (UNESCO Constitution), 16 November 1945, in force 4 November 1946, 4 UNTS 275.
2 Agreement Establishing the World Trade Organization (Marrakesh Agreement), 15 April 1994, 1867 UNTS 154; 33 ILM 1144 (1994).
3 A couple of ICJ cases referred to local fishing customs or practices. One could contend that fishing practices constitute *economic* and *subsistence* activities rather than *cultural practices*. Therefore, such cases would not focus on intangible cultural heritage *per se*. Yet, human rights courts have acknowledged that hunting and fishing practices can have a cultural component, contributing to the identity of given local communities. In any case these cases show the willingness of the ICJ to take into account local practices of fishing communities. See e.g. *Pulp Mills on the River Uruguay (Argentina v. Uruguay)*, Judgment of 20 April 2010, ICJ Reports 2010, p. 14 para 171 (referring to "pre-existing uses of the river"); *Case Concerning Maritime Delimitation in the Area between Greenland and Jan Mayen (Denmark v. Norway)* Judgment of 14 June 1993, ICJ Reports 1993, p. 38, para 73 (referring to the local fishing practices of migratory stocks).
4 Recommendation on the Safeguarding of Traditional Culture and Folklore, 15 November 1989, available at http://portal.unesco.org/en/ev.php-URL_ID=13141& URL_DO=DO_TOPIC&URL_SECTION=201.html (accessed on 15 May 2015).
5 For instance, Mexico proclaimed the Indigenous Day of the Dead (*el Día de los Muertos*) as a masterpiece of its intangible heritage. The festivity takes place each year at the end of October to the beginning of November to commemorate the transitory return to Earth of deceased relatives and loved ones and mark the completion of the annual cycle of cultivation of maize, the country's predominant food crop. Locals prepare the deceased's favourite dishes, believing that the dead can bring prosperity (e.g. an abundant maize harvest) depending on how well the rituals are performed (UNESCO 2005: 66).
6 Convention for the Safeguarding of the Intangible Cultural Heritage (CSICH), 17 October 2003, 2368 UNTS.
7 CSICH, Article 2(1).
8 Ibid.
9 Ibid.
10 Ibid., Article 16.
11 Ibid., Article 17.
12 UNESCO doc. ITH/10/5.COM/CONF.202/Decisions, 19 November 2010, p. 21.
13 Ibid., p. 45.
14 UNESCO, Decision 8, COM8.10.
15 Ibid., COM8.17.
16 Ibid., COM.
17 Convention Concerning the Protection of the World Cultural and Natural Heritage. Paris, 16 November 1972. 1037 UNTS 151, 11 ILM 1358.
18 Agreement on Trade-Related Aspects of Intellectual Property Rights (TRIPS Agreement), 15 April 1994, Marrakesh Agreement Establishing the World Trade Organization, Annex 1C, 1869 UNTS 299, 33 ILM 1197 (1994).
19 CSICH, Article 3(b).
20 TRIPS Agreement, Article 22.

21 Ibid., Article 24.
22 *European Communities–Measures Prohibiting the Importation and Marketing of Seal Products*, Reports of the Panel, 25 November 2013, WT/DS400/R, WT/DS401/R. *European Communities–Measures Prohibiting the Importation and Marketing of Seal Products*, Reports of the Appellate Body, 22 May 2014.
23 Regulation (EC) 1007/2009 of the European Parliament and of the Council of 16 September 2009 on Trade in Seal Products, 2009 OJ (L. 286) 36.
24 General Agreement on Tariffs and Trade 1994, 15 April 1994, Marrakesh Agreement Establishing the World Trade Organization, Annex 1A, 1867 UNTS 187.
25 Agreement on Technical Barriers to Trade, 15 April 1994, Marrakesh Agreement Establishing the World Trade Organization, Annex 1A, 1868 UNTS 120.
26 *European Communities – Measures Prohibiting the Importation and Marketing of Seal Products*, Reports of the Panel, para. 7.2.
27 Ibid., paras 7.161 and 7.164.
28 Ibid., para. 8(2).
29 Ibid., para. 7.292.
30 United Nations Declaration on the Rights of Indigenous Peoples, A/RES/61/295, (2007).
31 *European Communities – Measures Prohibiting the Importation and Marketing of Seal Products*, Reports of the Panel, para. 7.317
32 Agreement on Agriculture, 15 April 1994, Marrakesh Agreement Establishing the World Trade Organization, Annex 1A, 1867 UNTS 410.
33 Agreement on the Application of Sanitary and Phytosanitary Measures, 15 April 1994, Marrakesh Agreement Establishing the World Trade Organization, Annex 1A, 1867 UNTS 493.

References

Aikawa-Faure, N. (2009) "From the Proclamation of Masterpieces to the Convention for the Safeguarding of Intangible Cultural Heritage". In L. Smith and N. Akagawa (eds), *Intangible Heritage*, Routledge, London, pp. 13–44.

Beghin, J. (2014) "The Protectionism of Food Safety Standards in International Agricultural Trade". *Agricultural Policy Review* 1: 7–9.

Blake, J. (2009) "UNESCO's 2003 Convention on Intangible Cultural Heritage: The Implication of Community Involvement in 'Safeguarding'". In L. Smith and N. Akagawa (eds), *Intangible Heritage*, Routledge, London, pp. 45–73.

Bonadio, E. (2015) "Why Europe and the US are Locked in a Food Fight over TTIP". *The Conversation*, 7 August, http://theconversation.com/why-europe-and-the-us-are-locked-in-a-food-fight-over-ttip-45279 (accessed 14 January 2016).

Borghini, A. (2014) "Geographical Indications, Food and Culture". In P. B. Thompson and D. M. Kaplan (eds), *Encyclopedia of Food and Agricultural Ethics*, Springer, Heidelberg.

Broude, T. (2015) "A Diet too Far? Intangible Cultural Heritage, Cultural Diversity, and Culinary Practices". In I. Calboli and S. Radavan (eds), *Protecting and Promoting Diversity with Intellectual Property Law*, Cambridge University Press, Cambridge.

Brown, M. F. (2005) "Heritage Trouble: Recent Work on the Protection of Intangible Cultural Property". *International Journal of Cultural Property* 12: 40–61.

CESCR (2009) "General Comment No. 21, Right of Everyone to Take Part in Cultural Life". Article 15, Para. 1(a), of the International Covenant on Economic, Social and Cultural Rights, UN doc. E/C.12/GC/21, 21 December, United Nations Committee on Economic, Social and Cultural Rights.

Cody, E. (2012) "An American Chateau? French Winemakers Say No". *Washington Post*, 24 September, available at www.washingtonpost.com/world/europe/an-american-chateau-french-winemakers-say-no/2012/09/23/a4b08432-03ee-11e2-8102-ebee9c66e 190_story.html (accessed 14 January 2016).

Di Giovine, M. A. and Brulotte, R. L. (2014) "Introduction: Food and Foodways as Cultural Heritage". In R. L. Brulotte and M. A. Di Giovine (eds), *Edible Identities: Food as Cultural Heritage*, Ashgate, Farnham, pp. 1–28.

Friedman, T. L. (1999) *The Lexus and the Olive Tree*, New York, Farrar.

Gangjee, D. S. (2015) "Geographical Indications and Cultural Rights: The Intangible Cultural Heritage Connection?", In C. Geiger (ed.), *Research Handbook on Human Rights and Intellectual Property*, Edward Elgar, Cheltenham, pp. 544–559.

Gervais, D. (2003–2004) "Spiritual but Not Intellectual – The Protection of Sacred Intangible Traditional Knowledge". *Cardozo Journal of International and Comparative Law* 11: 633–670.

Hauter, W. (2007) "The Limits of International Human Rights Law and the Role of Food Sovereignty in Protecting People from Further Trade Liberalization under the Doha Round Negotiations". *Vanderbilt Journal of Transnational Law* 40: 1071–1098.

ILC (2006) *Report of the Study Group of the International Law Commission, Fragmentation of International Law: Difficulties Arising from the Diversification and Expansion of International Law*, UN Doc. A/CN.4/L.682, 13 April, International Law Commission, Geneva.

Kearney, A. (2009) "Intangible Cultural Heritage (Global Awareness and Local Interest)". In L. Smith and N. Akagawa (eds), *Intangible Heritage*, Routledge, London, pp. 209–224.

Kirshenblatt-Gimblett, B. (2004) "Intangible Heritage as Metacultural Production". *Museum International* 56: 52–65.

Korthals, M. (2008) "Ethics and Politics of Food: Toward a Deliberative Perspective". *Journal of Social Philosophy* 39: 445–463.

Kurin, R. (2004) "Safeguarding Intangible Cultural Heritage in the 2003 UNESCO Convention: a Critical Appraisal". *Museum International* 56: 66–77.

Lauter, D. (2012) "French Winemakers Concerned over 'Chateau' Change". *The Telegraph*, 16 September, available at www.telegraph.co.uk/foodanddrink/wine/9546440/French-winemakers-concerned-over-chateau-change.html (accessed 14 January 2016).

Lubina, K. (2009) "Protection and Preservation of Cultural Heritage in the Netherlands in the 21st Century". *Electronic Journal of Comparative Law* 13: 1–62.

Maffei, M. C. (2012) "Food as a Cultural Choice: A Human Right to Be Protected?". In S. Borelli and F. Lenzerini (eds), *Cultural Heritage, Cultural Rights, Cultural Diversity*, Martinus Nijhoff, Leiden.

Pentassuglia, G. (2009) *Minority Groups and Judicial Discourse in International Law: A Comparative Perspective*, Martinus Nijhoff Publishers, Leiden.

Scovazzi, T. (2008) "La notion de patrimoine culturel de l'humanité dans les instruments internationaux". In J. A. R. Nafziger and T. Scovazzi (eds), *Le patrimoine culturel de l'humanité/The Cultural Heritage of Mankind*, Brill, Leiden.

Searles, E. (2002) "Food and the Making of Modern Inuit Identities". *Food and Foodways* 10: 55–78.

Shimura, K. (2010) "How to Cut the Cheese: Homonymous Names of Registered Geographic Indicators of Foodstuffs in Regulation 510/2006". *Boston College International and Comparative Law Review* 33(1): 129–152.

Smith, F. (2009) *Agriculture and the WTO: Towards a New Theory of International Agricultural Trade Regulation*, Cheltenham, Edward Elgar.

Smith, F. (2011) "Indigenous Farmers' Rights, International Agricultural Trade and the WTO". *Journal of Human Rights and the Environment* 2: 157–177.

UNESCO (2005) *Masterpieces of the Oral and Intangible Heritage of Humanity – Proclamations 2001, 2003, 2005*, United Nations, Paris.

Vadi, V. (2012) "The Protection of Cultural Landscapes and Indigenous Heritage in International Investment Law". In L. Westra, C. Soskolne and D. Spady (eds), *Human Health and Ecological Integrity: Ethics, Law and Human Rights*, Earthscan, London, pp. 250–261.

Woods, M. (2002–2003) "Food for Thought: The Biopiracy of Jasmine and Basmati Rice". *Albany Law Journal of Science and Technology* 13: 123–144.

5 The right to food between food security and food sovereignty

Different perspectives of the battle against genetically modified organisms

Mery Ciacci

Introduction

In the past fifteen years, the use of genetically modified (GM) crops in agriculture and the production of genetically modified food have considerably increased. In spite of this, mistrust and resistance to consume GM food has also spread in different corners of the globe.[1] The introduction of the use of biotechnology[2] in agriculture was welcomed as a revolutionary remedy for solving the crisis of the agricultural sector in developing countries, stopping starvation in the world and promoting sustainable development. Yet, today the impressive number of people still suffering hunger and the scientific uncertainty about likely impacts of genetically modified organisms (GMOs) on the environment and human health are quite disappointing about such expectations. Further, the lack of appropriate global governance for biotech (BT) agriculture addressing economic, social and cultural impacts that may derive from the current diffusion of BT agriculture – and may, in particular, affect local and indigenous communities primarily thriving upon traditional agriculture – is leading to controversial consequences which threaten the fulfilment of the right to food, development and other fundamental freedoms of such communities.

The opposition to use, consume and produce GM food can be seen as a way to express and realize citizens' choice to their right to access, produce and consume their food, in line with their economic needs, socio-cultural preferences and their freedom to choose. Since the late '90s, the idea that people as producers and consumers of food should control the mechanisms and policies of food production and distribution has gained ground both at the national and international level. This concept, first defined as *food sovereignty* by the Via Campesina movement, is being increasingly recognized at different stages and opposes the idea that the adequate supply of nutrition – or, in other words *food security* – should be ensured through global imports and in the name of enhanced productivity. In the last years, an increased number of Governments, especially in Latin America, integrated food sovereignty as a goal to be achieved within their national constitutions or laws. On the basis of this, several States have banned the cultivation and the import of genetically modified organisms. On the other hand, supporters of the concept of food security criticize such a choice and argue that the spreading of GMOs would be beneficial for those countries where hunger and starvation remain a plague.

In this chapter I argue that the ongoing battle against GMOs carried on by several States, movements and groups of stakeholders around the world is a significant manifestation of the will to own at local and national level the control on the food chain, so contributing to the realization of food sovereignty. In order to do so, I will first provide an overview of the overall context concerning the introduction of GMOs in agriculture; secondly, I will introduce the notions of *food sovereignty* and *food security*. Then, looking at the debate on GMOs through the lenses of these notions, I will look at some examples of existing constitutional provisions banning GMOs in different Latin American countries. Finally, I will argue that the interpretation of the right to food provided by the UN Committee Economic, Social and Cultural Rights in its General Comment no. 12 should not be read as merely supporting food security, but as an official recognition of the need to move towards food sovereignty.

The context: the overall debate about genetically modified organisms and the right to food

Biotech agriculture: future expectations and actual impacts

Supporters of the production and consumption of GMOs claim that biotechnology in agriculture has been introduced in order to increase the production of foods, both in terms of quantity and quality, so as to solve the problem of hunger in the world. Indeed, BT agriculture is often associated with the fulfilment of the right to food (Mechlem and Raney 2007). Promoters of agricultural biotechnologies affirm that genetically engineered (GE) crops would increase yields, produce more food, control pests and weeds and raise worldwide food security. They would resist environmental challenges, such as drought, frost and soil salinity and provide resilience to climate change. Further, GE crops would also contribute to improve environmental quality by reducing pesticides and other chemicals in agriculture. According to these ambitious expectations, by boosting food production at the global level, agricultural biotech would lead to greater social gains, namely the eradication of poverty and alleviation of disparities by granting everyone the right to food (Shiva et al. 2001).

In the last twenty years, the cultivation of transgenic crops and the commercialization of biotech (BT) products have rapidly spread (FAO 2014). In spite of this fact, according to *The State of Food Insecurity in the World 2015*, still some 795 million people continue to suffer from hunger (FAO, IFAD and WFP 2015). Further, the expectations concerning improvements to environmental quality and economic wealth have not been met. So far, it could be asserted that biotechnological agriculture shares several traits with the Green Revolution. First, it embraces the same logic underpinning the Green Revolution, namely the logic of boosting food production to fight hunger in the world. The Green Revolution was a terrific success in terms of numbers: food production in developing countries doubled in only a few years (Gonzalez 2006: 378). However, this did not eradicate starvation, but contributed to exacerbate existing socio-economic disparities and trade distortions in developing countries. Indeed, the Green Revolution mostly favoured large and wealthy farmers capable of purchasing chemical products and technology, obliging small farmers to indebt themselves or to abandon their activities. This contributed to concentrate the market power in the hands of a few agrochemical companies and large-scale producers and the superabundance of food brought down the price of agricultural products at the global level to the detriment of small farmers. Today, the growing expansion of agricultural biotechnology risks entailing similar consequences. In fact, also BT agriculture privileges extensive monocultures and large-scale farms and, consequently, a great variety of seeds is no longer cultivated. This reduces the diversity of foods consumed, with negative consequences for human nutrition and the risk of famine in case of catastrophic failure of the monoculture, as well as the loss of cultural diversity. Extensive monocultural agriculture needs extensive acreage to be cultivated. Therefore, uncontaminated acres of land and forests are destroyed to leave the place to cultivations, entailing unpleasant consequences, such as the fact that local and indigenous communities living in these areas are often forced to move away and change their typical livelihoods and that the conservation of biodiversity is deeply endangered.

Several studies demonstrate, today, that additional negative impacts on the environment may derive from the use of GM crops: for instance, GM crops created super-pests and super-weeds, so increasing the use of herbicides (Shiva et al. 2001: 12).[3] This not only hinders the quality of the environment, but also threatens the survival of several beneficial insects, such as bees, and soil microorganisms. Further, GM crops claim to be climate change resilient, yet there is no scientific proof of such stronger resilience than other conventional crops (Shiva et al.: 17).[4] A new threat caused by the spreading of GM crops is the risk of genetic pollution. In countries that have allowed GM cultivations, like Canada, Brazil, Argentina, etc., several cases of genes' contamination from GM crops to conventional crops have been reported (Gonzalez 2007: 608; Shiva et al. 2001: 21). Seeds contamination cannot be controlled and there is no scientific agreement on the gravity of its consequences. However, it has been observed that when GM crops contaminate conventional crops, these last ones do not resist and stronger weed-killers have to be applied (Shiva et al. 2001).[5]

Another issue concerning the use of GMOs deals with their possible impacts on human health. Although today the likely effects on human health remain mostly unknown, several voices from the scientific community and public opinion highlight the need to "thoroughly reconsider all aspects of the safety of plant biotechnology", since there is increasing evidence that several elements used to improve the GM crops' resistance can be probable or possible human carcinogens (Landrigan and Benbrook 2015).

Genetically modified organisms and the patentability of seeds

Finally, BT agriculture can deeply affect the preservation of sustainable traditional agriculture of local and indigenous communities. First of all, traditional agriculture's maintenance is tightly linked to the conservation of biodiversity, which can be deeply affected by the extensive use of monoculture, pesticides and the risk of GM crops' contamination. Second, BT products are regulated by intellectual property law. On this point, the most relevant provisions are those contained in the Agreement on Trade-Related Aspects of Intellectual Property Rights (TRIPS), which reinforced the protection of IP rights by establishing a framework of minimum standards binding all WTO members. Under article 27 of the TRIPS, States should grant patent protection for innovations in all fields of technology implying novelty, an innovative step and capable of industrial application.[6] Paragraph 3(b) of Article 27 establishes the possibility for States to exclude some kinds of inventions from patenting, such as plants, animals and essentially biological processes for the production of plants or animals, other than non-biological and microbiological processes. This exclusion is rather narrow: in fact, micro-organisms, and non-biological and microbiological processes are eligible for patents. As for plant varieties, some types of intellectual property eligibility need to exist, either through patent protection, or a *sui generis* system created specifically for the purpose, or a combination of the two.[7] Therefore, the choices are rather restricted, given that "some kind of protection" for plant variety is generally required by the provision, and micro-organisms, like the ones used today in biotechnology, cannot be exempted. In addition, even when the State can claim that plants and animals already exist in nature and no inventor can be identified, this is not enough to adequately protect the traditional knowledge and practices of indigenous and local peoples from the impacts of BT agriculture (Footer and Awuku 2005: 250).

Once seeds are patented, they only belong to the patent holder, who usually then sells them on the market. Intuitively, only wealthy farmers have the possibility to purchase patented seeds, and all the fertilizers, herbicides and pesticides necessary to raise GM crops. Consequentially, small-scale farmers are marginalized, and often obliged to renounce their activities or indebt themselves to be able to buy the patented seeds (Shiva et al. 2001).[8] In addition, patented seeds cannot be saved, according to laws pertaining to royalties. Considering that traditional agriculture mostly relies on informal supply seed systems, where farmers' traditional practices to collect, select, save and exchange seeds provide for a large amount of the seeds/propagating material used (Correa et al. 2015), many traditional activities are disappearing and unemployment is thus growing (due also to the fact that GM crops decrease the need for manual labour).

Although the application of IP law to biogenetic resources raises ethical and moral concerns and its compatibility with the respect of social, economic and cultural rights is highly debated,[9] the TRIPS rules allowing the patentability of genetic resources keep being applied at the global level. Several countries, and in particular developing countries, receive a lot of pressure from developed countries to comply with TRIPS obligations; choices concerning patent law and GMOs are often a response to such a pressure imposed in free trade agreements (FTAs) (Correa et al. 2015). Similar choices do not only hinder the full implementation of the right to food, but also threaten the environment and the socio-cultural integrity of vulnerable communities, such as local and indigenous communities.

Defining concepts: food security and food sovereignty

Food security

In the light of the above frame, it seems quite controversial to assert that GMOs may be a valid contribution to fulfil the right to food and promote sustainable development. Such an assumption seems not to take into account either the economic, social, cultural and environmental impacts deriving from the current GMO governance, nor the evidence that the quantity of food produced in the world is already enough to cover the global population. The real core of the hunger problem is rather about access to food and access to the means of production.

The approach relying on the need to boost food production in order to fight hunger in the world seems to be in line with the more classical concept of "food security". Food security is a technical concept referring to the supply of food and individuals' access to it. The term began to be used in the 1970s, and the 1974 World Food Conference defined "food security" as "the availability at all times of adequate world food supplies of basic foodstuffs to sustain a steady expansion of food consumption and to offset fluctuations in production and prices" (FAO 2003). Such a definition focused on the quantity rather than quality of food: in the aftermath of the world food price crises in the early '70s, emphasis was given to strengthening food production so as to increase both the availability of basic foodstuffs and ensure the stability of prices. National food security's discourse gave priorities to aspects such as grain reserves, import and export quotas, food aid, and agricultural techniques to increase production (Windfuhr and Jonsén 2005). In 1981 Amartya Sen's essay *Poverty and Famines*, by challenging the traditional analysis of famines concentrating on food supply, highlighted the need to entitle individuals and groups to access food. The debate questioned the adequacy of the production-oriented approach to solve hunger and malnutrition and gradually switched from the overall availability of food to people's empowerment, or in other words, to access to food. Under this influence, FAO started to expand its definition of food security (FAO 1996) and in 2001 "food security" was defined as "a situation that exists when all people, at all times, have physical, social and economic access to sufficient, safe and nutritious food that meets their dietary needs and food preferences for an active and healthy life" (FAO 2002). Thus, today the concept of food security has been broadened so as to include also food quality aspects and not merely refer to sufficient, but also safe and nutritious food.

In the current definition, the accent on the need to ensure physical, social and economic access derives from the inputs given by Sen's critique to the production-oriented quantity, which showed that individual food security can be severely constrained despite sufficient national supplies. That said, one may question whether in the overall context of global trade rules and policy constraints, such as the one concerning GMOs above described, access to food is properly ensured to all groups and individuals. The answer seems to be negative. As Sen highlighted, it is important to establish entitlement to enough food through production-based, labour-based, trade-based, transfer-based, or other entitlement relationships. In other words, access to food can be properly enacted if a complementary set of appropriate policies dealing with the whole food chain is also foreseen. Therefore, the argument that GMOs may contribute to achieve food security in the world does not seem to be a valid one, since it reduces the whole hunger problem to a question of "increasing and improving quantity" without taking the aspects related to access to food into proper account.

Food sovereignty

It is in order to counteract the food security discourse that the concept of "food sovereignty" was raised and gained a place in the political debate at national and international level. The *Via Campesina* movement was among the first voices to talk about this idea in the '90s. Since then, "food sovereignty" has been defined as "the right of peoples to healthy and culturally appropriate food produced through ecologically sound and sustainable methods, and their right to define their own food and agriculture systems"[10] or "the right of peoples and sovereign states to democratically determine their own agricultural and food policies" (IAASTD 2009).

"Food sovereignty" is a political concept and goes beyond food security by invoking rights to take decisions about food production, as to what is produced and how and at what scale (Desmarais et al. 2009; Patel 2010). It advocates for building a process of local ownership of the food chain whose ultimate goal would be to counterbalance the disparities and inequalities created by international trade rules, corporations and market institutions dominating the global food system. In order to reach this goal, "food sovereignty" calls for a complementary set of policy decisions concerning land reforms and biodiversity policies to favour access to land, water, seeds and livestock breeds and credit, localizing agricultural production; making effective the right of consumers to decide what they consume through appropriate information; linking agricultural prices to production costs and to stop all forms of dumping; enhancing the populations' participation in agricultural policy decision-making; and promoting ecologically sound agriculture. The realization of food sovereignty as a national priority is, therefore, highly connected to other policies, such as environmental protection or land property law. For instance, the choice to discourage monoculture and enhance the protection of autochthone biological diversity may be seen as a step towards the realization of food sovereignty. It is through these lenses that we can look at the decision taken by several countries, in particular in Latin America, to limit or ban the use of GMOs, as well as the call for an effective revision of TRIPS rules on patents' obligations (namely article 27.3(b) of TRIPS).

National laws and policies on genetically modified organisms in Latin America: towards national food sovereignty?

Still a great number of States, belonging to different regions of the world, fear the introduction of GM crops. For instance, the EU and its Member States have, until now, adopted a precautionary approach to limit the introduction and commercialization of GMOs in the common market. The choice is mostly based on the reluctance of European citizens to consume GM food without having enough scientific evidence of its impact on human health. In addition, there are cultural reasons underlying such opposition: typical and diverse food is part of the regional culinary traditions of Europe and EU citizens do not seem to be willing to lose such a typical element of their cultural identity. Nonetheless, WTO obligations also put pressure on the EU, and the Commission recently adopted a new directive which may put at risk the preservation of such cultural integrity.[11]

In less developed and developing countries, the introduction of GMOs in agriculture is often perceived as a threat to local economies and social, cultural and environmental integrity: on these premises, many Latin American, Asian and African countries adopted restrictions on GMO imports and/or ban the cultivation of GM crops.

Several of these countries have incorporated food sovereignty as a national priority into their national constitutions and laws, like for instance Venezuela, Bolivia, Ecuador, Nepal, Mali and Senegal. In particular, in most Latin American countries, the adoption of rules banning or limiting GMOs seems to be directly connected to the realization of food sovereignty.

For instance, Article 255 of Bolivia's Constitution prohibits the importation, production, and commercialization of GMOs and Article 309 lists food sovereignty as a national objective. It is interesting to highlight that, in line with this choice, in 2010 Bolivia raised a significant claim in a Communication to the Council for TRIPS.[12] Bolivia asked for a ban on the patenting of all life forms, including gene sequences and microorganisms as well as all biological, microbiological and non-biological processes, in order to ensure "the protection of the innovations of indigenous and local farming communities and the continuation of the traditional farming practices including the right to save and exchange seeds, and sell their harvest" and the "protection for the rights of indigenous communities and prevent any private monopolistic intellectual property claims over their traditional knowledge".[13] Bolivia, besides referring to the United Nations Declaration on the Rights of Indigenous Peoples (UNDRIP), called upon the Bolivian Constitution which expressly recognizes indigenous Cosmovision and local traditional knowledge as the common heritage and expression of the identity of the State.[14] Since Bolivia considers the regime set by article 27.3(b) inconsistent with its constitutional provisions and the UNDRIP, and a threat to the sovereignty of people regarding their own resources, the choice to ban GM products is clearly a political choice. Unfortunately, such a decision makes the country irregular in terms of compliance with WTO commitments. Indeed, in October 2001 Bolivia had to revoke the ban, allegedly due to pressure by the Argentinian soy corporate sector under the WTO frame (Baumüller 2003).

Another example of limitations of GM crops as a way to express food sovereignty is contained in Ecuador's Constitution. Article 281 affirms that "Food sovereignty is a strategic objective and an obligation of the State to guarantee that individuals, communities, towns and nationalities achieve permanent self-sufficiency with foods that are healthy and culturally appropriate." In line with this, Article 401 of the Constitution declares Ecuador free of transgenic crops and seeds; while Article 402 forbids the granting of rights, including intellectual property rights, to by-products or synthetics obtained from collective knowledge associated with national biodiversity. These provisions seems to implement Article 13 of Ecuador's Constitution, stressing the need that "individuals and communities have . . . safe and permanent access to healthy, sufficient and nutritious food, preferably produced locally and in accordance with their different identities and cultural traditions".

Unfortunately, Ecuador is not free from imports and also must comply with WTO obligations. The country is thus exposed to pressure to open its market to GMOs; further, by importing foodstuffs from countries such as Argentina, Brazil and the US, the risk of introducing GM products is highly concrete.

In search of a legal basis to support food sovereignty under current international trade law: the right to food

The ban on or limitations to GMOs imports as a way to achieve food sovereignty in the above mentioned examples are the outcome of a political choice. Anyway, such a choice is challenged by States' obligations undertaken under the WTO frame. Is there any legal basis to support such a political choice vis-à-vis the obligations arising under current international trade law?

In my opinion, the Right to Adequate Food, as interpreted by the UN Committee on Economic, Social and Cultural Rights in its General Comment No 12, can offer a valuable legal basis to support States' efforts to adopt and implement national measures promoting food sovereignty. The UN CESCR points out that the right to food is fulfilled when "every man, woman and child, alone or in community with others, has *physical* and *economic access* at all times to adequate food or means for its procurement" (emphasis added).[15] The UN CESCR also clarifies that "[t]he right to adequate food shall therefore not be interpreted in a narrow or restrictive sense which equates it with a minimum package of calories, proteins and other specific nutrients. The right to adequate food will have to be *realized progressively*" (emphasis added).

According to this interpretation, the right to food is fulfilled not only when food is available in terms of quantity, but also when:

- food quality satisfies the dietary needs of individuals;
- food is secure (free from adverse substances);
- food is culturally acceptable and links with cultural heritage/identity of peoples; and
- access to food is physically and economically sustainable.

In order to realize these conditions, States should comply with the three following obligations (Söllner 2007). First, the obligation to respect: States should refrain from interfering directly or indirectly with the enjoyment of the right to food, for instance they should refrain from denying or limiting access to food or interfering arbitrarily with existing arrangements (e.g. by destroying existing functioning market systems). But can we limit the interpretation of such a duty to these examples? In my opinion, opening the market to GMO products in order to comply with WTO commitments to further liberalization may enshrine a way to indirectly interfere with the full enjoyment of the right to food, especially in those contexts where vulnerable communities exist. Second, the obligation to protect requires States to take measures to ensure that third parties such as individuals, groups, corporations, or other entities do not interfere in any way with the enjoyment of the right. So, measures banning or limiting GMOs can be seen as a concrete expression of this obligation. Third, the right to fulfil: States must take positive measures to facilitate and provide for individuals' enjoyment of their rights, such as the development of comprehensive national right-to-food strategies and the development of policies, as well as the development of an enabling framework in which as many individuals as possible can provide for their own food. In this sense, the set of constitutional provisions aiming to realize food sovereignty by means of empowering individuals and groups to access food – for instance by ensuring access to land, protecting traditional knowledge and biodiversity – could be interpreted as a way to enact the right to fulfil.

To conclude, the full realization of the right to food cannot be reduced to a means to achieve food security. A right to access "safe" and "nutritious" food might be interpreted as not merely encompassing a right to *sufficient* food. The right to adequate food seems to go beyond the earlier more instrumentalist view of the right to food as a means to achieve food security and to move towards the support of a broader frame of action for access to food, such as food sovereignty can be.

Conclusive remarks

So far biotechnological agriculture and the increased spreading of GMOs in agriculture have brought about detrimental consequences for the environmental, social and cultural integrity of societies, especially the more vulnerable ones such as local and indigenous communities in developing and less developed countries. Further, rather than solving the problem of hunger, it may exacerbate the disparities created by the current food system and undermine the process of affirmation of food sovereignty. The ongoing battle against GMOs carried on by several States, movements and groups of stakeholders around the world is a significant manifestation of the will to own at local and national level the control of the food chain, so contributing to the realization of the right to food and food sovereignty. Constitutional provisions banning or limiting the use and commercialization of GMOs in some Latin American constitutions should be read through this lens. Unfortunately, the political choice to connect these measures with the stated constitutional goal to achieve food sovereignty does not seem to be powerful enough to counter WTO obligations to further liberalize the GMO market or grant patent protection to new seeds variety.

Although the realization of food sovereignty seems to be quite ambitious and raises some doubts (Peña 2013), the food sovereignty discourse has the added value of involving what is not quantifiable in the food security paradigm, namely matters of culture, biodiversity and traditional knowledge. Therefore, framing GMO related provisions into a broader set of complementary measures such as those concerning access to land, seeds and water can be a valuable contribution to the realization of the right to adequate food, as interpreted in the General Comment no. 12. On the other hand, the interpretation of the right to adequate food can be read in such a way to support the affirmation of food sovereignty.

Notes

1 See, for instance, Prupis (2015).
2 Article 2 of the Convention on Biological Diversity (CBD) defines biotechnology as "any technological application that uses biological systems, living organisms, or derivatives thereof, to make or modify products or processes for specific use".
3 Among the several cases quoted in the full report mentioned, I recall here the example of the Roundup Ready cotton, a herbicide-resistant crop that can create the risk of herbicide resistant "superweeds" by transferring the herbicide resistance to weeds. This has also been admitted by Monsanto in the case of a notorious Australian weed, rye grass, which developed tolerance to its herbicide Roundup.
4 It is interesting to mention that some Indian farmers showed that through traditional ways they have been able to select and collect drought tolerant varieties of seeds, already available in nature.
5 Paradoxically, in many of these situations, the polluted farmers were sued by producers of GM crops for stealing his/her BT products.
6 Art. 21(1) of the TRIPS Agreement.
7 An example of *sui generis* protection for plant varieties in the multilateral framework of agreements is the International Convention for the Protection of New Varieties of Plants (UPOV), 1961.
8 I recall as an example the numerous cases of suicides of small farmers in India, related to the debt accumulated to buy BT products.
9 Conflicts between the TRIPS Agreement and human rights have been addressed in several UN documents. See: Sub-Commission on Human Rights, "Intellectual Property and Human Rights", Resolution 2000/7 and Resolution 2001/21, and High Commissioner for Human Rights, "The Impact of the Agreement on Trade Related Aspects of Intellectual Property Rights on Human Rights", UN Doc. E/CN.4/Sub.2/2001/13.
10 Declaration of Nyéléni, adopted on 27 February 2007, by the Forum for Food Sovereignty in, Sélingué, Mali.

11 Directive (EU) 2015/412 of the European Parliament and of the Council of 11 March 2015 amending Directive 2001/18/EC as regards the possibility for the Member States to restrict or prohibit the cultivation of genetically modified organisms (GMOs) in their territory.

12 Communication from Bolivia, IP/C/W/545, 26 February 2010.

13 *Ibid.*, Point 30(a)(b)(c)(d).

14 Bolivia quotes Article 100 of the Constitution, which "recognizes the Cosmovision, myths, oral history, dances and cultural practices, traditional knowledge and technologies of indigenous peoples and peasants as their heritage [and that] this heritage is part of the expression and identity of the State", and Article 382, that states: "it is the competence and duty of the State to defend, recover and protect biological material coming from natural resources, ancestral knowledge and anything else that originate in the territory".

15 General Comment No 12, *The right to adequate food (Art.11)*, 12/05/1999, E/C.12/1999/5, para 6.

References

Baumüller, H. (2003) *Domestic Import Regulations for Genetically Modified Organisms and their Compatibility with WTO Rules: Some Key Issues*, International Institute for Sustainable Development, Winnipeg.

Correa, C. M., et al. (2015) *Plant Variety Protection in Developing Countries: A Tool for Designing a Sui Generis Plant Variety Protection System: An Alternative to UPOV 1991*, APREBES.

Desmarais, A., Wiebe, N. and Wittman, H. (2009) "The Origins and Potential of Food Sovereignty". In H. Wittman, A. Desmarais and N. Wiebe (eds), *Food Sovereignty: Reconnecting Food, Nature and Community*, Fernwood Publishing and FoodFirst Books.

FAO (1996) *Rome Declaration on World Food Security and World Food Summit Plan of Action*, World Food Summit, Rome, 13–17 November.

FAO (2002) *The State of Food Insecurity in the World 2001*, FAO, Rome.

FAO (2003) "Trade Reforms and Food Security: Conceptualizing the Linkages". Available at www.fao.org/docrep/005/y4671e/y4671e06.htm.

FAO (2014) *Low Levels of GM Crops in International Food and Feed Trade: FAO International Survey and Economic Analysis*, Technical Background Paper 2, March, FAO, Rome.

FAO, IFAD and WFP (2015) *The State of Food Insecurity in the World 2015: Meeting the 2015 International Hunger Targets: Taking Stock of Uneven Progress*, FAO, Rome.

Footer, M. and Awuku, E. O. (2005) "Sustainable Agricultural Resources and Food Security: the Seed Treaty and Equitable Benefit Sharing". In M. C. Cordonier Segger and C. G. Weeramanatry (eds), *Sustainable Justice: Reconciling Economic, Social and Environmental Law*, Martinus Nijhoff Publishers, Leiden.

Gonzalez, C. G. (2006) "Markets, Monocultures, and Malnutrition: Agricultural Trade Policy through an Environmental Justice Lens". *Michigan State Journal of International Law* 14: 345.

Gonzalez, C. G. (2007) "Genetically Modified Organisms and Justice". *Georgetown International Environmental Law Journal* 19: 583–642.

IAASTD (2009) "Global Summary for Decision Makers". International Assessment of Agricultural Knowledge, Science and Technology for Development, available at www.unep.org/dewa/Assessments/Ecosystems/IAASTD/tabid/105853/Defa (accessed 15 January 2016).

Landrigan, P. J. and Benbrook, C. (2015) "GMOs, Herbicides, and Public Health". *The New England Journal of Medicine* 373(8): 693–695.

Mechlem, K. and Raney, T. (2007) "Agricultural Biotechnologies and the Right to Food". In F. Francioni (ed.), *Biotechnologies and International Human Rights*, Hart Publishing, Oxford, p. 131.

Patel, R. (2010) "What Does Food Sovereignty Look Like?" In H. Wittman, A. A. Desmarais and N. Wiebe (eds), *Food Sovereignty: Reconnecting Food, Nature and Community*, Fernwood Publishing and FoodFirst Books.

Peña, K. (2013) "Institutionalizing Food Sovereignty in Ecuador". Paper presented at Food Sovereignty: A Critical Dialogue, International Conference, 14–15 September.

Prupis, N. (2015) "Answering 'Resistance From All Sides', Germany Moves to Ban GMO Crops". 25 August, available at www.commondreams.org/news/2015/08/25/answering-resistance-all-sides-germany-moves-ban-gmo-crops.

Sen, A. (1981) *Poverty and Famines: An Essay on Entitlement and Deprivation*, Clarendon Press, Oxford.

Shiva, V., et al. (2001) *The GMO Emperor Has No Clothes: A Global Citizens Report on the State of GMOs – False Promises, Failed Technologies*, report coordinated by Navdanya and Navdanya International, the International Commission on the Future of Food and Agriculture, with the participation of The Center for Food Safety (CFS), available at www.navdanyainternational.it.

Söllner, S. (2007), "The 'Breakthrough' of the Right to Food: The Meaning of General Comment No. 12 and the Voluntary Guidelines for the Interpretation of the Human Right to Food". In A. von Bogdandy and R. Wolfrum (eds), *Max Planck Yearbook of United Nations Law*, vol. 11, Brill, Leiden, pp. 391–415.

Windfuhr, M. and Jonsén, J. (2005) *Food Sovereignty Towards Democracy in Localized Food Systems*, ITDG Publishing, Bradford.

6 Genetically modified crops and their impact on the environment

Creating a win–win for science and nature with a deontological legal framework

Ngozi Stewart

It is difficult to make a general judgment about Genetic Modification (GM) . . . The risks involved are not due to the techniques used, but rather to their improper or excessive application . . .

(Watson 2015)

Introduction

The impact of climate change is beginning to be felt in unprecedented ways. It has become a constant threat to food security, the global economy, coastal areas, arable land and freshwater resources. This chapter will focus on food security and the controversy over how genetically modified crops (GMCs) can be an avenue for alleviating it.

It will show the importance of foundational (ethical) thresholds in determining the extent of human beings' interference with nature. Two ethical thresholds will be considered in this research – the deontological and utilitarian thresholds. They are both anthropocentric theories – a theory on which the principle of sustainable development is based. However, utilitarianism prioritizes function over value while deontology prioritizes value over function. This chapter will anchor on the deontological theory because its objective is to preserve the value of the environment (nature) in an era of rapid development (GMC production). It is a theory that claims that the ethical assessment of actions must not be based on their consequences alone. In other words, certain acts are wrong in themselves, that is, irrespective of their consequences. Such acts are prohibited even if they increase or maximize net benefit. One potent way of establishing a deontological threshold is by promoting nature's right to exist as a standard for the exercise of property rights of any kind.

This chapter is divided into eight parts including the introduction. It will begin by discussing the effects of climate change on food security; and in part three points in the direction of how food security can be enhanced through the production of GMCs but balances this discussion with the divergent views on the production of GMCs. It proceeds, in part four, to explore the possibility of using ecological thresholds for tempering the impacts of GMCs on the environment. Part five proves that without an ethical reference system, regulating the impact of GMCs on the environment will be illusive; hence the argument for an effective ethical reference system for a regulatory framework in part six; and an actual attempt to draft the proposed core ethical provision in part seven. The chapter is concluded in the eighth part.

The effect of climate change on food security

The Food and Agricultural Organization affirms that food security exists when all people, at all times, have physical, social and economic access to sufficient, safe and nutritious food that meets their dietary needs and food preferences for an active and healthy life (FAO 2012). The four pillars of food security are availability, access, utilization and stability (ibid.). Article 11(1–2) of the International Covenant on Economic, Social and Cultural Rights (ICESCR; 1966) also addresses food security by its definition of the "Right to Adequate Food".

The world's population is estimated at over 7 billion, and the growth rate is increasing, indicating that it will reach about 8.1 billion by 2030 (United Nations 1999). According to FAO figures, there are 815 million undernourished people in the world (FAO 2012). There is therefore a heavy burden on the world to provide food security to all the countries but the question is, by which technology can food security be best tackled?

Food security is diminished when food systems are negatively impacted by a number of environmental and socio-economic factors *in addition* to climate change such as population growth and unsustainable use of natural resources. The Intergovernmental Panel on Climate Change (IPCC)'s Fifth Assessment Report has announced some challenges for global agriculture: yields are expected to decline by 2 per cent per decade due to climate change, while the demand for food is expected to increase by 14 per cent per decade (IPCC 2013).

Climate change affects food security in four different ways (Harris 2009):

- *Temperature increase:* Plants get stressed from higher temperatures – evaporation is increased and, consequently, fertility rate and productivity are reduced.
- *Changing patterns:* Seasons and ecological dynamics are becoming more and more unpredictable. The result of this is more uncertainty for farmers especially with respect to when to plant crops.
- *Rising sea levels:* Increase in the saline water of the seas compromises freshwater aquifers and even predisposes island areas to storm surges.
- *Water:* The resultant water shortage from climate change leads to widespread food shortages and other related issues.

Researchers are agreed on the fact that climate change will increase the risk of reduced crop productivity associated with heat and drought stress (Gornall et al. 2010: 2779; Ayinde and Olatunju 2011: 189–194; Lobell and Gourdji 2012: 1686; Sengar and Sengar 2014). Negative impacts in average crop and pasture yield will likely be clearly visible by 2030. For example, in places like Brazil, rice and wheat yields could decline by 14 per cent (Kloeke 2015). Intensification of food production must therefore be accompanied by concerted action to reduce greenhouse gas emissions from agriculture to avoid further acceleration of climate change and avert threats to the long term viability of global agriculture (ibid.).

The next section will examine biotechnology (precisely, the production of genetically modified crops) as an adaptation strategy to the impact of climate change on food security.

Enhancing food security through genetically modified crops: divergent narratives

Eradicating hunger is a central part of the United Nations' Millennium Development Goals (United Nations 2010). However, the means of achieving this goal remain controversial. Some see the development and use of GM crops as key to reduce hunger, while others consider this technology as a further risk to food security. Solid empirical evidence to support either of these views however remains negligible.

GM crops are the crops whose DNA has been modified by using genetic engineering techniques, with the aim to introduce a new trait to the plant which does not occur naturally in the species (Lamichhane 2014: 43, 45). Specific changes are made in the DNA of these crops by genetic engineering techniques that encourage extra nutrients to be produced, faster growth and ability to resist diseases and other purposes. Virtually every area in the food production market is genetically modifying foods for the purpose of better taste, faster growth, disease resistance and improvement of nutrients (Green and Owen 2011: 5819–5829).

Corn is a good example of a crop that has been genetically modified. It has got the gene which is insect resistant and due to this, the farmers do not have to spray pesticides that are harmful to the soil as well as the crop. Soybean on the other hand which is very much used in every altered form is also being produced genetically, so that the farmers do not have to spray insecticides or pesticides (Pua and Davey 2007: 73). They have been genetically modified to offer improved oil profiles for processing or for healthier edible oils. Tomatoes have been modified in order to increase their shelf life. Tomatoes were genetically modified to prevent their from rotting. Canola oil has been genetically altered for resistance against pesticides (ibid.).

Proponents of GMC production argue that there are three possible pathways through which GMCs can impact on food security. First, GM crops could contribute to food production increases and thus improve the availability of food at global and local levels. Second, GM crops could affect food safety and food quality. Third, GM crops could influence the economic and social situation of farmers, thus improving or worsening their economic access to food (Moellenbeck 2001: 668–672; Dahleen 2001: 627–628; World Bank 2007; Harris-Lovett 2015).

On the other hand, environmental activists, religious organizations, public interest groups, professional associations and other scientists and government officials, for example, have raised concerns about GM foods. A cursory assessment of these opinions reveals a (legitimate) concern not about the way the crops are modified but their impacts after modification. In June 2015, Pope Francis made a profound comment on the emerging trend of GMCs. In the words of Pope Francis:

> It is difficult to make a general judgment about Genetic Modification (GM) whether vegetable or animal, medical or agricultural, since these vary greatly among themselves and call for specific considerations. *The risks involved are not due to the techniques used, but rather to their improper or excessive application* . . .
>
> (Watson 2015)

Most concerns about GM foods have fallen into three categories: environmental hazards, human health risks, and economic concerns (Helmuth 2000: 782–783).

The focus of this chapter is on the environmental impacts of the production of GMCs; hence the dedication of the subsequent paragraphs to a detailed discussion on the subject matter.

The introduction of genetically modified plants into the environment may have devastating effects on biodiversity. Birds, insects, and other animals which are dependent on certain crops for survival may find themselves unable to eat the genetically engineered crops due to the introduced gene or modification (Crawley et al. 2001: 682–683). They may be allergic to the new traits, or find them poisonous. Also, if they fed on the organisms which were once pests to the crop, then they may not have a source of food, as the pests would no longer be in the crop. Therefore, these animals would have to find other sources of food, or face starvation. This would impact the entire food chain and the predator-prey relationships (Brookes and Barfoot 2010).

The introduction of a modified organism into the environment may also displace indigenous fauna and flora. If the new strain is superior to the parent strain, it may take over the habitat or eliminate the wild strain (Crawley et al. 2001: 682–683; Altieri 2015).

Growing genetically modified or conventional plants in the field has raised concern for the potential transfer of genes from cultivated species to their wild relatives. However, many food plants are not native to the areas in which they are grown. Locally, they may have no wild relatives to which genes could flow.

Moreover, if gene flow occurs, it is unlikely that the hybrid plants would thrive in the wild, because they would have characteristics that are advantageous in agricultural environments only (Ammann 2009: 240–264).

Genetically modified crops may have indirect environmental effects as a result of changing agricultural or environmental practices. However, it remains controversial whether the net effect of these changes will be positive or negative for the environment. For example, the use of genetically modified insect-resistant Bacillus Thuringiensis (Bt) crops is reducing the volume and frequency of insecticide use on maize, cotton and soybean. Yet the extensive use of herbicide and insect-resistant crops could result in the emergence of resistant weeds and insects (Pua and Davey 2007: 73).

Most of the impacts discussed in the above paragraphs remain speculations. The broad consensus is that the environmental effects of genetically modified plants should be evaluated using science-based assessment procedures, considering each crop individually in comparison to its conventional counterparts. When considering the risks and benefits – both direct and indirect – of a new technology, they should be compared against existing alternatives. Any ecological (or health) impacts of a GM crop should be balanced against the impacts of the agricultural practices the GM crops would replace (Tencalla et al. 2009: 61–73).

Arguably, environmental impacts of GMCs can be analysed using a framework based on the direct drivers of change in ecosystems and biodiversity (discussed further below). Unfortunately, this type of framework is lacking as a guide to the regulation of GMC production. The next section will therefore begin a pathway to the creation of a proposed framework.

This framework is crucial in the light of the undeniable reality of the existence of GMCs in our world. The encyclical of Pope Francis could not have been less apt considering the proven (dastardly) effects of GMCs on the environment and human health so far (see KREM 2015). However, holding fastidiously to absolute views will first be a manifest denial of the increasingly strong (global) presence of GMCs despite the continuing debate on biotechnology and GM crops; and second ridicule the role of law as a medium of regulation.

Impact of genetically modified crops on the environment: exploring ecological thresholds

In the previous section a need for comparison of two methods of crop production was seen as a way of determining the environmental impact of GMC production. This section will explore that comparison by engaging the use of ecological thresholds. The term "threshold" in ecology is closely linked to resilience: time needed for a system to return to a global equilibrium after perturbation (Groffman et al. 2006: 1–13).

An ecological threshold can be described as a point or a zone where there is a relatively rapid change from one ecological condition to another (Huggett 2005: 301–310; Luck 2005: 299–300). The question as to what defines a desired state is subjective – a question which cannot be answered from a purely scientific and objective perspective (Muradian 2001: 7–24; Scheffer and Carpenter 2003: 648–656). Thus one of the major factors militating against the use of ecological thresholds in environmental management is the lack of general principles for applying these concepts to different kinds of response variables and ecosystems (Groffman et al. 2006: 1–13). In nature, populations usually fluctuate around some trend or stable average and ecosystems are assumed to respond smoothly to gradual change in external conditions. Occasionally, however, such a scenario is interrupted by an abrupt shift to a dramatically different regime. Dramatic regime shifts are known for a range of ecosystems including lakes, coral reefs, oceans, forests and arid land (Scheffer and Carpenter 2003: 648–656). Hence, each stressor and ecosystem response must be evaluated independently, a process that is usually not appropriate for regulatory decision-making as it requires years of site-specific research. Ecological sciences are currently more able to predict the magnitude of change (that is, the possible alternative state) than the precise threshold value (Muradian 2001: 7–24). It must be noted however that for most ecological indicators, fully operational thresholds will probably rarely be available. This inevitably challenges policy-makers as they cannot precisely rely on defined ecological thresholds that would facilitate decision-making processes.

The imperative of an ethical reference system

In the absence of thresholds, damage could be defined by using a baseline approach. In principle, the baseline approach helps to determine when a change has to be regarded a damage as the definition of damage is not dependent on a precise threshold, but on the comparison of two different states. The first state (that is, the status quo) is thereby indicating how things usually are. Damage occurs if the difference between the status quo and the second state is judged to be too sufficiently large to be adverse. Unfortunately, the baseline approach is not recommended to gauge environmental damage because the "status quo" method which it employs is fundamentally flawed in the sense that the *base line* state may (and usually will) be a state in which the environment is usually greatly impacted. What therefore should be the criteria for acceptability?

Ecology as a science needs to rely on ethics if it strives to evaluate possible and real impacts of GMCs on biodiversity because concepts such as *damage* and *risk*, which play a pivotal role in evaluation, are inherently value laden. Thus the normativism of ethics can be used to guide the description of science.

While there is agreement among ethicists that biodiversity is valuable, there is no agreement on the importance of this value (compared to other relevant values) and whether it is an inherent or just an instrumental value. It is this value that determines the ecological protection goals which become codified as norms. The role of the law is therefore to determine whether these norms have been violated or not. Following, the next section will examine existing ethics with a view to arguing for an effective ethic for regulating the impact of GMC production on the environment.

Striking a balance with a foundational threshold: deontological versus utilitarian

Ethical reference systems (ERS) usually swing between deontological and utilitarian. Under the deontological ERS, exposing another to risk (without their consent) is only admissible if all precautionary measures have been taken to reduce the risk to a point where the occurrence of harm can be deemed unlikely. Risk thresholds then serve to determine how far a risk must be reduced in order to be acceptable. Therefore from a deontological perspective, it is easier to distinguish what is *acceptable* from what is *accepted*. Deontology therefore creates a duty of care that limits risks to a particular level; moral right to life must be respected (Ross 1930: ch 1; Freeman 1994: 313–349; Scanlon 1998). Under the utilitarian system, on the other hand, decisions are based on highest expected benefits; duty to maximize utility.

According to deontological approaches decisions are based on an absolute normative threshold defining a limit that no risk may exceed. The main idea underlying the deontological criterion is that there is a duty of care according to which risks should be reduced to a point where the occurrence of harm is not to be expected. Requiring zero risk, however, cannot be justified since this would make social life impossible. That is why a normative threshold is introduced. Risks exceeding this threshold are prohibited, irrespective of the chances associated with them, while risks below the threshold are acceptable.

If the risk of an option exceeds the defined threshold, it has to be reduced by appropriate risk management options to possibly remain below the threshold. Within the acceptable options (that is, those options where the risk remains below the given threshold), the one bearing the lowest risk or impact may be preferred.

The main idea underlying the utilitarian criterion is that there is only one moral duty: the duty to maximize expected net benefit. If this goal can only be reached by exposing other individuals or populations to certain risks, this must be done, even if these risks are very high. For each option, the total chance is compared to the total risk. There is an obligation to choose the option with the highest expected benefit (total chance minus total risk). When following a utilitarian approach, there is no normative threshold (Jacobson 2008: 159–191; Conway and Gawronski 2013: 216–235).

Within the context of the core issue in this paper – regulating the impact of GMC production on the environment – an ethical reference system that is deontological (largely precautionary) is preferred. This is because, whether for or against, the claims about the impacts of GMC production have not been overwhelmingly substantiated by science; and waiting for the conclusions of science before precautions are taken is manifest evidence of a feigned commitment to sustainable development. The next section therefore comprises an attempt to create a deontological framework that should effectively regulate the impact of GMC production on the environment.

A compromise legal framework: an attempt

So far, the existing regulations on GMCs provide (in different ways) for *what* the developers of GMCs should avoid, but not *how* it should be avoided. The *how* is what is crucial to effective management of the environment.

An essential means for ensuring that the trans-boundary movement of GMOs takes place within a framework of both national and international regulations is the Cartagena Protocol on Biosafety. Launched in 2000, the Protocol entered into force on 11 September 2003 and to date has 130 Parties. Although the Cartagena Protocol incorporates the Precautionary Principle (see article 11(8)), it is however neutral on the topic of whether GMOs should be introduced. Rather, it is designed to increase public confidence in the safety of any proposed introductions and marketed products, while providing the public and private sectors involved in the biotechnology industry and any farmers that use GMOs with a commercially valuable legal right to import, introduce, transport, or develop GMOs. Although this Protocol does not specifically address the impacts of GMCs on the environment, it recognizes that the precautionary principle is crucial in biotechnology-related issues. Flowing from this, this paper therefore argues that placing the preservation of biodiversity in a relational context has an important consequence for environmental preservation because it contextualizes and places limits on property rights like the genetic modification of crops. It will be implicit in such a framework that property owners (scientists, in this case) will look to (the continuity) of biodiversity as the standard or measure for their action. Thus a regulation on GMCs should have the following form:

The genetic modification of crops must be within the context of careful consideration and knowledge of the needs of the environment and its unique functions:

- The exploitation of biodiversity shall therefore be acceptable for the purpose of genetically modifying crops, to the extent that these vital functions are maintained.
- "Acceptable" shall denote that all practicable diligence has been taken to prevent harm to biodiversity.

The use of the word "acceptable" in the draft above is crucial. The word is used instead of "accepted" indicating that there is a standard for acceptability and that (objective) standard must be that while the GMCs are being produced and explored, nature should ALSO be able to maintain her vital functions; and due diligence must be employed to ensure that this is so.

Conclusion

Humanitarian, environmental and global security concerns demand a global commitment to improve the lot of the large proportion of the human population that is currently food insecure or vulnerable to food insecurity. This requires that we must build resilience to climate shocks and food price volatility, halt degradation, and boost productive assets and infrastructure. GMCs have the potential to solve much of the world's hunger and malnutrition problems, and to help protect and preserve the environment by increasing yield and reducing reliance upon chemical pesticides and herbicides. Yet there are challenges ahead for governments, especially in the areas of safety testing, regulation, international policy and food labelling. We must proceed with caution to avoid causing unintended harm to human health and the environment. This chapter is not advocating that every single crop and plant be genetically modified; rather it cautions that where there must be a modification, then it should be undertaken with all the precautionary measures that ensure nature's vital functions are maintained in the process; and where such precautionary measures have not been engaged, then the modification process should not proceed.

References

Altieri, M. (2015) "The Ecological Impacts of Agricultural Biotechnology". Available at www.actionbioscience.org/biotechnology/altieri.html·imer (accessed 16 October, 2015).

Ammann, K. (2009) "Biodiversity and Genetically Modified Crops". In N. Ferry and A. Gatehouse (eds), *Environmental Impact of Genetically Modified Crops*, CABI, Wallingford, pp. 240–264.

Ayinde, O. and Olatunju, G. (2011) "Effect of Climate Change on Agricultural Productivity in Nigeria: A Co-Integration Model Approach". *Journal of Human Ecology* 35(3): 189–194.

Brookes, G. and Barfoot, P. (2010) *GM Crops: Global Socio-Economic and Environmental Impacts 1996–2008*, PG Economics, Dorchester.

Conway, P. and Gawronski, B. (2013) "Deontological and Utilitarian Inclinations in Moral Decision Making: A Process Dissociation Approach". *Journal of Personality and Social Psychology* 104(2): 216–235.

Crawley, M. et al. (2001) "Biotechnology: Transgenic Crops in Natural Habitats". *Nature* 409: 682–683.

Dahleen, L. (2001) "Transgenic Approaches to Combat Fusarium Head Blight in Wheat and Barley". *Crop Science* 41(3): 627–628.

FAO (2012) *The State of Food Insecurity in the World*, Food and Agriculture Organization of the United Nations, Rome.

Freeman, S. (1994) "Utilitarianism, Deontology, and the Priority of Right". *Philosophy and Public Affairs* 23: 313–349.

Gornall, G. et al. (2010) "Implications of Climate Change for Agricultural Productivity in the Early 21st Century". *Philosophical Transactions of the Royal Society B* 35(1554): 2779.

Green, J. and Owen, M. (2011) "Herbicide Resistant Crops: Utilities and Limitations for Herbicide Resistant Weed Management". *Journal of Agricultural and Food Chemistry* 59(11): 5819–5829.

Groffman, P. et al. (2006) "Ecological Thresholds: The Key to Successful Environmental Management or an Important Concept With no Practical Application?" *Ecosystems* 9(1): 1–13.

Harris, B. (2009) "On People and Planet". Available at http://climateandcapitalism.com/2009/01/22/how-climate-change-threatens-food-security (accessed 16 October 2015).

Harris-Lovett, S. (2015) "GMO Rice Could Reduce Green House Gas Emissions". *Los Angeles Times*, 22 July, available at www.latimes.com/science/sciencenow/la-sci-sn-gmo-rice-methane-emissions-20150722-story.html (accessed 16 October 2015).

Helmuth, J. (2000) "Biotechnology: Both Sides Claim Victory". *Science* 287: 782–783.

Huggett, A. (2005) "The Concept and Utility of 'Ecological Thresholds'". *Biological Conservation* 124: 301–310.

IPCC (2013) *Climate Change: The Physical Science Basis*, Cambridge University Press, Cambridge.

Jacobson, D. (2008) "Utilitarianism Without Consequentialism: The Case of J. S. Mill". *Philosophical Review* 117(2): 159–191.

Kloeke, E. "How Will Climate Change Affect Food Security?" Available at www.elsevier. com/connect/how-will-climate-change-affect-food-security (accessed 16 October 2015).

KREM (2015) "Monsanto Denies Liability in Spokane River Pollution Suit". August 5, available at www.krem.com/story/news/local/spokane-county/2015/08/05/monsanto-denies-liability—spokane-pollution-suit/31178273 (accessed 16 October 2015).

Lamichhane, S. (2014) "Genetically Modified Foods – Solution for Food Security". *International Journal of Genetics* 5(1): 43, 45.

Lobell, D. and Gourdji, S. (2012) "The Influence of Climate Change on Global Crop Productivity". *Plant Physiology* 160(4): 1686.

Luck, G. (2005) "An Introduction to Ecological Thresholds". *Biological Conservation* 124: 299–300.

Moellenbeck, D. (2001) "Insecticidal Proteins From *Bacillus Thuringiensis* Protect Corn From Corn Rootworms". *Nature Biotechnology* 9(7): 668–672.

Muradian, R. (2001) "Ecological Thresholds: a Survey". *Ecological Economics* 38(1): 7–24.

Pua, E. and Davey, M. (eds) (2007) *Biotechnology in Agriculture and Forestry 59: Transgenic Crops IV*, Springer, Berlin.

Ross, W. (1930) *The Right and the Good*, Clarendon Press, Oxford.

Scanlon, T. (1998) *What We Owe to Each Other*, Harvard University Press, Cambridge, MA.

Scheffer, M and Carpenter, R. (2003) "Catastrophic Regime Shifts in Ecosystems: Linking Theory to Observation". *Trends in Ecology and Evolution* 18(12): 648–656.

Sengar, R. and Sengar, K. (2014) *Climate Change Effect on Crop Productivity*, CRC Press, Boca Raton, FL.

Tencalla, F., Nickson, T. and Garcia-Alonso, M. (2009) "Environmental Impact Assessment". In N. Ferry and A. Gatehouse (eds), *Environmental Impact of Genetically Modified Crops*, CABI, Wallingford, pp. 61–73.

United Nations (1999) *Human Development* Report, Oxford University Press, Oxford.

United Nations (2010) "Millennium Development Goals Fact Sheet". Available at www. un.org/millenniumgoals/pdf/MDG_FS_1_EN.pdf (accessed 16 October 2015).

Watson, E. (2015) "Pope Francis Weighs Into the GMO Debate". Available at www. foodnavigator-usa.com/People/Pope-Francis-weighs-into-the-GMO-debate? utm_source=copyright&utm_medium=OnSite&utm_campaign=copyright (accessed 16 October 2015).

World Bank (2007) *World Development Report 2008: Agriculture for Development*, World Bank, Washington, DC.

7 The water–energy nexus

The role of water law

Joseph W. Dellapenna

Introduction

Beneath our canopy of air, there are two essential resources for our lives – water and energy. We, and most of what we consume, depend on adequate supplies of water: We are but walking water. Some 60 per cent or more of our bodies are water (Guyton 1991: 274). As Tom Robbins wrote:

> Water—the ace of elements. Water dives from clouds without parachute, wings, or safety net. Water runs over the steepest precipice and blinks not a lash. Water is buried and rises again; water walks on fire and fire gets the blisters. Stylishly composed in any situation—solid, gas, or liquid—speaking in penetrating dialects understood by all things—animal, vegetable, or mineral—water travels intrepidly through four dimensions, *sustaining* (Kick a lettuce in the field and it will yell "Water!"), *destroying* (The Dutch boy's finger remembered the view from Ararat), and *creating* (It has even been said that human beings were invented by water as a device for transporting itself from one place to another, but that's another story). Always in motion, ever flowing (whether at steam rate or glacier speed), rhythmic, dynamic, ubiquitous, changing, and working its changes, a mathematics turned wrong side out, a philosophy in reverse, the ongoing odyssey of water is virtually irresistible.
>
> (Robbins 1976: 1–2)

Energy like water is central to our existence. Biologic processes depend upon energy and also create or release surplus energy (Scilli Staff 2009). With sufficient energy we can solve many problems. We can manufacture what we need, we can travel the universe, and we can manage, or even create, water. There is increasing recognition of inter-connection between water and energy (Healy et al. 2015). Little can be done with water (at least above the smallest scale) without significant inputs of energy; and little can be done to generate or harness energy without large amounts of water.

Consider the processes by which we manage water. We build dams, we pump water to abstract it and move it from place to place, and we purify water before and after use. We even "produce" water ancillary to the extraction of oil or natural gas (Engel et al. 2014). We do not undertake all of these processes for every use of water, but we undertake some of these processes for each use of water other than navigation – and, of course, navigation today is likely to involve oil, gas, or other thermal fuels, or at least harnessing the wind.

Producing energy also requires large amounts of water. Besides the use of water in hydroelectric generation, producing energy in a thermal power plant uses lots of water (Sovacool *et al.* 2014). We boil water to make steam (or at least pressurized fluids) to turn turbines (Elliott *et al.* 1997; US EPA 1997), and in nuclear plants we used controlled nuclear reactions to boil water (or some other liquid) (Cooke 2009). And in both, we are likely to use water to cool the system (US EPA 1997: 79). Geothermal (Bertani 2007) or tidal (Lewis *et al.* 2015) electricity generation are just new ways of exploiting water to make energy. While internal combustion engines do not exploit water to make energy, they often are water cooled. One of the most important problems with fracking is that it requires enormous amounts of water to produce gas or oil – water that competes with other potential uses and often is rendered useless for other users because of its contamination in the fracking process (Gaba 2014; Gallegos *et al.* 2015; Gray 2014, 2016). While wind and solar may appear to be exceptions to the use of water, water is consumed in producing wind turbines (Mortholst *et al.* 2002) and solar cells (IEA 2014). And this does not even take into account the water use or impacts of mining coal and other hard minerals necessary for energy production nor the water necessary to grow or process biofuels (Burr *et al.* 2012).

Although thermoelectric power generation is commonly said to use by far the largest volume of water in the United States today (even more than irrigation; see Figure 7.1), this statement is somewhat misleading. It underestimates total use and does not indicate (and therefore overestimates) how much of the water is consumed. Underestimating arises because some of the water used for biofuels and manufacturing other energy equipment, and even cooling internal combustion engines are subsumed under the headings of "irrigation", "industrial" or "domestic" use. Overestimating arises because significant amounts of the water used in thermoelectric power generation are returned to a water source, allowing reuse. In fact, according the US Geological Service, only about 4 per cent of the water used in generating thermoelectric power is not recycled for other uses (Kenny *et al.* 2009).

The relation of water and energy is changing because of growing populations, changing technologies for exploiting water and energy, and changing levels and types of economic activity. We also live on a planet undergoing disruptive climate change (IPCC 2013). Without attempting precise predictions, we can foresee climate disruption changing the timing and nature of precipitation throughout much of the planet (IPCC 2014: ch. 3). Hotter temperatures and drier air will cause higher rates of evapotranspiration and drier soils, both less supportive of plant life without irrigation (ibid.: 241–243). Arid regions will become wider (Montenegro and Ragab 2010). The melting of glaciers and mountain snowpack is reducing or destroying the storage capacity of these immense reservoirs of fresh water, which sustain rivers during dry months (IPCC 2014: 236–243, 253, 256). Climate disruption will bring more extreme events – droughts and floods – at more frequent intervals (ibid.: 246–248, 252–253). Rising sea levels will disrupt

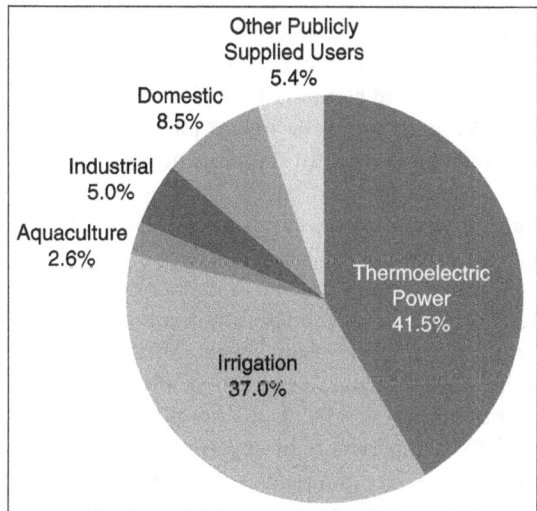

Figure 7.1 US freshwater withdrawals (2005)

Source: data from Kenny *et al.* (2009).

Notes: Livestock and mining combined use approximately 1 per cent of total use and are not included.

exiting water and energy infrastructure (Ayyub *et al.* 2012). In coastal areas, salt water will intrude into fresh waters upon which humans today depend (Ranjan 2012). And in many areas, the availability of water will decrease substantially (IPCC 2014: ch. 3).

Demand for energy will increase dramatically. If the available water declines, while populations grow or remains stable and economic activity also either grows or remains stable, more energy will be needed to abstract, store, treat, move, and use water. If regions become hotter, demand for energy for air conditioning will increase. One prediction even suggests that the Persian Gulf region will become so hot as to be nearly unliveable (Pal *et al.* 2015). Yet most energy production will be highly sensitive to declines in available water. These changes allow us to predict that climate disruption will cause more stress for already stressed water management regimes, rendering obsolete existing arrangements for water management even in regions where water has historically been plentiful (Dellapenna 2010). This leads to the question of just what role can water law play in crafting successful responses to these challenges.

What role can water law play?

Pressure for water law reform at the national and international levels has been growing across the world because of population growth and changing patterns of use (Dellapenna 1997). The pressure intensifies with climate disruption. In designing legal reforms, the problem is to balance competing metrics – the conflict between the public and the private interest in water, and the need for flexibility to meet our needs and for adequate certainty to enable investment in water facilities and infrastructure. We must build the institutional competence necessary for proper water management and we must create greater capacity for cooperation across agency boundaries and between public agencies and private entities. And we must develop funding for all of this in an increasingly neoliberal world were taxes are anathema and markets reign supreme.

Despite water's ubiquity, usable water is a scarce and valuable resource. The United States provides an interesting case study of the possibilities because water allocation law in the United States remains state law. There are three distinct bodies of water allocation law in the United States, each with characteristics that render it more or less capable of providing suitably adaptive responses to global climate disruption (Dellapenna 2008). Riparian rights evolved to the east of Kansas City as a body of law that treats water as a species of common property. Appropriative rights emerged to the west of Kansas City as a body of law that treats water as a species of private property. Finally, in the second half of the twentieth century, about half of the states that had applied riparian rights developed a third model, regulated riparianism, predicated on treating water as public property (Dellapenna 2015: §6.01(b)(1)). The correspondence between forms of American water law and the several types of property is more than a simple curiosity. It enables comparison to other forms of water law across the globe, and it enables prediction of whether existing forms are adaptable to changing circumstances, or whether an entirely new form must be substituted when water demand or supply changes dramatically. Each model must be evaluated for its ability to cope with global climate disruption both in terms of the water economy and on the energy economy, and regarding the reforms that might allow appropriate responses.

Water is a common pool resource, one that inevitably passes into and out of one's control; in this sense, water cannot be owned. Water, in some settings, is a private good (Gleick *et al.* 2002). We have all bought bottled water. Few things in this world are strictly indivisible and public. Yet the peculiar characteristics of water – its critical importance, its ubiquity, its heterogeneity, its renewability, its commonality, and its vulnerability – combine to make markets unworkable for bulk water in its natural state (Dellapenna 2000b). Bulk water in its natural state is the quintessential "public good". Even market fundamentalists use water metaphors when discussing what even they concede are public goods: "common pool resource"; "spillover effects"; and so on. From a legal point of view, the most central managerial problem regarding public goods is: How can we recover the cost of maintaining or enhancing the good when a significantly large group of people have access to, and the legal right to use, it without direct charge?

Partly because of water's public, ambulatory nature, water has been treated as a free good. Users seldom pay for water when they remove it from it a natural source, and even public delivery systems seldom charge for the water, as opposed to the services of extracting, purifying, delivering, and carrying away water after its use. From this arises the complex difficulties of treating water as property: to some meaningful extent it is "non-excludable" – water is either available (in quantity and quality terms) to a large number of users drawing from a common pool or it is available to no one (Kaul *et al.* 1999). It also means that investments in developing and protecting the resource are subject to free riders, and markets are difficult to establish, maintain, or operate (Dellapenna 2000b). Given the difficulties then with markets as a primary management tool, here I will not address the possibility of markets in detail.

Riparian rights (common property)

The system of "riparian rights" take its name from the Latin word *ripa*, meaning a riverbank (Dellapenna 2015). Rights pertaining to lakes, seas, and other large bodies of water, historically called "littoral rights" (from the Latin *litus*, meaning "shore"), today are included in riparian rights. Riparian rights provide a rule for the allocation of water to particular uses (ibid.: ch. 7). Under riparian rights, the right to use water is a natural attribute of the land, dependent on the natural availability of water to riparian land (land abutting or underlying water sources). In 2015, about half of the states east of Kansas still apply classic riparian rights to disputes over the allocation of water.

Many scholars assert that riparian rights once required protection of the "natural flow" of a watercourse, with only domestic uses of water (used to meet immediate human survival needs) being allowed to alter materially a water source in quantity or quality. In fact, courts never enforced a "natural flow rule" (ibid.: §7.02(c)). Instead, courts embraced the "reasonable use rule", under which riparian landowners can use water as they choose so long as each does not transgress the equal right of other riparians (ibid.: ch. 7). "Natural flow" language continues to appear in riparian rights cases, but courts always apply the reasonable use theory (ibid.: §7.02(c)).

The reasonable use rule is a common property system, under which owners of land contiguous to a watercourse are equal co-owners of the right to use the water. As co-owners, they exercise their individual judgment on whether, when, and how to use it. A court intervenes only when a use by one co-owner interferes directly with a use by another co-owner. This cannot be a simple inquiry of whether one water use is interfering with or harming another for each use necessarily interferes with the other; whichever prevails necessarily destroys or diminishes the other (Coase 1960). The result is a relational test that involves weighing the value of competing uses against each other to determine which is more socially valuable (Dellapenna 2015: §7.02(d)(2)).

Courts often discuss this balancing in vague terms. Yet a few things are clear. Besides not simply protecting natural flows, courts do not necessarily protect uses that began earlier (ibid.: §7.03(d)). Temporal priority has seldom, if ever, been relevant to riparian rights. Some have argued that the economic value of the competing activities explains the decisions, which would require a court to reconsider the result whenever product values changed significantly. Courts seem to give only minimal, if any, attention to such non-economic variables as the natural characteristics of the stream, general social concerns, or abstract justice (ibid.: §7.03(d)(3)). If these or other variables did affect the decision, then a change in these circumstances would also require reconsideration of the reasonableness of the competing uses.

Courts applying reasonable use sometimes try to avoid problems through *pro rata* sharing among competing users (ibid.: §7.03(c)(1)). That cannot work when a *pro rata* share is too small for anyone's use or for widely differing uses. Yet applying the reasonable use theory runs into serious difficulties, including vagueness, instability, unpredictability, the lack of processes for coping with extreme shortages or for protecting public values (ibid.: §7.05(a)), and systematic bias favouring large users (ibid.: §7.03(d)(3)). The only firm rule is that use on non-riparian land is *per se* unreasonable (ibid.: §7.02(d)(1)). These serious difficulties were thought sufficient to justify the shift from riparian rights to appropriative rights in the west, and the shift to regulated riparianism in the east (ibid.: chs 8, 9), and make riparian rights unlikely to cope with the problems that arise in meeting future energy needs or otherwise that arise because of global climate disruption.

The foregoing difficulties derive from the nature of riparian rights as common property. The result of such a common property system will be the "tragedy of the commons" (Hardin 1968). The tragedy arises because a common property system – a system of open access with individual decision making – can only function successfully when the common pool resource is available in much greater supply than the demand for the resource. Because each common owner can decide for herself whether to increase use regardless of the effect on other common owners (except for direct interference), each owner is free to appropriate the whole of each additional increment of use, but the entire group shares the cost imposed on the common resource. If we decide only by an individualized cost-benefit analysis, we will continue to grab until the resource is exhausted.

Commons have functioned successfully over extended periods even when close to their carrying capacity through informal rules enforced by small communities (Benkler 2006; Ostrom 1990). Such examples are irrelevant for a larger society where most persons are strangers to each other, informal sanctions are not effective, and formal law imposes no real limits on exploiting the commons (Dellapenna 2000a: 860–876). Under these circumstances, typical of states applying riparian rights, the tragedy of the commons will play out. This is more than a theoretical model. Common pool resources have been destroyed repeatedly in the past century when common property was not displaced by a different rule (e.g. Wyman 2008).

Appeals to moderation, ethics, and morality are self-defeating. Those who respond to such appeals simply leave the field to other common owners who continue to increase their exploitation to the point of exhausting the resource. Responsible users soon realize that heeding moral appeals reduces their gains with little or no benefit to the common resource. Many who agree with the moral appeal then join in grabbing whatever they can before it is all gone.

When exploitation of a common resource requires significant capital investment, the inability of investors to keep others from pre-empting their uses could cause underinvestment in the resource (Rose 1986). Fear of just such problems, deriving from the pervasive uncertainty under classic riparian rights, was used to justify rejection of riparian rights in the drier, western third of the United States in favour of a private property system (Dellapenna 2015: §8.02).

The reasonable use theory provides the flexibility necessary for adaptation to climate disruption, but the slow, laborious process of litigation between two individuals at a time is ill-suited to adapting water use to changing supply or demand. Decisions are made by judges who are not versed in hydrology, meteorology, or economics. Nor would a reversion to the natural flow theory help – it would only freeze out consumptive uses just when communities need new water sources. The tragedy of the commons has already led about half of the eastern states to abandon riparian rights in favour of regulated riparianism (Dellapenna 2015: ch. 9). Climate disruption most likely will only accelerate as the effects of climate disruption become more widely felt.

Appropriative rights (private property)

European settlers in the region west of Kansas City needed water for mining, irrigation, and later industrial and municipal uses, but they could not satisfy their needs under riparian rights (Dellapenna 2015: ch. 8). The "Anglo" intruders swept away Spanish-Mexican law and law that could have been derived from aboriginal practices (ibid.: §8.01). Regarding both land and water, the early prospectors were trespassers who took what water they needed, enforced by "vigilance committees" elected by the prospectors themselves. The resulting vigilante law was based upon the most elementary notion of justice – first in time, first in right (Hundley 1992: 67–73). The later organized governments had little choice but to follow the customs of the miners. After 150 years, the miners' rule has been developed into a complex system of water administration in every appropriative rights state. Eventually, everywhere across the West, appropriative rights displaced riparian rights (Dellapenna 2015: §§8.02–8.04).

Appropriative rights basically is a private property approach to water allocation in which water rights are defined as to quantity, time, place, and manner of use, and most importantly according to temporal priority relative to other uses. Like riparian rights, appropriative rights have serious problems even without reference to climate disruption (Gaffney 1969). Appropriative rights are more uncertain than the governing principle – first in time, first in right – suggests, if only because the earliest priority dates predate the modern administrative machinery. Despite statutes and legal proceedings designed to put these claims on record, the earliest (and hence most valuable) water rights still have not been quantified on some watercourses, while prescriptive, abandoned or forfeited rights also create gaps in the official record (Goplerud 2015).

There is a more serious problem than gaps in the records. Appropriative rights actually encourage waste because the "first in time, first in right" principle rewards "jumping the gun" to use water before it is needed, or even if it will never actually be needed (Neuman 1998). Diverting water, a cost to society, is a private gain to an appropriator, creating a claim to water in the future, but one must invest real social capital to divert, store, and apply water. Capital is switched from productive uses to the capturing of sub-marginal resources. As a result, excessive diversions are the rule under appropriative rights, yet for most appropriations there are inadequate investments in the post-diversionary aspects of development, especially to save water. Appropriators thus live in a setting where it is smart to waste water.

Even if all appropriators were careful only to appropriate as much water as they actually need, the legal regime violates two fundamental economic principles: (1) marginal productivity; and (2) pooling of risk (Malloy 1990: 20–33). Under the rule of "first in time, first in right", when there is not enough water to satisfy all water rights, a junior appropriator must drop out first and lose everything before the next senior appropriator loses anything. Senior appropriators are protected by exaggerating the risk to junior appropriators. A junior appropriator may lose marginal units of higher productivity than the least valuable marginal units of a senior appropriator. There is no pooling of risk whatsoever. Appropriative rights thus introduce changes in the aggregate variability of supply beyond the natural variability and distribute these risks unequally. When administered by large irrigation districts, appropriated water may be applied to rationally defined service areas. When claimed by individuals or small districts, a further inefficiency arises because areas served from a single source are generally scattered. The farther one is from a source, and the more convenient it is to others, the greater the incentive to be first. So typically, the first claimants on a source are scattered and soon the supply is fully claimed. The included dry lands can never get water from the closest source. They can, however, find other, more remote, sources. The results need not be imagined, they may be seen throughout the West.

Even more than for riparian rights, appropriative rights favour large users. To make an appropriation, it must be used "beneficially" (Neuman 1998). The amount of water that one can use beneficially is a function of the amount of land or the size of the factory that one owns. Appropriative rights thus tend to distribute what are defined in law as publicly owned waters to those who already own the most. No effort was made to protect the public interest in the waters or to distribute the fruits of those waters among the disadvantaged of society (Gaffney 1969: 138). Many appropriative rights states have now enacted statutes requiring consideration of the public interest in evaluating applications to make a new appropriation, but such statutes do not preempt existing water rights and thus have little practical effect in any water basin in which most or all available water has already been appropriated. Whether today we should favour protecting endangered species or other public values rather than irrigation is a hotly debated question, but a debate that almost cannot occur under appropriative rights or other private property regimes.

These rather serious problems could perhaps be dealt with fairly easily if markets worked under appropriative rights, but, for reasons that I have addressed at length elsewhere (Dellapenna 2000b), there never has been a market for appropriative rights to any significant extent. Appropriative rights thus create a rather peculiar form of private property, one that rather than ensuring free transferability and efficient use of the resource, effectively freezes uses in place unless the state intervenes directly to transfer the water to other uses (Dellapenna 2015: §§6.01(b)(2), (b)(3)). Small-scale transfers of water rights among users who are making similar uses at more or less the same place are the only ones that regularly occur under appropriative rights without major state intervention to dictate the transactions and their terms.

These serious problems make appropriative rights even less able to respond to global climate disruption. A rigid freezing of water use in wasteful patterns that disregard fundamental principles of marginal utility is the very opposite of what will be necessary when water use patterns need to be adjusted to a new reality. Sustaining appropriative rights where they already exist will be difficult or impossible, and major efforts are underway to subvert them through a pretence of markets that are not really markets (ibid.: §§6.01(b)(2), (b)(3)). Replacing riparian rights with appropriative rights would face additional severe impediments in states with numerous and well-developed water uses (ibid.: §8.05).

Regulated riparianism (public property)

Hawaii and more than half of the states east of Kansas City now use administrative permit systems to replace traditional riparian rights, or to administer groundwater, or both (Dellapenna 2015: ch. 9). Rather than importing appropriative rights, these states have developed a system of water administration called "regulated riparianism" because it is based on riparian principles (ibid.: §6.01(b)(1)). This really is a system of public property. The transition from limited regulatory intervention to more or less comprehensive regulation often occurred incrementally rather than from conscious design. As a result, it is debatable whether certain states have moved from relying largely on common law riparian rights to regulated riparianism. The following summary description of regulated riparianism is based upon its common core of principles as found in actual regulated riparian statutes and articulated in the *Regulated Riparian Model Water Code* (ASCE 2003). No state has a system precisely like the one described here or in the Model Code, although several come very close.

Regulated riparianism fundamentally departs from classic riparian rights in requiring that no water is to be withdrawn from a water source without a time-limited permit from the state where the withdrawal occurs (Dellapenna 2015: §9.03(a)). The permits determine the water right, not the riparian nature of the use, yet the new system remains within the riparian tradition because the criterion for deciding on permits is whether the proposed use is "reasonable" (ibid.: §9.03(b)). The criterion of "reasonableness" is applied very differently than at common law. The administering agency decides before a use begins whether the use is reasonable, both in terms of general social concerns and in terms of effects on other permitted uses (ibid.: §§9.03(a)(5)(A), 9.03(b)). Water users gain a significant advantage because they know for the duration of the permit whether their use is reasonable; they cannot be caught unaware by a judicial decision that wipes out their investment without a penny of compensation. Permits allow potential investors to gauge whether their investments can be profitable and inform potential investors about the proper scale of the investment.

The administering agency imposes conditions designed to protect other lawful users and public values (ibid.: §§9.03(a)(5)(A), 9.05). The statutes often provide preferences for certain classes of uses (ibid.: §§9.03(a)(3), 9.05(c)). Unlike appropriative rights, temporal priority has a strictly limited role (ibid.: §9.03(b) (3)). Uses on non-riparian land are no longer unreasonable *per se* (ibid.: §9.03(2) (2)). Finally, permits generally are issued only for a specific period of time (ranging one to twenty years) (ibid.: §9.03(a)(4)). Upon expiration of a permit, the continued reasonableness of a use is re-examined, introducing a desirable flexibility.

While users are sometimes made to pay recurring fees for permits, the fees are not payment for the water itself (ibid.: §9.03(a)(5)(c)). Statutes that set a uniform fee regardless of the amount of water used clearly do not charge for the water used. Even if the fee is variable, it generally is set according to the presumed ability of the user to pay, rather than according to the value of the water or its use. Charging based on water use could provide incentives for efficient use of water in a setting where markets are not likely to work (ASCE 2003: §4R-1-08).

Today, the main quantity and quality threats to the availability of water in eastern states are not pollution or withdrawal, but manmade physical and ecological transformations of water sources and the lands on or in which the sources are found. Global climate disruption is the most extreme example. Regulated riparian statutes address these problems through mandated planning and protections for the public interest, including requirements for the agency to define and protect a minimum flow (Dellapenna 2015: §9.03(b)), specific protections for the public interest in water resources (ibid.: §§9.05(b)(3) to 9.05(d)), and long-term planning (ibid.: §9.05(a)). The administering agency also usually has broad discretion to plan for and to deal with extreme water shortages, including suspending permits in whole or in part (ibid.: §9.05(d)). The agency can incorporate permit conditions based on its plans or as necessary to protect the public interest.

Fear of the political (and legal) repercussions of a transformation of traditional water rights has led many state legislatures to exempt from the permit requirement large classes of users (usually agricultural) who were using water when the new statute came into effect (ibid.: §9.03(a)(3)). This introduces a significant temporal element. A more sophisticated solution would be to guarantee existing users an initial permit, but subjecting renewal to the same process as any other permit, limiting temporal preferences to a single permit cycle (ibid.: §9.03(b) (3)). Existing users who decline to apply for a permit within a short period of time after the new regulated riparian system comes into effect can be conclusively presumed to have abandoned their water right. There is some evidence, however, that administering agencies prefer to use temporal priority or *pro rata* sharing as the allocation method least likely to provoke litigation or other difficulties for the agency (ibid.: §9.053(f)). This sabotages the whole scheme of regulated riparianism. No clear way to prevent such sabotage has emerged.

Regulated riparianism scores well for introducing flexibility and for protecting public values compared to appropriative rights, important dimensions of adaptation to global climate disruption. Regulated riparian statutes usually do not provide for the transfer of water rights between potential users (ibid.: §9.03(d)). Even if this were done, there seems to be little reason to think a market would function better under regulated riparianism than under appropriative or riparian rights (Dellapenna 2000b). In practice, the agencies free up far less water through the renewal process than theory suggests because the agencies prefer to tighten conditions on existing uses rather than to deny renewal outright (Dellapenna 2015: §9.03(a)(4)). Non-renewal of permits probably will remain an infrequent and cumbersome device unless the state is willing to create a good deal of investment insecurity and political controversy.

Another problem that could impact on adaptation to global climate disruption would be if regulated riparian statutes do not provide adequate security for investment in water facilities. If the duration of a permit is too short, leaving too little time to recover the initial cost of a project, private actors will be reluctant to invest in water facilities. Additional uncertainty could arise from the administering agency's power to modify permits during water emergencies (unforeseen water shortages). In the actual operation of regulated riparian systems, however, neither investment nor other insecurities seem to have caused difficulty (ibid.: §9.03(a)(4)). Administering agencies have been, if anything, too sensitive to large institutional investors in water. The cost of imposing an elaborate administrative system might also be substantial (ibid.: §9.05(a)(5)(D)). Similar problems could be anticipated in other public property systems.

Conclusion

Law can either impede or facilitate adaptation to global climate disruption. The three models of water law found in the United States can be taken as exemplars of the basic approaches to water law found in many countries. None of these three models of water law – riparian rights, appropriative rights, and regulated riparianism – guarantees that societies will strike a good balance between the public interest and private need, between flexibility and certainty, that will enable optimal adaptation. Yet given the increasing failure of traditional riparian rights (a common property system), and the only slightly better performance of appropriative rights (as close to a private property system as we are likely to achieve for water), and the general failure of markets as a basic water management tool (as opposed to state intervention masquerading as a market), there seems little choice but to move to a public property (regulated riparian) system. Regulated riparianism is not a perfect system, but it appears so far to be the best suited to coping with the enormous challenges that the world will face during the coming century. This conclusion, if valid, would have important implications for energy policy as well. Regulated riparianism would enable more thorough and effective planning both for energy consumption in water management and for water use in energy production. In other words, this newest form of water law (at least in the United States) deserves careful consideration in any effort to reform water law to respond to the ongoing global climate disruption.

References

ASCE (2003) *The Regulated Riparian Model Water Code*, ASCE Stdd. 40–03, ed. J. W. Dellapenna, American Society of Civil Engineers, Reston, VA.

Ayyub, Bilal M. and Michael S. Kearney (2012) *Sea Level Rise and Coastal Infrastructure: Prediction, Risks, and Solutions*, American Society of Civil Engineers, Reston, VA.

Benkler, Yochai (2006) *The Wealth of Networks: How Social Production Transforms Markets and Freedom*, Yale University Press, New Haven, CT.

Bertani, Ruggero (2007) "World Geothermal Generation in 2007". *Geo-Heat Centre Quarterly Bulletin* 28(3): 8–19.

Burr, Carolyn F., Rebecca W. Watson, and Chelsea Huffman (2012) "Water: The Fuel for Colorado Energy". *University of Denver Water Law Review* 15: 275–327.

Coase, Ronald H. (1960) "The Problem of Social Cost". *Journal of Law and Economics* 3: 1–44.

Cooke, Stephanie (2009) *Mortal Hands: A Cautionary History of the Nuclear Age*, Bloomsbury, New York.

Dellapenna, Joseph W. (1997) "Population and Water in the Middle East: The Challenge and Opportunity for Law". *International Journal of Environment and Pollution* 7: 72–110.

Dellapenna, Joseph W. (2000a) "Law in a Shrinking World: The Interaction of Science and Technology with International Law". *Kentucky Law Journal* 88: 809–883.

Dellapenna, Joseph W. (2000b) "The Importance of Getting Names Right: The Myth of Markets for Water". *William and Mary Environmental Law and Policy Review* 25: 317–377.

Dellapenna, Joseph W. (2008) "Climate Disruption, the Washington Consensus, and Water Law Reform". *Temple Law Review* 81: 383–432.

Dellapenna, Joseph W. (2010) "Global Climate Disruption and Water Law Reform". *Widener Law Review* 15: 409–445.

Dellapenna, Joseph W. (2015) "Riparianism". In Amy Kelly and Robert Beck (eds), *Waters and Water Rights* 1: chs. 6–9, OverDrive, Cleveland, OH.

Elliott, Thomas C., Kao Chen, and Robert C. Swanekamp (1997). *Standard Handbook of Powerplant Engineering*, 2nd edn, McGraw Hill, New York.

Engel, Mark A., Isabelle M. Cozzarelli, and Bruce D. Smith (2014) "USGS Investigations of Water Produced during Hydrocarbon Reservoir Development". USGS Survey Fact Sheet 2014–3104, 16 July, available at http://dx.doi.org/10.3133/fs20143104.

Gaba, Jeffrey M. (2014) "Flowback: Federal Regulation of Wastewater from Hydraulic Fracturing". *Columbia Journal of Environmental Law* 39: 251–318.

Gaffney, M. Mason (1969) "Economic Aspects of Water Resources Policy". *American Journal of Economics and Sociology* 28: 131–144.

Gallegos, Tanya J. *et al.* (2015) "Hydraulic Fracturing Water Use Variability in the United States and Potential Environmental Implication". *Water Resources Research* 51: 5839–5845.

Gleick, Peter H., Gary Wolff, Elizabeth L. Chalecki, and Rachel Reyes (2002) *The New Economy of Water: The Risks and Benefits of Globalization and Privatization of Freshwater*, Pacific Institute, Oakland, CA.

Goplerud, Peter III (2015) "Adjudication of Water Rights". In Amy Kelly and Robert Beck (eds), *Waters and Water Rights* 1: ch. 16.

Gray, Janice S. (2014) "Frack Off! Law, Policy, Social Resistance, Coal Seam Gas Mining and the Earth Charter". In L. Westra and M. Vilela (eds), *The Earth Charter, Ecological Integrity and Social Movements*, Routledge, Abingdon, pp. 129–147.

Gray, Janice S. (2016) "Trans-jurisdictional Water Governance in the Context of Unconventional Gas Mining: The Australian Experience". In J. Gray, C. Holley, and R. Rayfuse (eds), *Trans-jurisdictional Water Law and Governance*, Routledge, Earthscan.

Guyton, Arthur C. (1991) *Textbook of Medical Physiology*, 8th edn, W. B. Saunders, Philadelphia, PA.

Hardin, Garrett (1968) "The Tragedy of the Commons". *Science* 162: 1243–1248.

Healy, Richard H. *et al.* (2015) *The Water-Energy Nexus: An Earth Sciences Perspective*, Washington, DC: USGS Geological Survey Circular 1407, available at http://dx.doi.org/10.3133/cir1407.

Hundley, Norris Jr. (1992) *The Great Thirst: Californians and their Water, 1770s–1990s*, University of California Press, Berkeley, CA.

IEA (2014) *Technology Roadmap: Solar Photovoltaic Energy*, International Energy Agency, Paris.

IPCC (2013) *Climate Change 2014: The Physical Science Basis*, Fifth Assessment Report, IPCC, Geneva, available at http://ipcc.ch/pdf/assessment-report/ar5/wg1/WG1AR5_ALL_FINAL.pdf.

IPCC (2014) *Climate Change 2014: Impacts, Adaptation, and Vulnerability*, Fifth Assessment Report, IPCC, Geneva, available at http://ipcc.ch/pdf/assessment-report/ar5/wg2/WGIIAR5-Chap3_FINAL.pdf.

Kaul, Inge, Isabelle Grunberg, and Marc Stern (eds) (1999) *Global Public Goods: International Cooperation in the 21st Century*, Oxford University Press, Oxford.

Kenny, J.F., N.L. Barber, S.S. Hutson, K.S. Linsey, J.K. Lovelace, and A. Maupin (2009) *Estimated Use of Water in the United States in 2005*, Circular 1407, USGS Geological Survey Washington DC, available at http://pubs.usgs.gov/circ/1344/pdf/c1344.pdf.

Lewis, M., S.P. Neill, P.E. Robins, and M.R. Hashemi (2015) "Resource Assessment for Future Generations of Tidal-Stream Energy Arrays". *Energy* 83: 403–415.

Malloy, Robin Paul 1990. *Law and Economics: A Comparative Approach to Theory and Practice*, West Publishing Co., St. Paul, MN.

Montenegro, A. and R. Ragab (2010) "Hydrologic Response of a Brazilian Semi-arid Catchment to Different Land-Use and Climate Change Scenarios: A Modeling Study". *Hydrological Processes* 24: 2705–2723.

Mortholst, Paul Erik, Robert Y. Redlinger, and Per Andersen (2002) *Wind Energy in the 21st Century: Economics, Policy, Technology, and the Changing Electricity Industry*, Palgrave Macmillan, Basingstoke.

Neuman, Janet C. (1998) "Beneficial Use, Waste, and Forfeiture: The Inefficient Search for Efficiency in Western Water Use". *Environmental Law* 28: 919–996.

Ostrom, Elinor (1990) *Governing the Commons: The Evolution of Institutions for Collective Action*, Cambridge University Press, Cambridge.

Pal, Jeremy, S. Elfatih, and A.B. Eltahir (2015) "Future Temperature in Southwest Asia Projected to Exceed a Threshold for Human Adaptability". *Nature Climate Change*, doi:10.1038/nclimate2833, available at www.nature.com/nclimate/journal/vaop/ncurrent/full/nclimate2833.html.

Ranjan, Priyantha (2012) "Effect of Climate Change and Land Use Change on Saltwater Intrusion". Available at www.eoearth.org/view/article/152361.

Robbins, Tom (1976) *Even Cowgirls Get the Blues*, Houghton Mifflin, Boston, MA.

Rose, Carol M. (1986) "The Comedy of the Commons: Custom, Commerce, and Inherently Public Property". *University of Chicago Law Review* 54: 711–781.

Scilli Staff (2009) "Harvesting Energy from Humans". *Popular Science*, 29 January, available at www.popsci.com/environment/article/2009-01/harvesting-energy-humans.

Sovacool, Benjamin K. and Alex Gilbert (2014) "Developing Adaptive and Integrative Strategies for the Electricity-Water Nexus". *University of Richmond Law Review* 48: 997–1032.

US EPA (1997) *Profile of the Fossil Fuel Electric Power Generation Industry*, doc. no. EPA/310-R-97-007, US Environmental Protection Agency, Washington, DC.

Wyman, Katryna M. (2008) "The Property Rights Challenge in Marine Fisheries". *Arizona Law Review* 50: 511–543.

Part II

The common good, climate change and the right to health

8 Lessons learned from climate change on the need to fix responsibility in governments, scientific environmental organizations and environmental NGOs for applied ethical analyses

Donald A. Brown

Introduction

Global environmental issues such as climate change raise numerous serious ethical issues that should guide policy-makers in formulating policy. Yet recent research concludes these ethical issues are largely being ignored by policy-makers, the media covering public debates about these issues and, surprisingly, often even environmental non-governmental organizations (NGOs) engaged in policy disputes. This chapter therefore argues that there is a need to fix responsibility in governments, scientific organizations (especially those that make policy recommendations on environmental issues) and environmental NGOs engaged in environmental policy debates to identify ethical issues that arise in policy formation, explain how these ethical issues were considered or ignored, and expand the work of NGOs working on environmental issues to include a much deeper applied ethics component of their work.

Lessons learned from climate change

As the international community approaches the twenty-first Conference of the Parties (COP-21) to the United Nations Framework Convention on Climate Change (UNFCCC), many nations are utterly failing to make commitments on reducing their greenhouse gas emissions reduction targets to levels needed to prevent dangerous climate change.

Because of this, there is also widespread agreement among observers of UNFCCC negotiations that there is little hope that the international community will develop an adequate international response to climate change unless nations increase their GHG emissions reductions commitments (i.e. INDCs) to levels that represent their fair share of safe global emissions. And so it is widely agreed that nations must base their INDCs both on achieving safe atmospheric GHG levels that will limit warming to tolerable levels and the nation's just percentage of global emissions that will achieve this level. These two issues, namely the issue of what is safe enough, and the issue of what is each nation's fair share, are ethical and moral issues at their core.

Research conducted by Widener University Commonwealth Law School and the University of Auckland concludes that the ethical issues in setting national commitments on GHG emissions are being largely ignored by governments in setting GHG emissions reduction commitments, by the press in covering national debates about climate policy, and even more surprisingly by domestic NGOs who are proponents of climate change policies.[1]

This is so despite the fact that:

- It is impossible for a nation to think clearly about climate policy until the nation takes a position on two ethical issues: (i) what warming limit the nation is seeking to achieve through its policy; and (ii) what is the nation's fair share of safe global emissions.
- Climate change policy-making raises numerous other ethical issues that arise in policy formulation (see below).
- Ethical arguments made in response to the arguments of climate change policy arguments are often the strongest arguments that can be made in response to the claims of climate policy opponents because most arguments made by opponents of climate policies fail to pass minimum ethical scrutiny.
- Climate change more than any other environmental problem has features that scream for attention to see it fundamentally as a moral, ethical, and justice issue. These features include: (i) It is a problem overwhelmingly caused by high-emitting nations and individuals that is putting poor people and nations who have done little to cause the problem at greatest risk; (ii) the harms to the victims are potentially catastrophic losses of life or the destruction of ecosystems on which life depends; (iii) those most at risk usually cannot petition their own governments for protection, their best hope is that high emitters of GHGs will respond to their moral obligations to not harm others; and (iv) any solution to the enormous threat of climate change requires high emitting nations to lower their GHG emissions to their fair share of safe global emissions, a classic problem of distributive justice.

The Widener/Auckland research identified above has also discovered that most participants in national debates about climate policies, including scientists, journalists, and NGOs around the world have largely ignored the numerous ethical issues that arise in climate policy formation and instead usually have narrowly responded to the arguments of the opponents of climate policy which have almost always been variations of claims that climate change policies should be opposed because: (a) they will harm national economic interests, or (b) there is too much scientific uncertainty to warrant action.

Yet numerous issues arise in climate change policy formation for which ethical and moral considerations are indispensable to resolve these issues and moral arguments about these issues are by far the strongest responses to arguments on these issues usually made by opponents of climate policies. These issues include, among many others:

- Can a nation justify its unwillingness to adopt climate change policies primarily on the basis of national economic interest alone?
- When is scientific uncertainty an ethically acceptable excuse for non-action for a potentially catastrophic problem like climate change given that waiting until the uncertainties are resolved makes the problem worse and more difficult to solve?
- Should proponents or opponents of climate change policies have the burden of proof to scientifically demonstrate that climate change is or is not a threat before climate change policies are enacted?
- What level of proof, such as, for instance, 95 per cent confidence levels or the balance of the evidence, is needed to demonstrate climate change is a threat that warrants policy responses?
- What amount of climate change harm is it ethically acceptable for a nation to impose on those nations or people outside their jurisdiction who will be harmed without their consent?
- To what extent does a nation's financial ability to reduce GHG emissions create an ethical obligation to do so?
- What are the rights of potential victims of climate change to consent to a nation's decision to delay national action on the basis of national cost or scientific uncertainty?
- Who gets to decide what amount of global warming is acceptable?
- Who should pay for reasonable adaptation needs of victims of climate change?
- Do high emitting nations and individuals have a moral responsibility to pay for losses and damages caused by climate change to people or nations who have done little to cause climate change?
- How should national GHG targets consider the per capita or historical emissions of the nation in establishing their national climate commitments?
- Do poor, low-emitting nations have any moral responsibility for climate change and what is it?
- When should a nation be bound by provisions of international law relevant to climate change that they agreed to including provisions in the United Nations Framework Convention on Climate Change such as the "no-harm", and "precautionary principle" and the duty of developed nations to take the lead on climate change?

The Widener/Auckland research mentioned above has also concluded that these ethical issues are mostly being ignored in national debates about climate policy while, for the most part, a narrow economic rationality is largely the actual basis for national climate change policy. This is so, despite the fact that in the international negotiations issues about the justice of national commitments on climate change both in regard to national INDCs, and national acceptance of responsibility for the costs of adaptation to climate change and damages and losses from climate change in poor countries that have done little to cause climate change are at the centre of the most contentious issues in the climate change negotiations. For this reason, the utter failure of national media to cover the ethics and justice issues at the centre of international climate change disputes is startling.

What is the cause of the failure to identify and consider the ethical issues entailed by policy-making on climate change?

As we have seen, most of the debate on climate change policy-making at the national level has been focused on responses to arguments made by opponents of climate change policies which have usually been claims that proposed climate change policies will impose unacceptable costs on national economies, or there is too much scientific uncertainty to warrant expensive national action on climate change. Such claims have both factual and normative assumptions. Citizens and environmental groups have unknowingly been tricked into responding to these arguments by making factual responses to these claims, such as climate change policies will increase jobs, despite the fact that each of these arguments contain hidden assumptions which clearly flunk minimum ethical scrutiny.

For example, opponents of climate change policies in the United States have frequently based their opposition to climate policies on the claim that climate change policies will destroy US jobs or the US economy.

The response of NGOs and citizens to this argument has largely been to assert that climate change policies will create jobs and boost the economy. Yet this response, unknowingly, implicitly supports the very dubious hidden normative assumption of the climate policy opponents' arguments, namely that the United States should not adopt climate policies if the policies will hurt the US economic interests despite the fact that this argument is obviously wrong when viewed through an ethical lens because polluters not only have economic interests, they clearly have moral responsibilities to not harm others. This claim that nations have duties to not harm others is a strong moral claim because almost all cultures agree with the Golden Rule which holds that someone should not be able to kill others because it would be costly to the killer to stop the killing behaviour.

Thus, the failure to respond to the arguments of the opponents of climate change policies arguments on moral grounds is an astonishing oversight in light of the fact that the moral objection is very strong to someone who claims that they can seriously harm others if their economic interests are threatened if they are required to limit their harmful activities. Such a claim violates the most non-controversial ethical rules including the Golden Rule and many well-accepted provisions of international law based on the Golden Rule such as a rule called the "no harm principle" which asserts that all nations have a legal duty to prevent their citizens from harming people outside their jurisdiction.

If citizens who support climate policies ignore the ethical problems with the arguments made by opponents of climate policies on the grounds that climate policies will impose costs on those who are harming others, they are playing into the hands of those responsible for putting the planet at risk from climate change.

There are also deeply problematic ethical assumptions that have remained largely unchallenged when the opponents of climate change policies have argued that a nation, such as the US, should not adopt climate change policies due to scientific uncertainty. [2]

And so, for 30 years, the opponents of climate change policies have succeeded in framing the climate change debate in a way that ignores obvious ethical and moral problems. Surprisingly both environmental organizations and the national press have also failed to bring attention to the obvious moral problems with the opponents of climate policies' arguments.

And so a major cause of the failure to consider ethical problems with the arguments of opponents of climate change policies is the successful framing by opponents of climate policies of issues to be considered in policy formation.

However, an equally important cause of the failure to expressly consider the ethical dimensions of environmental policy is attributable to two problems.

First, most employees of environmental policy offices are technically trained in science or economics. As a result, they are often very poor in spotting the ethical problems with arguments made about policy. In fact they are often expected to perform their policy analyses exclusively through the lens of science and economics, disciplines which pretend to be "value-free" yet often hide very controversial normative assumptions.

Second, higher education is largely failing to train those engaged in environmental issues to spot ethical questions. Although many schools of higher education teach environmental ethics usually as an elective, most students enrolled in courses in environmental economics or sciences have no exposure to these ethics courses.

In addition, courses on environmental ethics frequently fail to include discussion of the ethical questions that arise when environmental economics and science are applied as prescriptive guides to public policy. This is so because the major focus of academic environmental ethics has been to explore ethical questions about human and environmental relationships, not ethical questions that frequently arise in policy formation such as the ethical limits of economic arguments, problems of procedural and distributive justice, or the ethical issues that arise when government officials must make decisions in the face of uncertainty.

Philosophers often categorize training about how to calculate something, the kind of training often provided in higher education in environmental science and economics, as "instrumental" rationality to be distinguished from "ends" rationality, or what are the right ends of society, the domain of ethics or political philosophy. Instrumental rationality is rationality about what means can be used to achieve certain ends where the ends are not in question. Instrumental rationality focuses on how to do something, not on why something should be done.

The kind of critical thinking usually taught in science and economics is most frequently "means" rationality, not "ends" rationality. This is so because most economic analyses applied in public policy assume that governments should maximize public welfare or efficiency goals of policy that are not often questioned by the economic analysis despite the fact that welfare maximization goals sometimes dramatically conflict with other valid societal goals such as distributive or procedural justice, guaranteeing human rights, or how the environment and humans should be valued.

Science training is often focused on knowledge of how nature works or how to search for answers about how nature works that are currently unknown; it is not concerned with ethical questions that frequently arise when science is applied to public policy such as who should have the burden of proof, what quantity of proof should satisfy the burden of proof, or who gets to decide about what should be done in response to uncertain harms when some people more than others are at risk.

What should be the goal of a good life or what is right or wrong are matters of ethics, questions about the "ends" of society. However, how to calculate costs and benefits or how to conduct experiments to achieve adequate levels of confidence are understood to be questions of "means" to achieve societal goals and therefore the domain of instrumental rationality.

Now, occasionally, environmental ethics literature has acknowledged problems with the almost exclusive focus on instrumental rationality that is the domain of science and economics when these disciplines are used to guide public policy. However, much of the environmental ethics literature ignores or minimizes many of the problems entailed by the dominance of instrumental rationality in science and economics, the disciplines which almost always frame environmental controversies and frequently the only disciplines that are taught in environmental policy courses.

Because certain value-neutral policy languages structure specific environmental controversies, and because the environmental ethics literature does not usually focus on ethical analyses of concrete problems, most environmental ethics literature is not relevant to some of the most frequent and crucial issues that arise in policy-making, namely economic and scientific arguments about whether to act or not to protect the environment.

The need to fix responsibility for applied ethics analysis

Given the extraordinary importance of the ethical dimensions of environmental policy and the utter failure of governments, scientific organizations, and NGOs to consider many ethical issues that should guide policy, there is a need for the following:

- Government environmental policy-makers should expressly create responsibility in an office or individual for identification, analyses, and responses to ethical issues raised by environmental policy-making.
- So that citizens understand and can respond to how ethical issues were responded to by government policy-making, governments should explain how ethical principles affected policy decisions.

- Scientific organizations that make recommendations on environmental policy options, such as IUCN, the Intergovernmental Panel on Climate Change (IPCC), the Academies of Sciences, the American Association for the Advancement of Science and similar professional science organizations should assign express responsibility for ethical issue spotting as part of their work to an office or individual in the organization and expressly identify in reports how ethical issues were considered, if at all, in policy recommendations.
- Institutions engaging in ethical analyses of environmental issues including academic environmental ethics programs and NGOs making recommendations on environmental policy, should become much more focused on applied ethical analyses of concrete issues that arise in environmental policy-making rather than abstract discussion of ethical principles. In this regard they should assume responsibility for educating citizens about the ethical issues that often arise in the "value-neutral" disciplines of science and economics.

The priority focus of an applied ethics responsibility

Those engaged in applied ethics work in government, scientific organizations that make policy recommendations, and NGOs should be mindful of the following:

- Ethics is understood to be the domain of inquiry that examines what is right or wrong, obligatory or non-obligatory, just or unjust, and when duties arise about human behaviour.
- Every claim about what should be done to protect nature usually has both a factual assumption (usually the domain of science) and normative claim (usually the domain of ethics and law). The normative assumption is usually not stated expressly. For instance, a national commitment on climate change may actually be based on protecting the coal industry but this assumption is not identified when decisions are made. To reveal actual normative assumptions often requires that policy-makers and governments respond to questions about the decision.
- When law is inadequate to achieve ethically acceptable outcomes, ethical inquiry can provide important advice on how the law should be amended or enlarged to achieve an ethically acceptable, just, and equitable legal regime.
- Because many of the normative assumptions embedded in claims about what should be done to protect natural resources are often hidden in what appear to be at first glance "value-neutral" languages of science and economics, those responsible for applied ethical analyses in organizations should make explicit the hidden ethical assumptions to enable adequate ethical and public scrutiny. This will sometimes require that proponents of policies be asked to state explicitly what the normative basis of their position is because the normative assumptions are usually not evident from reading arguments

for or against programs that will protect natural resources. For instance, a government may claim they took justice into account when they made commitments to reduce GHG emissions yet the actual level of commitments were based on limiting economic costs to economy.

- Because many policy-makers and leaders of environmental organizations are not trained in spotting ethical issues that often arise in the policy languages of science, economics, and law, those responsible for applied ethical analysis should identify the ethical issues that arise in policy formation.
- Because ethical analysis of natural resource issues can lead to: (a) agreement among ethicists on what ethics requires, (b) disagreement among ethicists on what ethics requires, (c) an overlapping consensus among ethicists that a position taken by individuals, organizations, or governments on these issues fails to pass ethical scrutiny despite disagreement on what ethics requires, the applied ethical analysis should identify ethical controversies raised by the ethical issues that must be faced in policy formation including the range of respectable ethical opinions when there is a reasonable disagreement among ethicists on what ethics requires while identifying, if appropriate, positions on these issues that have been taken that fail to withstand ethical scrutiny.

Notes

1 See Nationalclimatejustice.org under "lessons learned".
2 See, for example, my article "The Ethical Duty to Reduce Greenhouse Gas Emissions in the Face of Scientific Uncertainty", available at http://ethicsandclimate.org/2008/05/19/the-ethical-duty-to-reduce-greenhouse-gas-emissions-in-the-face-of-scientific-uncertainty (accessed 13 November 2015).

9 Climate change challenges in law and ethics

Can individual and collective rights be protected?

Kathleen Mahoney

Climate change is probably the greatest moral and ethical question of this era, but the legal treatment of climate change grounded in morals and ethics is almost non-existent, especially in the policy making of provincial and federal governments in Canada. A legal theoretical void exists because when climate change is discussed at all, is almost always in terms of economic or technology-based benefits, technical legal defences, statutory interpretation or issues of scientifically precise causation rather than on the moral and ethical issues and fundamental human rights.

We ignore the many moral, ethical and human rights dimensions of climate change at our peril. Consideration of these dimensions must play a central role in legal and policy decisions about climate change if peace, economic stability and even the survival of the planet are to be assured for the future generations. In this chapter I discuss what I believe are some of the considerations and challenges that must be a part of the legal thinking, policy decisions and judicial decision-making going forward.

The ethical, moral and human rights concerns raised by climate change are those that inform universally recognized human rights as well as the very nature of justice and equity. Distributive justice, compensatory justice, procedural justice and human rights are all implicated.

There are three major moral imperatives that should influence legal thinking for both legislation and jurisprudence:

- Responsibility to future generations of humanity.
- Responsibility to different populations around the world.
- Responsibility to the natural world and its natural state.

Future generations and generational equity

With respect to the first imperative, the effects of climate change on future generations whose rights are at the mercy of those living in the present must be considered. Aboriginal elders echo this point when they say Aboriginal peoples have the moral duty to steward the environment for the benefit of seven generations to come.

Philosophers and economists support this view.[1] They say it is ethically indefensible to give more weight to the welfare of current generations than to that of future generations just because they were born sooner in time. If we wish to create intergenerational equality, the only morally defensible approach to any intergenerational event is to treat the generations equally. This moral imperative only intensifies when one generation causes the harms by their acts or failure to act, subjecting subsequent generations to foreseeable harm to their resources or opportunities. But the responsibility to mitigate climate change for future generations is difficult to place in conventional legal discourse because it requires legal tools we presently do not have.[2]

Foreseeing harms to climate for future generations that will accumulate and persist for a very long time, even centuries or millennia, is difficult enough. The fact of the irreversibility of the effects of global warming such as the melting of the polar ice cap or the extinction of species, adds a further layer of complexity to legal or philosophical analyses to make decisions regarding responsibility for harms caused, degree of mitigation expenditures and when and on what they should be spent. For example, investments in infrastructure may make mitigation cheaper for future generations but if future generations are already stuck with irreversible, catastrophic damage that can no longer be prevented, mitigation efforts in infrastructure may be futile. Policy making in this scenario is very difficult indeed and underscores the urgency of stopping or slowing climate change as soon as possible.

It is precisely the long-term, possibly preventative nature of climate change responses that makes it a moral and ethical issue. The mismatch between the need to act now for benefits that will accrue for future generations explains why so many politicians, economists, industrialists, judges, and other decision-makers find it so difficult to engage and respond to climate change. Their interests are usually focused on immediate evidence of present harm, election cycles or quarterly returns. This short-term thinking or generational buck-passing is immoral and unethical.

UNESCO suggests that instead of thinking about an infinite number of future generations, it would make more sense legally to think about the next 100 years. Although the full impacts of climate change may not be known in 100 years, at least it is possible for lawyers and judges to think of three future generations in terms of rights, needs, interests, and harms that we can imagine, put a value on, and plan for in the present. Although this approach falls short of the First Nations imperative to steward the environment seven generations into the future, it at least would illuminate wider concerns than those that affect the present generation alone and provide a point of departure for future generations in their struggles to diminish the impacts of climate change.

Poorer populations and procedural and distributive justice requirements

For the second imperative, that of moral responsibility to poorer populations around the world most affected by climate change, ethical challenges abound. Very little has been done by legal scholars to articulate relevant conceptual issues, as well as what public policy should look like in practice.

The fact that climate change is a global phenomenon is the starting point. Once emitted, greenhouse gas emissions can have climate effects anywhere on the planet, regardless of the source. While all countries would like to see global emissions curtailed, none wish to experience the economic consequences of doing so individually.[3] At the same time, many of the countries and populations most vulnerable to climate change are those who have emitted or are emitting the least amount of greenhouse gasses. People living in the Arctic, in desert or semi-arid or equatorial regions, large river deltas or coastal and island regions, the elderly and children, all fall into categories of those most vulnerable to climate change.[4] In the Canadian context, those most affected right now by climate change are the First Nations and Inuit peoples.[5] They are bearing a disproportionate brunt of resource development as their treaty rights, food sources, water and cultures are compromised by climate change and environmental damage it causes.

Professor Simon Caney's research[6] has contributed to a growing consensus that climate policies ought to be guided by ethical ones and that emission rights must be looked at holistically in conjunction with human rights. He argues that it is wrong to isolate climate change from human rights, poverty and health.

Procedural justice requires that the most vulnerable groups must participate in the decision-making about measures to prevent, mitigate or adapt to climate change. In order to meet this requirement, mechanisms need to be put in place to ensure this dialogue can take place and that local and traditional knowledge are integrated in the decision-making processes.

Poorer nations who have not emitted comparable amounts of greenhouse gasses believe it unfair to demand that they minimize their emissions for the sake of future generations, especially if future people are likely to be better off, and if many present high emitters are already much richer than the future poor are likely to be. This skewed vulnerability in both cause and effect can result in serious unfairness between countries and regions. This will require the global community to cooperate in limiting emissions. Principles of distributive and procedural justice must then inform the decision-making process. Decision-makers largely ignore these age-old principles.

Unfairness in the distribution of the negative consequences of climate change is not the only distributive justice issue. The unfairness in the distribution of the benefits derived from actions that cause climate change should also be considered. For example, the Chipewyan First Nation that is down river from the oil sands and is adversely impacted by the environmental degradation caused by the mining of the bitumen is not allowed any share in the revenues from resource development on their lands. Chief Alan Adam has stated that if the nation had a fair share of the oil sands revenues, they could use them for programs and services to ameliorate some of the impacts of the harms caused by the oil sands development.[7] From a distributive justice standpoint, the Chief is correct. The nation should not only have their share of the revenues from the development, they should have a say in how the development is done and how the impacts should be ameliorated.

On a broader scale, distributive justice principles would dictate that developed countries should be permitted fewer emissions than developing countries because developed countries are responsible for most of the cumulative emissions contributing to climate change and are more able to bear the burden of reducing emissions. Nonetheless, developed countries favour the immoral argument that they should not be responsible for emissions prior to 1990 because of ignorance about the climatic effects of these emissions, and that present inhabitants of developed countries should not be held responsible for past emitters now dead. This tension and a lack of consensus need to be resolved.

Other issues such as vulnerability to climate affected by previous wrongs, such as the legacy of colonization, slavery, and economic exploitation, compound injustices and increase feelings of alienation and anger which we see played out in the Canadian courts where First Nations feel their treaty and aboriginal rights are disrespected by resource developers and governments.[8]

Human rights and compensation issues

Climate change scientists say it is possible that some impacts such as sea level rise will threaten the very existence of entire coastal countries and communities.[9] Displaced populations and climate change refugees who will be seeking asylum in other countries will invoke the Universal Declaration of Human Rights[10] and other applicable human rights conventions, which may be reluctant to receive them. Others will claim that their individual life, liberty and security of the person rights are violated as well as their rights to use their property to enhance their well-being and choose their own way of life.

The right to life and the right to health are violated when people are killed by violent storms, droughts and intense heat waves and heat waves, droughts, floods and storms intensify diseases. There is a widespread view that oil sands development strips away the rights of First Nations and affected communities to protect their children, land and water from being poisoned.[11] How will the rights of the most vulnerable be validated? Will human rights bodies even recognize these rights or will the rights themselves be put under threat by climate change?

International law of environmental human rights supports Charter claims for protection from environmental degradation arising from resource development. The Supreme Court consistently prefers statutory interpretations that align with international legal obligations.[12] In the realm of Charter interpretation, international human rights law has played a particularly important role.[13] As Lamer CJ pointed out in *Re BC Motor Vehicle Act*:

[Principles of fundamental justice] represent principles which have been recognized by the common law, the international conventions and by the very fact of entrenchment in the Charter, as essential elements of a system for the administration of justice which is founded upon a belief in the dignity and worth of the human person and the rule of law.[14]

As such, the international law of environmental human rights is likely to be an important consideration in environmental Charter claims.

In *Millership v Kamloops (City)*[15] Millership brought an action against the defendants for damages for personal injuries as a result of public fluoridation of the water in Kamloops. Millership argued, "Public water fluoridation is in breach of a large range of federal and provincial statutes and the Charter of Rights and Freedoms, as well as a number of international treaties."[16] The Court dismissed the case on evidentiary basis.

Two Alberta cases, *Domke v Alberta*[17] and *Kelly v Alberta*[18], provide a glimpse into judicial thinking on the right to health versus economic development. Landowners in both cases objected to the issuance of approvals permitting the drilling of sour oil wells in the vicinity of their homes, citing concerns regarding the potential release of toxic gases and the negative effects on their health.

In *Domke*, the plaintiffs appealed a decision by the Energy and Utilities Board ("EUB") that granted two sour well licences to Highpine Oil. They argued that their Charter rights to life, liberty and security of the person were violated when the EUB allowed the oil company to refuse to provide information to the plaintiff about production results of a new well. EUB concluded that the company met the relevant regulatory standards and that had sufficiently addressed the citizens' concerns. It found no breach of Charter rights, as the risk that the proposed wells posed was minimal.

On appeal, in considering whether there has been an infringement of the homeowners' rights to life, liberty and security of the person the court said the Board had correctly applied the law to the facts, stating it was not prepared to second guess the Board's factual findings.[19]

The plaintiffs also challenged the constitutionality of the regulations that allowed producers to withhold the information. They argued that the period of secrecy could result in exposure to higher levels of gaseous emissions than had been projected for the purposes of obtaining the drilling permit. The Court disagreed using astounding circular reasoning, saying that the claim was speculative as there was no evidence to support the notion that actual emission levels might be higher than projected.

In the *Kelly* case,[20] the Applicants alleged violations of their life, liberty and security of the person rights from an EUB decision that would have put the homeowners at risk of an inherent hazard during the drilling and completion of the wells.

The case failed because the court said the homeowners had to prove direct causation of actual or threatened harm sufficient to qualify as a deprivation of their life, liberty or security of the person.[21] The Court refused to consider the broader ecological context and the effects of cumulative environmental damage. If it had done so, it would have recognized that the simple "but for" test to prove legal causation is inappropriate when the problem involves protecting people from environmental degradation[22] because the requirements of the test ensure they will never be able to succeed. "But for" causation requirements in the environmental context are impossible to prove because complexities of the causal interactions between humans and their environment make the data insufficient.

A review of the cases brought so far reveals that they fail not on legal applications of the Charter, but on evidentiary basis. Courts strictly apply the law of causation to these claims such that plaintiffs must prove on a balance of probabilities that but for the activities being challenged, the violations of human rights would not have occurred. This high level of proof requires clear scientific evidence.

The Harper government, however, not only inhibited the availability of this kind of evidence but also went so far as to stop the production of environmental data altogether, making it very difficult for court actions to succeed. Chris Turner, in his book *The War on Science*,[23] says when the lead researcher on bitumen mining and freshwater pollution found indisputable proof that they were linked and that Alberta's oil industry had created an unacceptable risk to public and ecological health, the response of the Harper government was to ensure that there is no new information, no field samples, no analysis, no conclusions to report to the press or be interpreted by experts. Instead of re-writing environmental regulations for the oil sands so that the oil industry operates within parameters that properly safeguarded the public good, the Harper government saw that the problem was in the very fact that the issue was investigated at all. Turner commented, "Only by destroying all the firearms, in other words, can you guarantee there will be no more smoking guns."[24]

Against the ever-tightening access to evidence and the lack of policy responses to proven harms, individuals seeking to protect their rights to life, liberty and security of the person under section 7 of the Charter face a very difficult task of meeting the legal requirements of a Charter challenge. Judges who take a very conservative approach to environmental claims compound this. With the election of a new liberal government in October 2015 that campaigned on respecting scientists and working co-operatively with the bureaucracy, there is some reason to be hopeful that Canadian public policy is undergoing significant change. Whether this will have an influence on the narrow approach the courts have taken, is less hopeful.

As rapid innovation continues to expand and become increasingly complex, it would seem reasonable for the courts to send a signal to producers that the negligent creation of risk will be penalized by tort law, even if data is insufficient to establish "but for" causation of injury in any given case. From the public health perspective, it is imperative that a meaningful price be placed on risk creation. If plaintiffs could succeed by establishing on a balance of probabilities that they have been harmed and if the burden of proof shifted to the industry to show that their harms were not created by their industrial activities, this would redress the present imbalance between the rights of citizens to their life and health and the rights of the oil industry to create profits. This is wholly consistent with tort's longstanding function of adjusting the law to deal with new social needs. Collins argues that there is little difference between an individual who is killed by state-permitted emissions and one who is shot by state police. Both should be protected from the deprivation of life.[25] As part of this approach, legal evidentiary demands must be made commensurate to the ability of an individual hurt by state-supported environmental damage to prove causation.[26]

In the Supreme Court decision in the *Operation Dismantle* case,[27] the appellants alleged that Canada's decision to allow the US to test cruise missiles in Canada violated §7 of the Charter. They argued that the tests increased the risk of nuclear war in Canada. They sought a declaratory relief, an injunction and damages. The Supreme Court of Canada dismissed the case, holding that the case turned on whether there would be an actual increase in the risk of war which the appellants could not prove; the Court admitted, however, that some kinds of government conduct resulting in the risk of physical injury could amount to violations of §7, leaving the door open for other Charter challenges.

In *Millership v British Columbia*[28] and *Locke v Calgary*,[29] the applicants alleged that fluoridation of water without prior consent violated §7 rights to liberty and security of the person. Both cases confirmed the applicability of §7 to environmental harm; however, both cases failed on the scientific evidence. Even if that evidence did exist, the court said the bylaw would be saved by the principles of fundamental justice which required that a fair balance be struck between the interests of a person who claimed that his security has been violated and those of society. The Court was convinced in this case, that such a fair balance was struck by the said bylaw because the benefits to public health and the health of children were sufficiently pressing objectives to uphold the practice.

The importance of the cases lies in the fact that in both *Millership* and *Locke* the court considered a §7 violation seriously and left little doubt that a remedy under §7 is available for environmental threats to health.

However, all of that may be reversed in the case of *Ernst v EnCana*[30] decided by the Alberta Court of Appeal, now on appeal to the Supreme Court of Canada. The case involves Jessica Ernst, a landowner who lives near Rosebud, Alberta. Her water is supplied by a private well that draws from the Rosebud Aquifer. Between 2001 and 2006, EnCana started "fracking" near Ernst's property. Ernst's water soon became severely contaminated with flammable levels of methane.

Ernst brought claims against EnCana, Energy Resources Conservation Board (ERCB) and Alberta regarding the severe contamination of her well water and other harms; lawsuit claims against EnCana (for negligence, nuisance and other torts); the ERCB (for breaches of Ms. Ernst's fundamental freedoms under the Charter of Rights and Freedoms and for the negligent failure to implement the ERCB's inspection scheme); and Alberta (for among other things, negligent investigation).[31]

There is a lot riding on this precedent-setting lawsuit. If Ernst wins damages under the Charter, then many other residents who have felt the effects of fracking on their water supply will follow suit. It could also establish a duty of care that companies would have to public resources such as water.

In response to Ernst's Charter claim, the ERCB argued that a generally worded statutory immunity clause of the ERCB bars Ernst's claims under the Charter as well as for damages for negligence.[32]

In deciding whether the Charter claim could go forward, the Court of Queen's Bench said no. The judge said the immunity clause was similar to statutes of limitations that do apply to Charter claims. He said: "I see commonalities between statutory immunity provisions and limitation periods of general application that apply to Charter claims for personal remedies. Both are statutory bars to claims that may otherwise have merit."[33] He went on to bolster his position by saying there are strong policy reasons for the application of immunity clauses to claims for personal remedies under the Charter. Policy considerations are given effect when the merits of a claim for a Charter breach are examined. In my view, these policy considerations also apply when determining whether a statutory immunity clause applies.[34]

Justice Whittman's analogy of the general immunity clause with statutes of limitations is not a good one in the author's opinion. Limitation periods are not adopted to bar a *claim* outright; instead, they simply provide rules regarding *how promptly* a claim must be made. What seems to be lost in the reasoning is that the very purpose of the Charter is to restrain government action.

Chief Justice Wittmann's decision in *Ernst v EnCana* and the Court of Appeal's affirming of it, could mark an ominous new direction for court cases involving personal rights against government-sponsored environmental damage. Before *EnCana*, Charter challenges were mostly deemed to be valid but ultimately came short on evidentiary grounds. If the Supreme Court accepts the reasoning of the Alberta courts, it will send the message that all legislatures have to do in the future is pass immunity legislation to be free from Charter claims by those whose basic human rights are violated by government action or inaction.

The Alberta Court's decisions fly in the face of Canada's international obligations. Canada is a signatory to international human rights treaties that require the state to protect the rights that climate change adversely affects. The Supreme Court of Canada regularly looks to international human rights conventions to interpret domestic legislation, so international treaties can have an impact domestically. The Canadian Charter of Rights and Freedoms in section 7 protects citizens from violations by the state of their security of the person.

If violations of human rights through climate change are recognized as compensable, numerous issues arise. Who should bear the costs of mass movements of peoples? How will causal responsibility be assigned? What mechanisms and structures will be required to address these claims? Who will be the beneficiaries of the compensation and how will it be distributed?

These questions of distributive justice, human rights and compensation will only be answered when the ethical discussion about climate change is put at the forefront of negotiation tables, economic forecasting, political debates and meetings of world leaders.

Pope Francis, arguably more than any other world leader, has set out the moral framework for the global response to climate change. Francis's vision for change is comprehensive. He addresses the challenges of food production due to uncontrolled fishing. He reminds readers of his encyclical that migrants are forced to flee poverty induced by environmental degradation but are not recognized internationally as refugees. He offers a corrective to past theological interpretations that say that God gave humanity dominion over the earth and challenges the idea that humanity should be the centre of concern when it comes to the earth's future.[35]

The natural world

The third theme is the moral and ethical duty to protect the natural world. Environmentalists tell us there is a need to develop a much stronger ethical and legal relationship between humans and the rest of nature. From a legal perspective it raises the philosophical and ethical question as to whether justice is meant to be not only for human beings but also for the world's other sentient creatures.

Climate change raises numerous new questions about the moral value of nonhuman nature including asking ourselves whether we have obligations to protect nonhuman animals, unique places, or nature as a whole, and if so, when and what form such obligations take?

In making his theological argument, Pope Francis starts from the ground up, saying as much as we need to protect poor or marginalized humans – the moral responsibility extends to protecting plankton in the ocean's food chain and worms in threatened ecosystems.

A number of influential studies place a third of the world's species on a path to climate-driven extinction as a result of global climate change.[36] There is scientific evidence that caribou populations in Canada have dramatically diminished in their natural, historic habitat which used to occupy the area of the oil sands development. Historically, philosophers, scientists, and advocates focused on establishing good reasons for caring about the plight of threatened species and their habitats. Now, because of the effects of global climate change, questions such as the following should be asked: Should species be protected at all? Should they be protected within the historical habitats in which they evolved? Is it appropriate to modify habitats through creating parks, refuges, and protected areas?

It is inevitable that conservation decisions in the future will require more interventionist conservation policies leading to debates regarding risks, benefit and of novel practices as well as consideration of impacts on basic human rights, distributive justice and procedural justice for those whose livelihoods, food supply and cultural practices depend upon the preservation of the natural world.

Conclusion

Climate change needs to be carefully considered and acted upon through the lens of ethics, morals and human rights if these matters are to get the full consideration they need. Pope Francis cautions politicians, law makers, business executives and others that more will be required of them to properly address the challenge. He writes in *Laudato Si'*:

> To take up these responsibilities and the costs they entail, politicians will inevitably clash with the mindset of short-term gain and results which dominates present-day economics and politics, but if they are courageous, they will attest to their God-given dignity and leave behind a testimony of selfless responsibility. A healthy politics is sorely needed, capable of reforming and coordinating institutions, promoting best practices and overcoming undue pressure and bureaucratic inertia.[37]

Climate change involves serious legal issues, especially in its global, intergenerational, human rights, aboriginal and treaty rights and ecological dimensions. Our theoretical tools are underdeveloped in many of the relevant areas. The need for work in moral, legal and political philosophy that articulates compelling reasons as to how and why we should address climate change is clear. Are judges, lawyers and legal academics sufficiently prepared to participate in these discussions? Is space being created for those who need to be heard? This is the challenge for us all.

Notes

1 John Rawls, *Theory of Justice* (Cambridge, MA: Harvard University Press, 1971); see also Edith Brown Weiss, "In Fairness To Future Generations and Sustainable Development", *American University International Law Review* 8(1) (1992): 19–26 for a discussion of Edmund Burke, who in 1790 referred in general terms to the idea of a partnership "Between those who are living, those who are dead, and those who are to be born". In 1690 Locke (First Treatise) referred to an idea of *joint possession at the overlap*. In 1789 Jefferson claimed that "The earth belongs in *usufruct* to the living"; and at the beginning of the twentieth century Jean Jaurès even worked out a concept of "everlasting *mortgage*".

2 In their article "Climate Change and Future Generations", Revesz and Shahabian discuss the futility of using traditional methods of calculating future losses when it comes to mitigating climate change. See Richard L. Revesz and Matthew R. Shahabian, "Climate Change and Future Generations", *Southern California Law Review* 84 (2011): 1097.

3 The Canadian former prime minister, Stephen Harper, has made this clear many times, as have leaders in many other countries. For example see the commentary here.

4 For a discussion of disproportionate impacts on the poorer nations see the Intergovernmental Panel on Impacts, Adaptation and Vulnerability at www.ipcc.ch/ipccreports/tar/wg2/index.php?idp=674.

5 Ashleigh Downing and Alain Cuerrier, "A Synthesis of the Impacts of Climate Change on the First Nations and Inuit of Canada", *Indian Journal of Traditional Knowledge* 10(1) (January 2011): 57–70.

6 Simon Caney, *Justice Beyond Borders: A Global Political Theory* (Oxford: Oxford University Press, 2005).
7 See commentary at www.ammsa.com/publications/alberta-sweetgrass/no-resource-revenue-sharing-says-alberta-government.
8 For a thorough discussion of case law on these issues see Bill Gallagher, *Resource Rulers: Fortunes and Folly on Canada's Road to Resources* (Waterloo, ON: Bill Gallagher, 2012).
9 Downing and Cuerrier, "A Synthesis of the Impacts of Climate Change on the First Nations and Inuit of Canada".
10 UN General Assembly, *Universal Declaration of Human Rights*, 10 December 1948, 217 A (III), available at www.refworld.org/docid/3ae6b3712c.html (accessed 1 December 2015).
11 For example see Bishop Desmond Tutu's position at www.tutufoundationusa.org.
12 Lynda M. Collins, "An Ecologically Literate Reading of the Canadian Charter of Rights and Freedoms", *Windsor Review of Legal and Social Issues* 26 (2009): 7.
13 Jutta Brunnée and Stephen J. Toope, "A Hesitant Embrace: The Application of International Law by Canadian Courts", *Canadian Yearbook of International Law* 40 (2002): 9.
14 Reference re s. 94(2) of Motor Vehicle Act (British Columbia), [1985] 2 S.C.R. 486 at para 71 [*Re BC Motor Vehicle Act*].
15 *Millership v. Kamloops (City)*, 2003 BCSC 730 at para 6.
16 Ibid.
17 *Domke v Alberta (Energy Resources Conservation Board)*, 2008 ABCA 232.
18 *Kelly v. Alberta (Energy Resources Conservation Board)*, 2009 ABCA 349.
19 The Court applied the test in *Blencoe v British Columbia (Human Rights Commission)*, 2000 SCC 44, [2000] 2 SCR. 307 at para. 47.
20 *Kelly v. Alberta*.
21 Collins, "An Ecologically Literate Reading of the Canadian Charter of Rights and Freedoms".
22 Ibid.
23 Chris Turner, *The War on Science: Muzzled Scientists and Wilful Blindness in Stephen Harper's Canada* (Vancouver: Greystone Books, 2013).
24 Ibid.: ch. 2.
25 Ibid.
26 The House of Lords decision in the case of *McGhee v National Coal Board*, [1972] 3 All ER 1008 recognized the high degree of difficulty plaintiffs face when having the burden to prove causation on a balance of probabilities when scientific evidence of causation of effects from harmful pollution is unclear. To prevent injustices to the victims of negligence they reversed the burden to the defendants requiring them to disprove causation.
27 *Operation Dismantle v the Queen* [1985] 1 SCR 441.
28 2003 BCSC 82 [*Millership*].
29 [1993], 44 ACWS (3d) 375 (ABQB) [*Locke*].
30 *Ernst v EnCana*, 2014 ABCA 285.
31 Appellant's Factum, at para 3.
32 43. No action or proceeding may be brought against the Board or a member of the Board or a person referred to in section 10 or 17(1) in respect of any act or thing done purportedly in pursuance of this Act, or any Act that the Board administers, the regulations under any of those Acts or a decision, order or direction of the Board.
33 *Ernst v EnCana* at para 84.
34 Ibid. at 83.
35 Pope Francis, "Encyclical Letter *Laudato Si'* of the Holy Father Francis on Care for Our Common Home", 24 May 2015, available at http://w2.vatican.va/content/

francesco/en/encyclicals/documents/papa-francesco_20150524_enciclica-laudato-si.
html.
36 C. D. Thomas et al., "Extinction Risk from Climate Change", *Nature* 427(6970):
145–148. See also Lee Hannah (ed.), *Saving a Million Species, Extinction From Climate
Change* (Washington, DC: Island Press).
37 Pope Francis, "Encyclical Letter *Laudato Si'* of the Holy Father Francis on Care for
Our Common Home".

10 David versus Goliath

Voluntary professional societies of epidemiology and the industrial juggernaut

Colin L. Soskolne

Sustaining counter-weights to moneyed influence

Epidemiology is the discipline that bridges laboratory sciences with the human experience. It is thus relied on as the evidentiary base for advancing knowledge about the relationship between cause and effect, informing health policy, and in litigation. There are many examples of the wilful manipulation of epidemiology from moneyed interests (i.e. the Goliaths) whose purpose is to derail the work of epidemiologists working in support of the public interest. The need for epidemiologists to focus on translating their work at the interface between research and policy thus was born in 2006 with the founding of the International Joint Policy Committee of the Societies of Epidemiology (IJPC-SE). The IJPC-SE was established with its mission to advance valid science. It develops position statements to distinguish valid from invalid (i.e. "junk") science, the latter which continues to infiltrate the scientific literature and decision-making bodies. As voluntary members of our profession (i.e. the Davids), we on the IJPC-SE serve the public interest by, among other things, being vigilant to both the misuse and abuse of our discipline in relation to understanding causality, advancing public policy, as well as in the courts in tort actions usually on behalf of plaintiffs. The great challenge is for the Davids to conduct their work without a secure funding base, in contrast to the Goliaths who contribute fortunes to the manufacture of doubt and the fomentation of uncertainty. Two different models are explored for their practicality, effectiveness, and sustainability – one "democratic" and the other "benevolent dictatorship" – in support of voluntary organizations working in the public interest.

Introduction to epidemiology

The discipline of epidemiology is defined as the study of the occurrence and distribution of health-related states (or, events) in specified populations. Included in the definition is the study of the determinants influencing such states, and the application of this knowledge to control any health problem (Porta 2008).

The focus in epidemiology is on community (or, population) health, and on the *prevention* of disease, disability and premature death in any group of people (or animal) sharing some common characteristic. Epidemiologists study health problems with a view to developing policy interventions to correct the problem. Effective policy interventions usually will incorporate incentive and disincentive regimes to change behaviours, whether in relation to a business practice or of a personal nature.

The nature of conflicting (or competing) interests

Conflicting interests can arise in the work of epidemiologists. This situation can be prompted through concern by stakeholders over the consequences of any new incentive and/or disincentive regime that could be instituted through policy interventions informed through epidemiological enquiry.

On the one hand, the self-interest of those with business investments trumps all other considerations because their primary consideration revolves around maintaining the *status quo* to protect their investments; business growth is the paradigm with which many business managers are aligned. The least thing desired by such managers is information that might call into question the nature of their industrial processes and products, particularly should that information require investment in eliminating harms associated with their business interests.

While business interests do dominate, other interests are at stake, including personal self-interest. The focus of this chapter is on professional integrity that can assume second place to our colleagues' obligation to protect the public interest. Personal self-interest arises when one is seduced by the lure of financial rewards, usually very lucrative payments, typically in the service of powerful moneyed interests. One mechanism to control such behaviour in science has been to require that scientists declare any conflicting interests associated with their work. The colleagues seduced by money often fail to disclose the fact of competing/conflicting interests in order to please their sponsor. And, ever more devious ways of avoiding such declarations are being seen through the motivation for litigation-driven research.

What is driving these behaviours? It is the current world view of the sustainability of limitless growth and wealth accumulation, whether from a business or personal perspective. Limitless growth has been deconstructed and debunked as a fallacy, revealing it a paradigm whose time has passed. While it dominates, however, more of the same behaviours will be seen and will thus need to be challenged. The tension between a sustainable future and one of short-term profit is what is at play. All of these behaviours flow from the neoliberal view of unfettered economic growth.

Also dominating are social, cultural and religious interests, which are not unrelated to interest in accumulating wealth. All constitute and, indeed, contribute to the dominant narrative that sustains social systems. It is unwise to separate the personal behaviours of scientists described above from the dominant paradigm. Epidemiologists may view the continued dominance of the neoliberal mantra of limitless growth as the most upstream determinant of disease.

In whose best interests do epidemiologists work?

According to the above definition of epidemiology, the work of the epidemiologist is to protect community/population health. Thus, by definition, the public interest should trump other interests. Some epidemiologists will focus their talents on local communities and others will address global health issues. Some will focus on populations of the elderly, while others will concern themselves with the newborn, and also on the unborn (i.e. the foetus). And, some will focus on future generations.

Regardless of the area of focus, the epidemiologist who best serves the public interest is the epidemiologist who is aware of the pressures that operate in the real world and that exercise conscious and subconscious influence on her/his research. Influence is of concern because it determines corruption of the scientific method and thus detracts from the validity of the research findings. Pressures can range from the very questions that are permitted to be asked, through the conduct of the research, to the dissemination of the findings. When research is funded by private funds as opposed to public funds, different biases and issues of partiality are of concern. There thus is a moral tension in choosing a public over a private/corporate source of funds to address a particular scientific question.

Can we deny that influence drives junk science?

There are many examples, some more and some less recent, of exposés of relevance to our topic and that we cannot deny. The problem of influence was first brought to this author's attention through the work of Epstein (1978), which demonstrated the tension between the generation of evidence and the political pressures from powerful interests with influence. In 1982, on the topic of corporate ethics and environmental pollution, Lord noted that "Corporations create 80% of our GNP. They, of all entities working, have the most potential for good or evil in our society" (Lord 1982). This was said to be the case in 1982; today, with ever-growing corporate wealth nationally and internationally, this number is surely more like 90 per cent.

To the present generation of scholars as well as the public in countries with industrialized economies, the tobacco example is likely best known. The story of disinformation, lies, manipulation and deceit emerged through access to information after more than 50 years of influence. This is a sordid story with multiple aspects from which the world could learn if future harms are to be prevented. Sadly, this is not proving to be the case in that the culture of management and legal expertise that cut its teeth on advancing tobacco interests, applied those skills to many other products, including asbestos and climate change (Michaels 2008; Oreskes and Conway 2010; Baur et al. 2015).

In a systematic review of the literature on the scope and impact of financial conflicts of interest, Bekelman et al. (2003) demonstrated a statistically significant association between industry sponsorship and pro-industry conclusions. They concluded that financial relationships among industry, scientific investigators and academic institutions are widespread with industry sponsorship being associated with restrictions on publication and data sharing.

In a study by Friedman and Richter (2004), authors of drug studies with conflicting interests are reported as being more likely to present studies with positive findings.

While the vast majority of scientists have agreed with the evidence that climate change is a real phenomenon, authors like Lomborg (2001) have denied both the fact of climate change and that it is associated with human activity in terms of greenhouse gas emissions. In this same vein, Koch Industries (USA) with powerful interests (in particular in oil) as well as in asbestos through one subsidiary (Georgia-Pacific), have provided vast amounts of money in support of at least one ideological neoliberal think tank based in Chicago, USA, the Heartland Institute (Edmonton Journal 2008). This Institute had the gall in seeing as its mandate to send some 11,000 brochures and video packages denying that human activity is the driving force behind global warming to all Canadian school libraries in 2008. The intent was to create doubt in the minds of Canadian children as to the real facts about the causes of climate change. Brown has provided much evidence as to the power of ideological think thanks in derailing the pursuit of truth (Brown 2008, 2015).

In 2013, the Director of the United States National Institute for Environmental Health Sciences made the point that "Industry attacks on Public Health research have become more strident" (Morris and Hamby 2013).

In 2014, internal documents revealed an industry "pattern of behaviour" on toxic chemicals which suggest a pattern of concealment from workers relating to benzene and worker cancers (Heath and Morris 2014; Morris and Hamby 2013).

Some have even concluded that independent, scientific organizations and publications concerned with public health have betrayed their mission and the public interest by allowing financial conflicting interests to influence their work and their policies. For example, a report and article (Simon 2015) focussed on the food industry and the question was asked as to whether the American Society for Nutrition has lost all credibility.

The role of the professional society

The role of any professional society is to:

- serve as a transparent voice for advancing the discipline by providing a forum to facilitate networking to maximize engagement at multiple levels and scales in the public interest;
- foster the development of uni-, multi- and trans-disciplinary research methods;

- incentivize personal and professional integrity in both research and practice by setting normative standards for ethics, peer oversight, and account-ability; and
- provide a public face.

To be a cohesive forum with a particular focus, each society will develop a mission statement in which core values will be articulated. Such statements provide the anchor for the group's activity and collective motivation. In epidemiology, the mission includes the need to "maintain, enhance, and promote health in communities worldwide; and to work to protect the public health interest above any other interest".

What group would be better positioned to have the skills to differentiate between the valid and invalid applications of the discipline?

As with any science, it is possible to manipulate experimental and control groups in ways that introduce bias (Soskolne 2015). The text by Cranor (2011) nicely explicates the several classical techniques that skew results, from biased methods to junk science. The manipulation of science in this way goes counter to the scientific method and thus fails in its service to protect the public interest through the pursuit of truth. To prevent such occurrences, ethical training and oversight are crucial additions to the core training of scientists, particularly in those disciplines where the stakes can be high and influence strong.

The role of the scientist is to do the best possible science so that, armed with valid information, the scientist is more securely positioned when truth needs to be spoken to power. Yet, all sorts of pressures operate on the applied health scientist and these have implications for policy. Thus, anything that tempts scientists away from the pursuit of truth warrants careful scrutiny if misconduct that can derail scientific discourse is to be prevented.

To contain the problem of temptation, Clayson and Halpern (1983) were among the first in public health to bring to attention the fact that "Industry's offensive against the regulation of health and safety hazards uses academics to downplay or deny the seriousness of the hazards." Despite this knowledge, the frailty of even the most accomplished of scientists to succumb to temptation remains real. The pervasiveness of the misuse of and inappropriate application of the scientific method has been extensively reported (Epstein 1978; Davis 2002, 2007, 2010; Michaels 2008; McCulloch and Tweedale 2008) and, from the most recent such text (Oreskes and Conway 2010), a full-feature docudrama was released in 2015. Each of these texts documents egregious examples of the manu-facture of doubt through the production of invalid (i.e. "junk") science whose role it is to confuse both policy-makers and the public across a broad range of exposure circumstances and industries.

The main purpose behind fomenting uncertainty through the production of "junk" science is to render the policy-maker incapable of making changes in health policy, if only because policy-makers prefer to make policy in the presence of the least amount of uncertainty. More and more, however, the production of invalid science is making its way into the literature and being used to defend the indefensible in litigation.

Human and system frailties and the need for oversight

Most people are naively positive about the good in people. Yet, evidence points to the need for less naivety. Each of us is human and thus at risk of being seduced into corrupting the scientific method to serve a sponsor. Fortunately, there are those among the professionals who take seriously the obligation to be vigilant and especially careful in peer review. This group of professionals has recognized the need for oversight (as in human research ethics boards/ IRBs), as well as for keeping ourselves on track with ethics guidelines and related activities. Above all, we need to be aware of forces at play that influence both science and policy. Great vigilance and personal integrity are required to counter the influence of financially interested parties and of corrupt/morally bankrupt governments.

The relentless pressure from vested interests results in so-called professionals manoeuvring their way onto review panels, influencing Boards of our professional associations, and infiltrating the literature with junk science. In particular, litigation-driven research will shape scientific understanding and thus public health policy because public health officials and bureaucrats are not aware of the substantial conflicting interests associated with the literature that makes its way into the public domain. Expert witness tensions arise between the plaintiff and defence sides of the argument in tort actions where the rubber hits the road concerning policy decisions. To try to contain this phenomenon, the current major initiative of the IJPC-SE is its Working Group on Conflict-of-Interest and Disclosure.

The IJPC-SE: its goal and approach

The International Joint Policy Committee of the Societies of Epidemiology (IJPC-SE) is a volunteer-driven, not-for-profit consortium, currently comprising 19 national and international member-professional societies/associations. The goal of the IJPC-SE is to serve the public interest by informing health policy and related areas of endeavour through its work at the nexus of research and policy. It coordinates inter-professional society activities that are related to research and practice in the impartial generation of evidence, as well as in evidence-based policy application, formulation, implementation and evaluation. It promotes epidemiological best practices to inform policy.

The IJPC-SE was formed in Seattle in 2006 at the second (5-yearly) North American Congress of Epidemiology. In 2016, it will be in its tenth year of operation. The first chair (2006–2007) was Roberta Ness; the second chair (2008–2009) was Susan Sacks. The third chair (2010–2014) was Stanley H. Weiss. The fourth chair is the present author, Colin L. Soskolne (2014–2016).

The IJPC-SE 2012 position statement on asbestos

A major initiative was the development in 2011 and release in 2012 of The IJPC-SE's Position Statement on Asbestos:

> This Statement was an important act of collaboration and leadership by societies of epidemiology in calling for national and international policy to be based on the scientific evidence. While the asbestos industry spends millions of dollars on marketing and political lobbying, they are losing the battle of credibility, thanks to organizations such as the IJPC-SE speaking up to defend epidemiologic evidence and public health policy. More organizations have since joined the IJPC-SE, in part, I believe, because they see that the IJPC-SE is playing a positive and meaningful role in serving the public good.
> (Kathleen Ruff, quoted in
> *The Epidemiology Monitor*, June 2015)

Until 2012, the IJPC-SE had operated without any real infrastructure. It was after the importance of the role of the Committee was realized with the uptake of the Position Statement on Asbestos that a website was launched. Then, founding bylaws and related policy documents were developed, and not-for-profit status was set in motion in 2014–2015. This was done because of the need for transparency given the impact that the Statement was having in the world.

Current IJPC-SE initiatives

On the work product side of its mission, a current initiative of the IJPC-SE is to create a Position Statement on Conflict-of-Interest and Disclosure. The need to address conflict-of-interest and disclosure issues more forthrightly was brought about by high-profile failures of epidemiologists to fulfil norms and expectations in these areas.

On the organizational side of its mission, there were two choices before the then-members of the IJPC-SE: one was for a model of benevolent dictatorship and the other a model of democracy. There is no question that the demands for transparency alone contributed heavily to the need for an organization with infrastructure including governance documents. A benevolent dictatorship would not require much infrastructure and no governance structure as compared with the needs of a democratic organization. Is one model more suited to any particular mission?

The *advantages* of each model were considered as follows:

- A *benevolent dictatorship* is driven by a single powerful leader and requires far less reliance on volunteer commitment. There is thus greater ease in decision-making.
- A *democracy* requires devolved roles to spread the load. It requires shared decision-making leading to greater buy-in and a broader base of support. If funding were to be sought, it would appeal more to funders in support of its mission.

The *disadvantages* of each model were viewed as follows:

- A *benevolent dictatorship* lacks continuity and likely would be unattractive to potential sponsors.
- A *democracy* requires adherence to processes defined in bylaws. It also requires volunteerism having limited time and energy to devote to its articulated mission.

The way forward

In this chapter, the case has been made for having an effective counter-balance to the forces that currently are undermining the pursuit of truth. In the absence of an effective voice of reason (David), the forces of malfeasance (Goliath) will dominate.

In a world that aspires to democratic values, true democracy can be attained only through a well-informed public, underscored by an improved government science, technology and innovation strategy. The world of today pits the Davids against the Goliaths in its inability to regulate that which could be moderated by regulation.

The public interest needs to be protected. In a context where influences are so strong as to be damaging to the public interest, a strategy is needed that will offer *incentives* to non-profit professional organizations (such as the IJPC-SE) in support of capacity-building to expose junk science, particularly where applied science works at the nexus of policy. Such policies would be strengthened if they were to introduce *disincentives* (i.e. regulatory penalties) for those engaging in producing junk science. The IJPC-SE could serve as a practical, effective and sustainable counter-balance for good, and for a mere fraction of the monies devoted by vested interests to causing harm. However, those monies needed to protect the public interest will need to come mainly from private and public sources. The democratic model adopted is one that should be self-sustaining and endure well beyond its current board.

Acknowledgment and disclosure

As a professional legacy, I have been chairing for the past two years, and bankrolling for the past three years, the International Joint Policy Committee of the Societies of Epidemiology (IJPC-SE), a voluntary global consortium of member professional societies of epidemiology working in the public interest. I do so in the hope that it will become self-sustaining and, on a relatively small annual budget, will endure as a counter-weight to the many heavily funded ideological, neoliberal think tanks. The latter have fomented uncertainty and manufactured doubt relating to the body of valid scientific evidence generated to advance science and inform policy.

References

Baur, X., Budnik, L. T., Ruff, K., Egilman, D. S., Lemen, R. A. and Soskolne, C. L. (2015) "Ethics, Morality, and Conflicting Interests: How Questionable Professional Integrity in Some Scientists Supports Global Corporate Influence in Public Health". *International Journal of Occupational Environmental Health* 21: 172–175.

Bekelman, J. E., Li, Y. and Gross, C.P. (2003) "Scope and Impact of Financial Conflicts of Interest in Biomedical Research: A Systematic Review". *JAMA* 289(4): 454–465.

Brown, D. A. (2008) "The Ominous Rise of Ideological Think Tanks in Environmental Policy-Making". In C. L. Soskolne (ed.), *Sustaining Life on Earth: Environmental and Human Health through Global Governance*, Lexington Books, Lanham, MD, pp. 243–256.

Brown, D. (2015) "The Seeds of the Corporate Funded Climate Disinformation Campaign, the 1971 Lewis Powell Memo". Ethics and Climate blog, posted 31 October, available at http://ethicsandclimate.org/2015/10/31/the-seeds-of-the-corporate-funded-climate-disinformation-campaign-the-1971-lewis-powell-memo (accessed 17 November 2015).

Clayson, Z. E. and Halpern, J. L. (1983) "Changes in the Workplace: Implications for Occupational Safety and Health". *Journal of Public Health Policy* 4: 279–297.

Cranor, C. F. (2011) *Legally Poisoned: How the Law Puts Us at Risk from Toxicants*, Harvard University Press, Cambridge, MA.

Davis, D. (2002) *When Smoke Ran like Water: Tales of Environmental Deception and the Battle against Pollution*, Basic Books, New York.

Davis, D. (2007) *The Secret History of the War on Cancer*, Basic Books, New York.

Davis, D. (2010) *Disconnect: The Truth about Cell Phone Radiation, What the Industry has Done to Hide It, and How to Protect your Family*, Dutton, New York.

Edmonton Journal (2008) "Climate Change Questioned in Schools Mailout". *Edmonton Journal* (5 May): A5.

Epstein, S. S. (1978) *The Politics of Cancer*, Sierra Club Books, San Francisco, CA.

Friedman, L. S. and Richter E. D. (2004) "Relationship Between Conflicts of Interest and Research Results". *Journal of General Internal Medicine* 19: 51–56.

Heath, D. and Morris, J. (2014) "Exposed: Decades of Denial on Poisons: Benzene and Worker Cancers: 'An American Tragedy'". Available at www.publicintegrity.org/2014/12/04/16330/internal-documents-reveal-industry-pattern-behaviour-toxic-chemicals (accessed 17 November 2015).

Lomborg, B. (2001) *The Skeptical Environmentalist: Measuring the Real State of the World*, Cambridge University Press, Cambridge.

Lord, M. W. (1982) "Keynote Address: Corporate Ethics and Environmental Pollution". In J. S. Lee and W. N. Rom (eds), *Legal and Ethical Dilemmas in Occupational Health*, Ann Arbor Science Publishing, Ann Arbor, MI, pp. 5–18.

McCulloch, J. and Tweedale, G. (2008) *Defending the Indefensible: The Global Asbestos Industry and its Fight for Survival*, Oxford University Press, New York.

Michaels, D. (2008) *Doubt is Their Product: How Industry's Assault on Science Threatens Your Health*, Oxford University Press, New York.

Morris, J. and Hamby, C. (2013) "Industry Muscle Targets Federal Report on Carcinogens". Center for Public Integrity, available at www.publicintegrity.org/2013/07/30/13068/industry-muscle-targets-federal-report-carcinogens (accessed 17 November 2015).

Orsekes, N. and Conway, E. M. (2010) *Merchants of Doubt: How a Handful of Scientists Obscured the Truth on Issues from Tobacco Smoke to Global Warming*, Bloomsbury, London.

Porta M. (ed.) (2008) A *Dictionary of Epidemiology*, 5th edn, Oxford University Press, New York.

Simon, M. (2015) "Nutrition Scientists on the Take from Big Food: Has the American Society for Nutrition Lost All Credibility?" Available at www.eatdrinkpolitics.com/wp-content/uploads/ASNReportFinal.pdf.

Soskolne, C. L. (2015) "Public Health and Environmental Health Risk Assessment: Which Paradigm and in Whose Best Interests?" In L. Westra, J. Gray and V. Karageorgou (eds), *Ecological Systems Integrity: Governance, Law and Human Rights*, Routledge, Abingdon, pp. 191–200.

11 Some considerations on the role of the Security Council in facing the Ebola outbreak

Sabrina Urbinati

Introduction

The outbreak of Ebola[1] in West Africa started in late 2013 in Guinea. Soon it spread to Liberia and Sierra Leone and went out of control. This outbreak was the largest and most complex since the discovery of the Ebola virus in 1976. In fact, there were more cases and deaths in the Ebola outbreak in West Africa than in all others combined.[2]

On 18 September 2014, during an emergency meeting, the Security Council (SC) adopted resolution 2177 (res. 2177), entitled *Peace and Security in Africa*, where it determined that "the unprecedented extent of the Ebola outbreak in Africa constitutes a threat to international peace and security".[3] This recognition was reiterated by the President of the Council in a statement made on 21 November 2014.[4]

The recognition of the Ebola outbreak in West Africa as a threat to international peace and security is an unprecedented step. In fact, even if the SC had already treated infectious diseases, it had never declared one of them a threat to international peace and security. This determination contributes to expanding the concept of "threat to international peace and security" and raises new issues about the powers and the role of the SC in dealing with this new kind of threat.

The aim of this contribution is to try to give greater conceptual coherence to this development. For this purpose, I will examine the reaction of the International Community to the Ebola outbreak; then, I will recall the progressive expansion of the concept of "threat to international peace and security" and demonstrate that, before the Ebola outbreak, infectious diseases were not completely unknown by the SC as a major international threat; finally, I will analyse the content of res. 2177.

The international community reaction to the Ebola outbreak in West Africa

Before and after res. 2177, not only the three most affected States, but many others,[5] as well as international organizations – within and outside the family of the United Nations (UN)[6] – non-governmental organizations (NGOs),[7] and other private actors[8] mobilized themselves increasingly to fight against this horrible infectious disease, providing financial resources, structures, humanitarian and military personnel, as well as medical equipment and personnel. For instance, the World Food Programme (WFP) and the Food and Agricultural Organization (FAO) provided food aid for millions of people, who were suffering the consequences of the Ebola outbreak, such as the increasingly limited food availability and rising prices due to the established quarantine regimes and loss of crops. Moreover, the World Bank (WB) and the International Monetary Fund (IMF) provided financial resources to carry out the activities necessary to put under control the Ebola outbreak. Finally, the African Union (AU) Peace and Security Council approved its first humanitarian mission in the region. Among these actors, and beside the SC, the work of the World Health Organization (WHO), the Secretary General (SG) and the General Assembly (GA) deserved to be mentioned.

The WHO, which received the notification concerning the Ebola outbreak in March 2014, responded both from a normative and operational perspective. From the normative one, the WHO applied articles 12 (Determination of a public health emergency of international concern)[9] and 15 (Temporary recommendations)[10] of the International Health Regulation (IHR).[11] Thus, on the 8 August 2014 the WHO declared that the Ebola outbreak in West Africa was a "public health emergency of international concern" and adopted temporary recommendations. These recommendations were addressed "partly to the affected countries and partly to third States and aimed at preventing a further spread of this disease, while avoiding over reactive measures or unnecessary isolation of the affected countries".[12] From the operational perspective, it is impossible to make an exhaustive list of all the activities carried out by the WHO during the fight against the Ebola outbreak in West Africa. From a general point of view, it is possible to group the WHO operational activities conducted in the affected countries in several clusters: supporting surveillance, community engagement, case management, laboratory services, contact tracing, infection control, logistical support and training and assistance with safe burial practices. Moreover, at the level of the international community, the WHO urged the provision of financial, logistic, sanitary and technical assistance and tried to deal with these aids.[13]

The SG played a facilitating role in relation to international assistance and cooperation. Also in this case it is impossible to list all the activities he carried out to fight against the Ebola outbreak. Some examples will be useful. On 12 August 2014, he appointed the UN System Senior Coordinator for Ebola Virus Disease to provide overall strategic direction and to assist governments in the region in addressing the crisis. On 5 September 2014, he activated the Crisis Response Mechanism of the UN and appointed the Deputy Ebola Coordinator and Emergency Crisis Manager. On 17 September 2014, the SG sent a letter to the Presidents of the SC and GA,[14] where he underlined the threat to security deriving from the Ebola outbreak and announced his will to create the UN Mission for Ebola Emergency Response (UNMEER).[15] On 19 September 2014, he established the UNMEER with the aim to ensure the rapid, effective, efficient and coherent response to the crisis.[16] After its establishment, the UNMEER coordinated all the activities of the other actors involved in the fight against the Ebola outbreak in West Africa, also those of the WHO. Having achieved its core objective, the UNMEER closed on 31 July 2015. Since 1 August 2015 the oversight and coordination of all the actors involved in the fight against the Ebola outbreak is led once again by the WHO.

The GA adopted on 19 September 2014 resolution 69/1 (res. 69/1),[17] where it gave full support to the SG's action and in particular to the establishment of the UNMEER. Moreover, in res. 69/1, the GA expressed its grave concern that the Ebola virus could be a threat to the post conflict recovery of the three affected States and highlighted the urgent need to contain such a virus owing to its possible grave humanitarian, economic and social consequences. In addition, the GA requested the SG to take such measures as may be necessary for the prompt execution of his intention.

The progressive expansion of the concept of "threat to international peace and security" in the UN practice

Since the end of the Cold War, the definition of "threat to international peace and security" has progressively been broadened. In fact, the SC has gradually included, under article 39 of the UN Charter,[18] massive human suffering and displacement arising from violations of human rights and humanitarian law, international terrorism, violent overthrow of democratic governments and illicit exploitation of natural resources, including diamonds and wildlife that can fuel violent conflict.

The starting point of this trend in the SC practice can be dated back to the statement adopted in 1992 by the then SC President, where he affirmed that "the absence of war and military conflicts among States does not in itself ensure international peace and security. The non-military sources of instability in the economic, social, humanitarian and ecological fields have become threats to peace and security."[19] Since this statement, the SC has discussed many times several issues concerning economic and social threats, such as climate change (in 2007, 2011, 2013 and 2015),[20] food crises (in 2002 and 2005)[21] and the connection between security and development (in 2011).[22] Nevertheless, it has never adopted a resolution affirming that one of these issues was a threat to international peace and security.

Also other UN bodies have confirmed the progressive broadening of the concept of "international peace and security". For instance, in 1992, the SG, in *An Agenda for Peace*,[23] introduces a broadened understanding of the categories of threats to international peace and security, by adding internal conflicts and economic and social issues. In 2000, the GA, in the *United Nations Millennium Declaration*,[24] added international organized crime to the concept. Nevertheless, the most important conceptual development came in 2004 with the Report of the High-Level Panel on Threats, Challenges, and Change.[25] This Report proposed that "Any event or process that leads to large-scale death or lessening of life chances and undermines States as the basic unit of the international system is a threat to international security"[26] and proposed as a separate cluster of threats "economic and social threats, including poverty, infectious diseases and environmental degradation".[27] In 2005, in *In a Larger Freedom: Towards Development, Security and Human Rights for All*,[28] the SG included, among the threats to peace and security in the twenty-first century, "not only international war and conflict but civil violence, transnational organized crime, terrorism and weapons of mass destruction".[29] He also included "poverty, deadly infectious disease and environmental degradation, since these can have equally catastrophic consequences".[30] All of these threats can cause death or lessen life on a large scale and all of them can undermine States as the basic unit of the international system. Thus the SG fully endorsed a broader notion of security including deadly infectious disease and environmental degradation.

As mentioned above, the adoption of res. 2177 was not the first time where the SC addressed the connection between security and infectious diseases. The main example, before the Ebola outbreak, is HIV/AIDS, which was discussed by the SC on 10 January 2000.[31] On that occasion the SC discussed the impact that HIV/AIDS had in Africa but did not adopt any resolution and concrete measures. Nevertheless, this was enough to integrate HIV/AIDS in subsequent SC's resolutions on peace-keeping, both from the perspective of protection military contingents as well as of including HIV awareness for civilian populations as part of the mandate of multi-dimensional peace-keeping. Thus, what it is really new in res. 2177 is the recognition that the Ebola outbreak was a threat to international peace and security.

The Security Council reaction against the Ebola outbreak in West Africa: resolution 2177

The SC expressed for the first time its concern about the Ebola outbreak in West Africa in a press statement on 8 July 2014. In this occasion it also conveyed to the international community the need to provide prompt assistance in order to prevent the spread of the virus.[32] Then, the SC manifested an increased concern when, on 15 September 2014, it adopted resolution 2176 renewing and modifying the United Nations Mission in Liberia (UNMIL) mandate.[33] Finally, as mentioned above, on 18 September 2014, the SC adopted res. 2177, where it qualified the Ebola outbreak in West Africa as a threat to international peace and security. The SC justify this qualification, recognizing "that the peacebuilding and development gains of the most affected countries concerned could be reversed in light of the Ebola outbreak" and underlining "that the outbreak is undermining the stability of the most affected countries concerned and, unless contained, may lead to further instances of civil unrest, social tensions and a deterioration of the political and security climate".[34] In fact, Liberia, Guinea and Sierra Leone, which were emerging with many difficulties from vicious civil wars, risked seeing their development and political gains reversed. Moreover, even if reasons about Ebola presenting a global security threat beyond the immediate affected region are not elaborated in detail, they are arguably linked to the risk of international spread of the disease.[35] Among the reasons, not explicitly mentioned, at the basis of the qualification of the Ebola outbreak in West Africa as a threat to international peace and security, it is possible to read another justification into the rationale for the adoption of res. 2177: when the SC decided to adopt it, the available information showed disappointing results from the assistance to the affected countries and an evident lack of coordination among the involved actors derived from the many difficulties and chaos in responding to the Ebola outbreak.[36]

The SC's qualification of the Ebola outbreak as a threat to international peace and security allowed this body to adopt measures under Chapter VII of the UN Charter (entitled: Action with Respect to Threats to the Peace, Breaches of the Peace and Act of Aggression). These measures are of several kinds: provisional measures (that can be both binding and not binding),[37] coercive measures[38] and measures involving the use of force[39] (that instead are binding). With regard to the aim of the present contribution, a brief description of the measures adopted by the SC to fight the Ebola outbreak is useful. First, the SC encouraged the affected States to establish domestic mechanisms for the early diagnosis and isolation of suspected cases and these were also useful in mitigating the effects of Ebola at the political, socioeconomic and humanitarian level. Second, the SC called on the Member States as well as airlines and shipping companies to lift any restrictive measure to transnational cross-border movements of people and trade, facilitate the delivery of humanitarian assistance, increase communication and provide for resources and assistance. Thus, the SC also urged international regional organizations, such as the African Union, the Economic Community of West African States and the European Union, to collaborate to this end. Third, the SC urged the Member States to implement the temporary measures that had been adopted by the WHO in conformity with the IHR. Moreover, the SC called on also the SG to ensure that every organization, within the family of the UN, sped up its own response to the Ebola epidemic and encouraged the WHO to continue to strengthen its technical leadership and operation support to governments and partners.[40] Finally, it has to be noted that in the preamble of res. 2177 the SC inserted another important measure, which unfortunately was not reiterated in its operative part. The SC stressed the crucial and immediate need for a coordinated international response to the Ebola outbreak to find a solution to the several lacks and disorganization.[41]

This description reveals that res. 2177 addresses humanitarian assistance as well as public health measures and concerns. The same can be said for the Presidential statement, mentioned in the introduction of this contribution, that goes in some detail into the necessary interventions for fighting the Ebola outbreak, such as medical evacuation and treatment capacities for first-line responders, the availability of treatment units, the deployment of vaccines and diagnostics, and the entrustment of the coordination of the international action to the UNMEER.

On the basis of this description it seems also possible to argue that the measures adopted by the SC to fight against the Ebola outbreak can be qualified as not binding provisional measures as provided in article 40 of the UN Charter. This provision establishes that the SC, before making a recommendation or adopting measures provided in articles 41 and 42, may decide provisional measures as it deems necessary or desirable. These measures aim at preventing the aggravation of the situation qualified, by the SC, as a threat to international peace and security, or breaches of the international peace and security, or act of aggression. Moreover the provisional measures shall be without prejudice to the rights, claims, or position of the parties concerned. Actually, as illustrated above, the measures stated in res. 2177, by their nature, aimed at preventing the deterioration of the Ebola outbreak as it stood on 18 September 2014 and they also are temporary, because they will end as soon as the outbreak is over.[42] Finally, they cannot be considered binding because of the language utilized by the SC and the absence of reference to Chapter VII or article 40 of the UN Charter. In fact, the SC utilized verbs such as "encourages", "calls on", "urges" and "request".[43]

Conclusion

Before res. 2177 the SC had never qualified an economic and social event as a concrete threat to international peace and security. Thus, as demonstrated above, res. 2177 is significant for two reasons.

First, it gives us an example of what powers the SC may utilize and what measures it may adopt in facing an infectious disease qualified as a threat to international peace and security. As illustrated, substantially the SC ratified all the measures already adopted by other actors such as and *in primis* by WHO. On the bases of article 40 of the UN Charter, these measures are not binding provisional measures. Nevertheless, as the records of the SC meetings tell us that Member States, taking part, did not requested stricter measures, it seems that the door was left open, in case the situation worsened, to the adoption of more severe measures, at least, under article 41 of the UN Charter, such as travel restrictions, quarantines, etc. On this basis, I would argue that the SC management of the Ebola outbreak served essentially to raise the pressure and the attention of the international community to strengthen and, above all, better coordinate its reaction. Actually, before the SC's engagement against the Ebola outbreak some criticisms were raised about the effectiveness of the international community response. Thus the SC intervention can be compared to a call to order because the international community can be expected to coordinate itself in a better way in such an emergency. Only the SC has such authority in the international system, having primary responsibility for responding to threats to international peace and security. This argument is confirmed by the fact that the SC held other meetings on the Ebola outbreak one and two months after res. 2177. The purpose of these meetings was to maintain the pressure and the attention on the issue. Then the attention of the SC diminished in the following months, when the situation provoked by the Ebola outbreak seemed improved.

Second, res. 2177 gives a criterion for the determination of the possible measures the SC could adopt to fight other economic and social threats: it could adopt technical measures suggested by or ratify measures already taken by technical actors, such as the WHO to fight the Ebola outbreak. To push this further, in the area of climate change, for instance, the SC could adopt measures suggested or already taken by the institutions created by the United Nations Framework Convention on Climate Change.

Notes

1 Ebola virus disease is a severe and often fatal illness in humans: the average of Ebola death case rate is around 50 per cent. This virus was discovered first in 1976 in two simultaneous outbreaks: one in Sudan and the other in Democratic Republic of Congo. Symptoms are: the sudden onset of fever fatigue, headache, sore throat, muscle pain, vomiting, diarrhoea, rash, impaired kidney and liver functions, and in some cases internal and external bleeding. Ebola spreads through human to human transmission via direct contact with the blood, secretions, organs or other bodily fluids of infected people, and with surfaces and materials contaminated with these fluids. The incubation period is 2 to 21 days. Humans are not infectious until the development of symptoms. Currently no licenced vaccines exist, but there are some candidates under evaluation. At this stage, to improve survival, only supportive care (rehydration with oral or intravenous fluids) and treatment of specific symptoms can be used. It can be difficult to distinguish Ebola from other infectious diseases such as malaria, typhoid fever and meningitis. See www.who.int/mediacentre/factsheets/fs103/en/# (accessed 30 September 2015).
2 On 4 October 2015 WHO registered 28,457 cases and 11,312 deaths (Ebola Situation Report, http://apps.who.int/ebola/current-situation/ebola-situation-report-30-september-2015, accessed 4 October 2015). The Ebola outbreak spread also in other countries such as Nigeria, Senegal, Mali, USA, Spain and Italy.
3 Peace and Security in Africa, 18 September 2014, UN doc. S/RES/2177(2014).
4 Statement by the President of the Security Council, 21 November 2014, UN doc. S/PRST/2014/24.
5 For example: Cuba, UK and USA.
6 For example: the World Health Organization (WHO), the World Bank (WB), the International Monetary Fund (IMF), the World Food Programme (WFP), the Food and Agricultural Organization (FAO), the United Nations Children Fund (UNICEF), the Secretary General (SG) and the General Assembly of the UN (GA) and the African Union (AU).
7 For example: Doctors without Borders and Emergency.
8 For example: the Pasteur Institute and the Centres for Diseases Control and Prevention (CDC).
9 "1. The Director-General shall determine, on the basis of the information received, in particular from the State Party within whose territory an event is occurring, whether an event constitutes a public health emergency of international concern in accordance with the criteria and the procedure set out in these Regulations. 2. If the Director-General considers, based on an assessment under these Regulations, that a public health emergency of international concern is occurring, the Director-General shall consult with the State Party in whose territory the event arises regarding this preliminary determination. If the Director-General and the State Party are in agreement regarding this determination, the Director-General shall, in accordance with the procedure set forth in Article 49, seek the views of the Committee established under Article 48 (hereinafter the "Emergency Committee") on appropriate temporary recommendations. 3. . . . . 4. In determining whether an event constitutes a public health emergency of international concern, the Director-General shall consider: (a) information provided by the State Party; (b) the decision instrument contained in

Annex 2; (c) the advice of the Emergency Committee; (d) scientific principles as well as the available scientific evidence and other relevant information; and (e) an assessment of the risk to human health, of the risk of international spread of disease and of the risk of interference with international traffic."

10 "1. If it has been determined in accordance with Article 12 that a public health emergency of international concern is occurring, the Director-General shall issue temporary recommendations in accordance with the procedure set out in Article 49. Such temporary recommendations may be modified or extended as appropriate . . . 2. Temporary recommendations may include health measures to be implemented by the State Party experiencing the public health emergency of international concern, or by other States Parties, regarding persons, baggage, cargo, containers, conveyances, goods and/or postal parcels to prevent or reduce the international spread of disease and avoid unnecessary interference with international traffic."

11 OMS, International Health Regulation (2005), 2008, http://whqlibdoc.who.int/publications/2008/9789241580410_eng.pdf, accessed 1 October 2015. Under article 2 of IHR, "[t]he purpose and scope of these Regulations are to prevent, protect against, control and provide a public health response to the international spread of disease in ways that are commensurate with and restricted to public health risks, and which avoid unnecessary interference with international traffic and trade." The IHR is not an international convention, but a technical instrument adopted by the WHO on the basis of article 21 of the WHO Constitution. The last version of the IHR was adopted in 2005 and entered into force in 2007. It is binding only for WHO Member States. See P. Acconci, "The Normative Authority of the World Health Organization", in *Public Health*, 2015, pp. 1–9; L. O. Gostin, D. Sridhar and D. Hougendobler, "The Normative Authority of the World Health Organization", *Public Health*, 2015; H. De Pooter, *Le droit international face aux pandémies: vers un système de sécurité sanitaire collective?*, Pedone, 2015; S. E. Davies and J. Youde, "The IHR (2005), Disease Surveillance, and the Individual in Global Health Politics", in *The International Journal of Human Rights*, 2013, pp. 133–151; M. Fleming, "Combating the Spread of Disease: the International Health Regulations", *Columbia Journal of Transnational Law*, 2012, pp. 805–825; P. Acconci, *Tutela della salute e diritto internazionale*, CEDAM, 2011; L. Boisson de Cahzournes, "Le pouvoir réglementaire de l'Organisation Mondiale de la Santé à l'aune de la santé mondiale: réflexions sur la portée et la nature du Règlement Sanitaire International de 2005", in *Droit du pouvoir, pouvoir du droit: Mélanges offerts à Jean Salmon*, Bruylant, 2007, pp. 1157–1181; P. D. Fidler, "From International Sanitary Conventions to Global Health Security: The New International Health Regulations", *Chinese Journal of International Law*, 2005, pp. 325–392.

12 G. L. Burci, "Ebola, the Security Council and the Securitization of Public Health", *QIL* 10/2014, pp. 28–29.

13 More details are available on the WHO website: www.who.int/csr/disease/ebola/en, accessed 2 October 2015. See also P. Acconci, "The Reaction to the Ebola Epidemic within the United Nations Framework: What Next for the World Health Organization?", *Max Planck Yearbook of United Nations Law*, 2014, pp. 405–424.

14 Identical letters dated 17 September 2014 from the Secretary-General to the President of the General Assembly and the President of the Security Council, UN doc. A/69/389 – S/2014/679, 18 September 2014.

15 See the UNMEER website: http://ebolaresponse.un.org/un-mission-ebola-emergency-response-unmeer, accessed 9 October 2015.

16 Taking into account the sanitary and humanitarian nature of its purpose and activities, the UNMEER has to be considered an *exemplaire unique* and not a peace-keeping or a peace-enforcement operation. One of the reasons at the basis of this distinction is the fact that peace-keeping and peace-enforcement operations are established by the SC and not by the SG, as it has been done for UNMEER.

17 Measures to Contain and Combat the Recent Ebola Outbreak in West Africa, UN doc. A/RES/69/1, 23 September 2014.

18 "The Security Council shall determine the existence of any threat to the peace, breach of the peace, or act of aggression and shall make recommendations, or decide what measures shall be taken in accordance with Articles 41 and 42, to maintain or restore international peace and security."

19 Note by the President of the Security Council, UN doc. S/23500, 31 January 1992, p. 3.

20 UN doc. S/PV. 5663, 17 April 2007, UN doc. S/PV. 6587, 20 July 2011, and UN doc. S/PRST/2011/15, 20 July 2011, "Arria Formula Meeting on Climate Change", in *What's in Blue*, 14 February 2013, www.whatsinblue.org/2013/02/arria-formula-meeting-on-climate-change.php, accessed 8 October 2015; UN doc. S/PRST/2013/4, 15 April 2013, UN doc. S/PV.6946, 15 April 2013, "The Secretary-General's 2013 Security Council Retreat", in *What's in Blue*, 22 April 2013, www.whatsinblue.org/2013/04/the-secretary-generals-2013-security-council-retreat.php, accessed 8 October 2015; "Open Debate on Conflict Prevention and Natural Resources", in *What's in Blue*, 18 June 2013, www.whatsinblue.org/2013/06/open-debate-on-conflict-prevention-and-natural-resources.php, accessed 8 October 2015; "Arria-Formula Meeting on Climate Change as a Threat Multiplier", in *What's in Blue*, 29 June 2015, www.whatsinblue.org/2015/06/arria-formula-meeting-on-climate-change-as-a-threat-multiplier.php, accessed 8 October 2015.

21 UN doc. S/PV. 4652, 3 December 2002, and UN doc. S/PV. 5220, 30 June 2005.

22 UN doc. S/PV. 6547, 11 February 2011 and, in the same date, UN doc. S/PRST/2011/4.

23 UN doc. A/47/277, S/24111, 17 June 1992, para. 15.

24 UN doc. A/RES/55/2, 18 September 2000, para. 9.

25 A More Secure World: Our Shared Responsibility: Report of the High-Level Panel on Threats, Challenges and Change, UN doc. A/59/565, 2 December 2004.

26 Ibid., p. 12.

27 Ibid., p. 12.

28 UN doc. A/59/2006, 21 March 2005, para. 78.

29 Ibid., para. 78.

30 Ibid., para. 78.

31 S/PV.4172, 17 July 2000.

32 Security Council Press Statement on United Nations Office for West Africa, press statement SC/11466-AFR/2930, 9 July 2014, www.un.org/press/en/2014/sc11466.doc. htm, accessed 5 March 2015.

33 The Situation in Liberia, UN doc. S/RES/2176 (2014), 15 September 2014, "*Expressing grave concern about the extent of the outbreak of the Ebola virus in West Africa, in particular in Liberia, Guinea and Sierra Leone, Affirming that the Government of Liberia bears primary responsibility for ensuring peace, stability and the protection of the civilian population.*" L. Balmond, "Le Conseil de sécurité et la crise d'Ebola: entre gestion de la paix et pilotage de la gouvernance globale", *QIL* 10/2014, p. 22. The SC took into consideration the Ebola outbreak in West Africa, and especially in Liberia, when it further modified the UNMIL mandate: see The Situation in Liberia, UN doc. S/RES/2190 (2014), 15 December 2014 and The Situation in Liberia, UN doc. S/RES/2215 (2015). Moreover, still concerning the Ebola outbreak in Liberia see Milestone Expected to Be Reached in Liberia's Fight against Ebola, Senior Officials Tell Security Council, press statement SC/11882, 5 May 2015, www.un.org/press/en/2015/sc11882.doc.htm, accessed 24 August 2015, and UN doc. S/PV.7438, 5 May 2015.

34 Peace and Security in Africa, 18 September 2014, UN doc. S/RES/2177(2014). See also "Security Council Meeting and Resolution on Ebola Crisis", in *What's in Blue*, 17 September 2014, www.whatsinblue.org/2014/09/security-council-meeting-and-resolution-on-ebola-crisis.php/page=all&print=true, accessed 23 February 2015.

35 See Ebola Situation Report, http://apps.who.int/ebola/current-situation/ebola-situation-report-30-september-2015, accessed 4 October 2015; G. L. Burci, "Ebola, the Security Council and the Securitization of Public Health", *QIL* 10/2014, p. 30.

36 The data available on the WHO website told that on 16 September 2014 the cases of infection were 4,985 and the deaths 2,461. See "Security Council Meeting and Resolution on Ebola Crisis", in *What's in Blue*, 17 September 2014, www.whatsinblue.org/2014/09/security-council-meeting-and-resolution-on-ebola-crisis.php?page=all&print=true, accessed 8 October 2015. See also www.theguardian.com/world/2014/oct/17/world-health-organisation-botched-ebola-outbreak and http://news.yahoo.com/un-botched-response-ebola-outbreak-134221982.html; WHO Response to the Ebola Interim Assessment Panel Report, press statement, 7 July 2015, www.who.int/mediacentre/news/statements/2015/ebola-panel-report/en; WHO, "Report of the Ebola Interim Assessment Panel", July 2015, and WHO, "WHO Secretariat Response to the Report of Ebola Interim Assessment Panel", August 2015, both from www.who.int/csr/resources/publications/ebola/ebola-panel-report/en, all these documents accessed 8 October 2015. See also: P. Acconci, "The Normative Authority of the World Health Organization", *Public Health*, 2015, pp. 1–9.

37 Art. 40: "In order to prevent an aggravation of the situation, the Security Council may, before making the recommendations or deciding upon the measures provided for in Article 39, call upon the parties concerned to comply with such provisional measures as it deems necessary or desirable. Such provisional measures shall be without prejudice to the rights, claims, or position of the parties concerned. The Security Council shall duly take account of failure to comply with such provisional measures." See N. Krisch, "Article 40", in B. Simma, D.-E. Khan, G. Nolte and A. Paulus (eds), *The Charter of the United Nations. A Commentary*, 3rd edn, vol. II, Oxford University Press, 2012, pp. 1297–1304; J.-M. Sorel, "Article 40", in J.-P. Cot, A. Pellet and M. Forteau, *La Charte des Nations Unies: Commentaire, article par article*, vol. I, Economica, 2005, pp. 1171–1194; M. Arcari, "L'articolo 40 della Carta delle Nazioni Unite e le misure provvisorie del Consiglio di Sicurezza", in *Studi di diritto internazionale in onore di Gaetano Arangio-Ruiz*, vol. II, Editoriale Scientifica, 2004, pp. 1469–1526.

38 Art. 41: "The Security Council may decide what measures not involving the use of armed force are to be employed to give effect to its decisions, and it may call upon the Members of the United Nations to apply such measures. These may include complete or partial interruption of economic relations and of rail, sea, air, postal, telegraphic, radio, and other means of communication, and the severance of diplomatic relations."

39 Art. 42: "Should the Security Council consider that measures provided for in Article 41 would be inadequate or have proved to be inadequate, it may take such action by air, sea, or land force as may be necessary to maintain or restore international peace and security. Such action may include demonstrations, blockade, and other operations by air, sea, or land forces of Members of the United Nations."

40 Peace and Security in Africa, 18 September 2014, UN doc. S/RES/2177(2014). P. Acconci, "The Reaction to the Ebola Epidemic within the United Nations Framework: Next for the World Health Organization?", *Max Planck Yearbook of United Nations Law*, 2014, pp. 405–424.

41 "*Stressing* the crucial and immediate need for a coordinated international response to the Ebola outbreak", Peace and Security in Africa, 18 September 2014, UN doc. S/RES/2177(2014).

42 For instance, having achieved its core objective of scaling up the response on the ground, UNMEER closed on 31 July 2015. As of 1 August, oversight of the UN system's Ebola emergency response is led by the WHO.

43 N. Krisch, "Chapter VII Action with Respect to Threats to the Peace, Breaches of the Peace, and Acts of Aggression", introduction to Chapter VII: The General Framework, in Simma et al., *The Charter of the United Nations*, p. 1265; N. Krisch, "Article 40"; M. Arcari, "L'articolo 40 della Carta delle Nazioni Unite e le misure provvisorie del Consiglio di Sicurezza", in *Studi di diritto internazionale in onore di Gaetano Arangio-Ruiz*, vol. II, Editoriale Scientifica, 2004, pp. 1516–1517. Contrary to the possibility of the binding nature of the provisional measures is B. Conforti, *Le Nazioni Unite*, CEDAM, 2012, p. 242.

12 Human security in conflict and disaster

Ukraine at war

John Quinn, Tomáš Zelený and Vladimír Bencko

> Whenever peace – conceived as the avoidance of war – has been the primary objective of a power or a group of powers, the international system has been at the mercy of the most ruthless member of the international community.
>
> (Henry Kissinger, 1964)

Introduction

Former Ukrainian president Yanakovich's refusal to sign an agreement in November 2013 bringing Ukraine economically closer to the European Union (EU) set off a political and social revolution. A pre-planned Russian annexation of the Ukrainian Crimean Peninsula ignited prolonged conflict, killing, and stripping away fragile human and health security for almost seven million Ukrainians, ethnic Russians and a smorgasbord of cultures, religions and ethnic backgrounds in a European State (Quinn 2015).

During this transitional phase of social unrest from late 2013 to early 2014 and into most of 2015, Russia deployed lethal hybrid warfare with proxy Russian backed terrorists throughout two large Eastern Ukrainian regions of Donetsk and Luhansk creating a "hellscape" of non-government controlled areas, and annexed the Crimean Peninsula without any international community response (Snyder and Sharifulin 2014; Borowski 2014). The repercussions of violence in these regions may affect infectious disease risk throughout Ukraine and the region and eliminates basic systems for non-communicable disease (NCD) and other primary healthcare programmes for millions. The other environmental threats of weapons, bombs and destroyed infrastructure exacerbate the tenuous health security portrait. These factors of state fragility, Russian weapon systems, control of violence by non-state actors and increased disease risk may exacerbate the already degraded environmental impact and fallout from recent fighting and conflict. This may lead to no health security for those at risk that may not be quantified for a generation.

Definitions

A failed state is a country whose political or economic system has become so weak and lacking in capacity that the government is no longer in control or offering any component of the social contract. Fragile states are those countries at risk of failing and who find it increasingly difficult to deal with external and internal economic, political, environmental, social and security shocks (Grimm et al. 2014). Ukraine is a fragile state, at risk of failure from current and continued Russian backed incursions onto its sovereign territory. This is exacerbated by: expanding numbers of displaced peoples (well over 5 million in 2014 and 2015 (ACAPS 2015); no access to emergency services in many regions; a growing at-risk population without adequate access to primary healthcare services or vaccines; an ever expanding displaced geriatric population without access to NCD medications; and finally, all other non-combatants caught in the middle of war and faced with violence and threats of violence throughout the country.

Warfare in Ukraine has escalated incrementally since 2014 and the resulting mental stress on its approximately 35 million people is likely to cause poor clinical outcomes for decades. These multifactorial risks, burdens and variables leave Ukrainians and internally displaced peoples (IDPs) with a marked decrease in food, health, economic and human security.

As of late 2015 hostilities continue revealing many violations of international human rights law and international humanitarian law and demonstrating the aggressors lacking accountability (UNHCHR 2015). These violations are evidenced by the Organization for Security and Cooperation in Europe (OSCE) hourly monitoring updates and multiple reports of Minsk I and Minsk II ceasefire violations (Baer 2015); the Minsk I and II ceasefire agreements being the only international agreements between Europe, North America and Russia detailing how to stop warfighting activity by armed groups and militaries.

"Frozen conflict"

A commonplace in geopolitical literature in post-Soviet states is the *frozen conflict* (Lynch 2002). The term "frozen conflict" is a misnomer, the threat of violence is still violence. Loaded weapons in rifles, strategically positioned larger weapons and soldiers ready to deploy on multiple sides from a potential further escalation does not engender human or health security in any setting. Despite these facts, *Frozen Conflict* has become a hallmark of Russian Federation foreign policy. By definition, active armed conflict has stopped, yet no peace treaty and no political or social framework has been agreed upon by armed groups, combatants or states taking part in the conflict (usually with the blocking of resolutions by a third party or third state) describes "frozen conflict" (Kennelly 2006). These post-Soviet non-resolved conflicts can reignite at any point, can encourage a militarization of an entire region or country and define insecurity for these countries affected. The instability in frozen conflicts may strip a region of health security for all inhabitants. These *frozen conflicts* lack clearly defined boundaries and are stateless; lawless and ungoverned spaces are littered throughout the post-Soviet landscape.

Examples of these illegitimate, non-states are easily seen in Transnistria, a hairline splinter on the map between Ukraine and Moldova where Russian troops are stationed at multiple bases. Other examples include:

- The Negorna–Karabakh region, an ethnically Armenian enclave within Azerbaijan where a dispute between Azeris and Armenians festers.
- The Republic of Georgia: notably, the Abkhaz–Georgian conflict and the Georgian–Ossetian conflict, which has arisen owing to two largely unrecognized regions within the internationally recognized Republic of Georgia coming out of war in the early 1990s. The conflict was further stoked by Russian invasion of Georgia in 2008 leaving the Russian backed Republic of South Ossetia and Republic of Abkhazia in *de facto* control of the South Ossetia and Abkhazia regions in north and northwest Georgia (a wasteland for human rights and civil liberties; Jones 2008).
- The 2014 Russian annexation of Crimea and the scaled invasion of the Donbass, where Lugansk and Donetsk have propped-up regional powers creating a "hellscape" of criminality, where the rule of law does not exist and basic services and healthcare infrastructure are in tatters (Orttung and Walker 2015).

These stateless non-states left in "frozen conflict" form a buffer zone with regional powers to promote Russian foreign policy and corrupt business practices at the cost of human life and human security (Nygren 2007; Trenin 2003; Donaldson et al. 2014). Other examples of frozen conflict not directly linked to the former Soviet Union arguably are found in Northern Cyprus, Israel and Palestine, Northern Ireland and Sri Lanka, among others.

Moscow has a toolbox of soft and hard power tactics at its disposal to exert regional influence (i.e. bribes, energy exports, trade ties, military to military collaboration and guarantees), however supporting separatist movements both financially and militarily remains its strongest, yet bluntest, weapon (Mankoff 2014). It is possible that this frozen conflict description may be exactly the goal and planned outcome of the 2013–2015 Russian military backed interventions throughout Ukraine. However, such recent foreign policy has reduced state stability, mitigated state sovereignty and in the process stripped human and health security for millions of Ukrainians and multiple bordering communities. The lack of: clearly defined borders, the public admission of Russian involvement and/or the international community's response beyond economic sanctions, means that the likelihood of increased conflict and increased fatalities is high.

Ukraine's institutions

Ukraine is a fragile state and extractive Soviet and post-Soviet governmental and business practices over the past two decades have given way to a pan-institutional decline and deep ministerial decay. In sum, Russia never left Ukrainian politics despite its relative sovereignty and independence, Russia and the elite in Ukraine willing to collaborate with Russian business practices continued to profit from a system of taking out of the Ukrainian economy, political infrastructure, military and health system, instead of building it up (Way 2005; Kupatadze 2015).

Accountability was absent, corruption and nepotism rife and capacity extremely limited in post-Soviet era Ukraine (Markus 2015). The repercussions of such lack in capacity are a decrease in human and health security where institutions are unable to weather any shocks in times of crisis and disaster (Quinn 2015). The humanitarian crisis that has erupted has proven how much human life can be lost in such a short period of inaction and non-intervention in global health. When the public sector fails, sometimes the private and humanitarian sector can pick up slack and provide health security throughout limited and temporary solutions. A case example of the public–private partnerships that can boost institutional capacity is found in Box 12.1.[1]

Box 12.1   Short case report: public–private partnerships and institutional capacity building

Humanitarian organizations are good at filling the gaps for the very short term but have many shortcomings with mandate, motivation and incentives; as well timelines and endpoints. Prevention is the best medicine and state institutional capacity, accountability and transparency will lead to best outcomes (Fox 2015).

The precarious security situation in the Donbass has led to stove-piping of programs and the presence of multiple actors who are unwilling to communicate and share information and data across organizational boundaries. This duplication of aid sees two or more organizations addressing the region's immediate needs based on information that may be dated or one-sided and politically or financially motivated.

WHO, supported by the UN based Office for the Coordination of Humanitarian Affairs (OCHA), established the Health Sector in February 2014, and since then has been providing leadership and coordination to support both national and local health authorities in Ukraine. In late September 2015, UN and all related agencies (WHO) and all international humanitarian organizations were kicked out by the de facto governing separatists.

Moreover, WHO has been coordinating the collection, analysis, dissemination and communication of essential information on health risks, health needs and health sector responses, gaps and performance. Despite there being more than 50 registered partners in the Health Cluster, only a limited number are actually providing assistance.

WHO provides technical assistance appropriate to the health needs of the emergency (including the provision of health policy and strategy advice; promotion of expert technical guidelines, standards and protocols; best practice advice and implementation; and strengthening of disease surveillance and disease early warning systems). In order to provide this, WHO/Health Cluster flagship interventions include the Mobile Emergency Primary Health Care Units (MEPUs), Emergency Primary Health Care Posts (EPPs) and the provision of medicines. Donor funds that are provided to WHO are divided among the partners (e.g. The Ukrainian Red Cross, IMC); as of mid-2015, humanitarian programs were less than 45 per cent funded.

With these above mentioned advancements in state institutions and their capacity, the fragile state of Ukraine will become more stable on the stability spectrum only through allowing these state institutions securing human and health security to lead, to organize and coordinate and to prac-tice evidence based health policy. This can be enhanced through the support, guidance and material support offered through NATO and NATO member states. With President Poroshenko declaring the new Ukraine Military Doctrine is based on the duration of threat from Russia and demands full compatibility of the Armed Forces with NATO standards, there is a clear mandate to move forward and bring the Ukraine armed forces medical system up to international standards. This case report focuses on how healthcare and battlefield medical support from private and humanitarian agencies can increase state capacity, rule of law that increase stability for the European nation of Ukraine. Transparency of state institu-tions and a policy of prevention may lead Ukraine into a more independent and stable sovereign state.

Basic health portrait and bordering nations

Ukraine has battled with NCDs through the post-Soviet and transitional period, as well as some infectious diseases (pertussis, hepatitis B and C and TB/HIV with at risk groups, among others; OCHA 2015; UNHCHR 2015). This lower middle income country has nearly 46 million people and a life expectancy of 71 years. Neighbouring Belarus boasts a smaller 9.5 million people and life expectancy of 72 years; Moldova with just under 3.5 million and a life expectancy of around 70 (varying at 66 for men and 75 for women); while Russia has over 140 million and a life expectancy the same as Ukraine. Bordering EU nations Poland, Hungary, Romania and Slovakia share a much better basic portrait health. Financing for health is also different as percentage of GDP that is spent on healthcare; health outcomes in these states show further inequality with that of Ukraine.

Violence and the war

It is well understood and reported in the literature that war increases the incidence of violence for those exposed to it and increases exposure to physical and mental trauma. Since the beginning of hostilities in 2014, approximately 9,200–10,100 people have been killed with an additional 17,000–20,000 injured

(OCHA 2015).[2] Detailed reporting from the Human Rights Monitoring Mission in Ukraine puts total casualties (civilian and military) from mid-April 2014 to 15 August 2015 of at least 25,493 (Ukrainian armed forces, civilians and members of the armed groups) that include at least 7,883 people killed and at least 17,610 injured in the conflict area of eastern Ukraine.[3] Furthermore, according to the representative of Ukraine to the Trilateral Contact Group, the Government estimated that as of 8 June 1,200 people were missing in the conflict zone. The fog of war reveals little else despite best efforts.

Freedom of movement is severely restricted, whether violence is increasing or in a lull; procedures for simple movements across towns and villages to major checkpoints are restricted and aggravated by the scattered access to social services and general economic disruption. Indeed, the principal driver of human security vulnerability for the at risk populations of eastern Ukraine and bordering regions is the unknown resumption and prolongation of escalating hostilities with newly added lethal modern Russian weapon systems.

Of the 1.4 to 5 million Ukrainian IDPs (contested data for IDPs) and 5–7 million people at threat of further human security deterioration, 60 per cent represent the geriatric population and roughly 13 per cent paediatric (OCHA 2015) – two very disparate and sometimes difficult groups of patients to treat properly and prevent further disease in the prehospital and primary healthcare setting.

Further, people living in the territories controlled by armed groups continue to face obstacles in exercising any type of rights: civil, political, economic, social or cultural. Such people experience particular problems in accessing quality medical services and social benefits (UNHCHR 2015).

Non-communicable diseases and the war

It is anticipated that non-communicable diseases will be the number one global health threat for the next half century. Diabetes, hypertension, cardiovascular disease, chronic respiratory diseases, cancer and obesity are key NCDs which require long term primary healthcare services, prevention measures and healthcare infrastructure. The war in Ukraine exacerbates NCDs because prevention and/or reducing the risk factors associated with these diseases may be completely halted in many affected areas at war. Low-cost prevention focuses on the reduction of the common modifiable risk factors (alcohol misuse, any tobacco use, poor diet and physical inactivity). Many primary healthcare clinics throughout contested areas of Ukraine cannot implement programs employing these methods of prevention or risk reduction. They cannot offer these basic services to patients and consequently health burden and health security projections will be low and their negative impacts will be felt for decades.

Infectious disease and conflict

Herd immunity to many vaccine-preventable diseases are at critically low rates and have been well under 40–50 per cent for some vaccine preventable illnesses in Ukraine (Holt 2015). There was a similar paediatric health portrait in Syria circa 2012 before its polio and measles outbreak; conflict is a great medium for infectious disease to grow and spread for at risk populations (Sharara and Kanj 2014). Dr Dorit Nazan, the WHO Country Director in Ukraine warned of a potential polio outbreak due to low compliance with immunization and diminished stockpiles of vaccines from the unpredictable environment of war.

So it was no surprise that in late 2015, policy supported by the World Health Assembly (WHA) promoted an urgent response to the Ukrainian polio outbreak with three large-scale supplementary immunization activities and an appropriate amount of oral polio vaccine, covering a target population of 2 million children aged less than five years, and public declaration of the outbreak as a national public health emergency (Mayor 2015). A policy of prevention is cheaper than treatment with infectious childhood diseases such as polio. Conflict highlights the risk of increased disease burden brought on by sustained conflict and violence.

Environmental degradation and warfighting activity

Explosive remnants of war (ERW) and unexploded ordnance (UXO) will be a major health and environmental concern for the next three to four decades in Ukraine; water contamination, deaths and other sequelae from these munitions will plague the landscape. More globally, Ukraine faces significant environmental challenges like the still present fallout from the Chernobyl nuclear power plant disaster in 1986, prolonged industrial pollution and poor waste management also exacerbated by the conflict (ACAPS 2015).

Ukraine ranks 20th in the world for greenhouse gas emissions and is among the European countries with the highest levels of energy consumption and water usage (International Business Publications 2013). The coal-burning industries of eastern Ukraine, which emit high levels of sulphur dioxide, hydrocarbons, and dust, have created severe air pollution throughout the region. Some are still in operation; air quality is particularly poor in the cities of Dnipropetrovsk, Kryvyy Rih and Zaporizhzhya (Encyclopædia Britannica 2015). Major rivers, including the Dnieper, Dniester, Inhul and Donets, are polluted with chemical fertilisers and pesticides from agricultural runoff and poorly treated or untreated sewage (ibid.). These environmental risks and hazards hinder human and health security in the immediate and distant future if not addressed through transparent and accountable public health institutions.

Mental health in conflict

Very limited psychosocial support or psychoeducation is offered to IDPs in collective centres. Psychologists and volunteers working in the field are in need of training in psychological first aid and emergency related mental health issues (Eurdolian and Porter 2015). Shortcomings are compounded with civilian injuries and returning warriors with traumatic brain injury and the complex and nuanced mental health challenges that accompany it.

The psychological and physical rehabilitation of at least 40,000 demobilized soldiers in the Government-controlled areas is still largely provided by volunteers; services for survivors of sexual and gender-based violence are not available in the areas controlled by the armed groups and are insufficient in the Government-controlled areas (UNHCHR 2015). For those patients in acute crisis, basic acute first line treatments such as sedatives and neuroleptics are non-existent due to supply issues and logistical concerns.

Food and health security

The World Food Programme (WFP) released a report in July 2015 noting a 42 per cent increase year-on-year while the Kiev International Institute of Sociology (KIIS) reported that food insecurity is more severe in non-government controlled areas. Food prices for a full basket in the NGCAs in July 2015 were reported to be 70 per cent higher than the current national average (OCHA 2015). Macroeconomically, Ukraine has tremendous agricultural potential and could play a critical role in contributing to global food security. However, Ukraine must start with feeding its own people affordably and consistently.

Migration and people flows

Global migration from war, climate change and economics is very dynamic. Since 2010, Ukraine has also been a destination country for a number of new asylum-seekers. In 2010 there were 1,500 applicants; 2011, 890 people; 2012, 1860 people; 2013, 1310 people; 2014 with over 1,170 people and in the first six months of 2015, over 630.[4] The majority of people applying for asylum are from Afghanistan, Syria, the Russian Federation, Somalia and Iraq. Not all applicants will be accepted; however, the external and internal movement stressors will prove to be major health challenges for Ukraine. These migration trends affect the capacity of Ukraine's state institutions to provide essential and basic services to its citizens and newly arrived guests and residents.

Financial instruments of war

Labelled by the World Bank (WB) as a lower middle income country, Ukraine posted zero economic growth from 2012–2013 and in 2014 although there has been abysmal data to arrive at this conclusion with the regions and accompanying economic sectors of Crimea being annexed and Lugansk and Donetsk no longer under *de facto* control or authority of the central government. Part of Ukraine's fragility is related to basic economic factors such as exchange rate policy, loose and unaccountable fiscal policy, unknown conflict, outbreaks of violence and fiscal subsidies. Some of these structural shortcomings are slowly being brought under control by the central government with the support and guidance of international organizations such as the International Monetary Fund (IMF), the World Bank (WB) and departments within the European Union (EU) and European Commission (EC).

Unemployment is still higher than targets and is off the scale for non-government controlled areas. A booming security and defence apparatus is a growing budgetary concern and destruction of critical infrastructure like roads and bridges and ports have sequentially shocked the Ukrainian economy since early spring 2014. Owing to all of the above, inflation and reduced purchasing power and significant price hikes have increased the economic vulnerability of Ukrainians to an environment that engenders health security. Since the beginning of the year 2015, real income dropped by 23.5 per cent, wage arrears reached UAH 1.9 billion (approximately USD 87 million), while prices for basic commodities have increased by 40.7 per cent (UNHCHR 2015).

In early March 2015, the WB approved a US$214.73 million loan for the Serving People, Improving Health Project; a project aimed at supporting the implementation of reforms and improving service delivery in Ukraine's health sector (WB 2015). The new WB five-year project will seek to develop medical infrastructure and improve the quality of health services with the help of a new funding mechanism within hospitals, as well as enhanced primary and secondary prevention, early detection, and treatment of cardiovascular diseases and cancer – the key NCDs that put many Ukrainians at great risk of health insecurity. It is also expected to improve some first aid and emergency services (ibid.).

In mid-February 2015, the International Monetary Fund (IMF) released plans for approximately US$17.5 billion from itself and the international community to be allocated to stabilizing the Ukrainian economy. The IMF's Director commented, "[t]he fiscal position is getting stronger, the foreign exchange market has stabilized, and the banking sector is being repaired so that banks are sounder and can start to provide credit again" (IMF 2015). This needed growth will also bolster human and health security for those presently caught in the middle of fighting or frozen zone of conflict.

Conclusion

Ukraine is a fragile nation at war and it has diminished human and health security. The fear of a "frozen conflict" will put more people in Ukraine at risk. Increasing transparent and accountable institutional capacity of healthcare infrastructure is needed. In the immediate term, ensuring access to medical services and social protection to victims of human rights violations and those subjected to violence, must be paramount. Change will come slowly as there are simply no proven methods for generating major social, political, economic or cultural change relatively quickly when governing institutions are weak, corrupt and lacking rule of law (Mazarr 2014).

Specifically, the following are policy recommendations for the international community to consider:

• patient, long-term mentoring and aid-based partnerships based on continued economic assistance;

- training for emergency and primary healthcare workers;
- professional exchanges;
- military-to-military partnership; and
- economic investment which collectively focuses on enhancing effective and stable governance may push Ukraine from a fragile to stable state, and bolster human and health security.

Notes

1 The box is loosely based on a section of Quinn (2015).
2 UN Human Rights Monitoring Mission in Ukraine and the World Health Organization's conservative estimates based on available official data. These totals include: Ukrainian armed forces casualties as reported by the Ukrainian authorities; 298 people from flight MH-17; casualties reported by civil medical establishments and local administrations of Donetsk and Luhansk regions; and civilians and some members of armed groups. OHCHR and WHO report fatalities may be higher.
3 This may be a conservative estimate of HRMMU based on available data. Some data are incomplete due to geographic areas and because of overall under-reporting, especially of military casualties. Recent reports from the UN put the body count closer to 8,000 (www.un.org.ua/en/information-centre/news/2023). The increases in the numbers of casualties between the different reporting dates do not necessarily mean that these casualties happened between these dates: they could have happened earlier, but were recorded by a certain reporting date (UNHCHR 2015).
4 Taken from UNHCR Office of Statistics, 23 September 2015, available at http://unhcr.org.ua/en/resources/statistics.

References

ACAPS (2015) "Country Profile: Ukraine". August, available at: www.acaps.org/resourcescats/download/ukraine_country_profile_28_august_2015/416 (accessed 23 September 2015).

Baer, D. B. (2015) "Russia Must Live Up to the Agreements it Signed in Minsk: Russia Must Act, Not Just Talk". Press conference by the Ambassador to the OSCE Permanent Council, Vienna, 21 May 21.

Borowski, A. (2014) "Confidence in Social Institutions in the Post-Communist Countries". *International Letters of Social and Humanistic Sciences* 14: 7–17.

Donaldson, R. H., Nogee, J. L. and Nadkarni, V. (2014) *The Foreign Policy of Russia: Changing Systems, Enduring Interests*, M. E. Sharpe, New York.

Encyclopædia Britannica (2015) "Ukraine". Available at www.britannica.com/place/Ukraine (accessed 23 September 2015).

Eurdolian, A. and Porter, Z. (2015) *Ukraine: Situation Report 28*. United Nations Office for the Coordination of Humanitarian Affairs, Geneva.

Fox, J. A. (2015) "Social Accountability: What Does the Evidence Really Say?" *World Development* 72 (August): 346–361.

Grimm, S., Lemay-Hébert, N. and Nay, O. (2014) "'Fragile States': Introducing a Political Concept". *Third World Quarterly* 35(2): 197–209.

Holt, E. (2015) "Health Care Collapsing Amid Fighting in East Ukraine". *The Lancet* 385(9967): 494.

IMF (2015) "IMF Managing Director Christine Lagarde Visits Ukraine". Available at www.imf.org/external/np/sec/pr/2015/pr15404.htm (accessed 17 January 2016).

International Business Publications (2013) *Ukraine Oil & Gas Sector Energy Policy, Laws and Regulations Handbook*, Volume 1: *Strategic Information and Regulations*, International Business Publications, Washington, DC.

Jones, S. (2008) "Clash in the Caucasus: Georgia, Russia and the Fate of South Ossetia". *Origins* 2 (November).

Kennelly, K. G. (2006) "The Role of NATO and the EU in Resolving Frozen Conflicts". Thesis, Naval Postgraduate School, Monterey California, December, available at file:///Users/johnmquinnv/Downloads/ADA462609.pdf (accessed 22 September 2015).

Kupatadze, A. (2015) "Political Corruption in Eurasia: Understanding Collusion between States, Organized Crime and Business". *Theoretical Criminology* 19(2): 198–215.

Lynch, D. (2002) "Separatist States and Post-Soviet Conflicts". *International Affairs* 78: 831–848.

Mankoff, J. (2014) "Russia's Latest Land Grab: How Putin Won Crimea and Lost Ukraine". *Foreign Affairs* (May/June).

Markus, S. (2015) "Sovereign Commitment and Property Rights: The Case of Ukraine's Orange Revolution". *Studies in Comparative International Development* doi:10.1007/s12116-015-9188-0.

Mayor, S. (2015) "Polio Outbreak in Ukraine Likely to Spread, WHO Warns". *British Medical Journal* 351: h4749.

Mazarr, M. J. (2014) "The Rise and Fall of the Failed-State Paradigm: Requiem for a Decade of Distraction". *Foreign Affairs* (January–February).

Nygren, B. (2007) *The Rebuilding of Greater Russia: Putin's Foreign Policy Towards the CIS*, Routledge, Abingdon.

OCHA (2015) *Humanitarian Bulletin Ukraine* 1 (1–31 August).

Orttung, R. and Walker, C. (2015) "Putin's Frozen Conflicts". *Foreign Policy* Democracy Lab (13 February).

Quinn, J. (2015) "Notes from the Field: The Humanitarian Crisis in Ukraine". *Journal of Human Security* 11(1): 27–33.

Sharara, S. L. and Kanj, S. S. (2014) "War and Infectious Diseases: Challenges of the Syrian Civil War". *PLoS Pathogens* 10(11): e1004438.

Snyder, T. and Sharifulin, V. (2014) "Fascism, Russia, and Ukraine". *New York Review of Books* (20 March 20).

Trenin, D. (2003) "Southern Watch: Russia's Policy in Central Asia". *Journal of International Affairs* 56(2) (spring): 119–131.

UNHCHR (2015) *Report on the Human Rights Situation in Ukraine*, 16 May to 15 August, Office of the United Nations High Commissioner for Human Rights, Geneva.

Way, L. A. (2005) "Authoritarian State Building and the Sources of Regime Competitiveness in the Fourth Wave: The Cases of Belarus, Moldova, Russia, and Ukraine". *World Politics* 57(2) (January): 231–261.

WB (2015) "World Bank Provides Support for a Healthier Ukraine". Available at www.worldbank.org/en/news/press-release/2015/03/04/world-bank-provides-support-for-a-healthier-ukraine.

Part III

The common good and democracy for environmental governance

13 War, militarism and climate change

Time to connect the dots

Sheila D. Collins

The spectre of thousands of refugees surging into Europe from the wars in Syria, Iraq, Afghanistan and parts of Africa has commonly been attributed to a variety of geopolitical factors: interethnic and interreligious power conflicts; imperialist intervention in the Middle East by the big powers; and the failure of Arab societies to modernize. Less well recognized and accepted is the role climate change has played as one of the causal factors of such social and political upheaval as well as the role war and militarism play in contributing to climate change.

World leaders seem oblivious to the connections between environmental destruction and war. While they meet to negotiate reductions in greenhouse gases, they ignore a major source of those greenhouse gases – the military machines that continue to wreak havoc on the environment. It is as if climate change and war were distinct ontological categories when in fact they are symbiotically related. This myopia was illustrated in mid-September, 2014, when, two days after the largest climate march in history made its way through New York City's midtown, President Obama announced that he had just launched a heavy airstrike against militants in Syria – in effect, plunging the United States further into an unending military quagmire in the Middle East. He then proceeded on the same day to go to the United Nations to claim that he was serious about tackling climate change. Again, on the eve of the Paris summit on climate change in December 2015, Obama announced a new approach to the deepening Syrian conflict. After the complete failure of his $500 million effort to train Syrian rebels,[1] his only fallback was to send weapons to the leaders of a fractured and amorphous group of rebel fighters, a policy that is certain to result in more deaths and greater ecological destruction, not to mention the likelihood that the weapons will fall into the hands of the very people they were intended to be used against.

Climate change as a cause of war

Why this failure to understand the connections between war, militarism and climate change? After all, competition over resources – land, water, minerals, energy – has for hundreds of years been the ground of conflicts within and between nations despite the fact that such conflicts have usually been attributed to ethnic, religious, ideological or national rivalries, or to supposedly more noble values such as civilizing or "Christianizing" undeveloped peoples, "making the world safe for democracy", ridding the world of those (often wrongly) held responsible for a nation's ills, or so-called "humanitarian intervention". The colonial conquest of the Americas was all about resources: gold, silver, tin, rubber. The modern wars in Africa have often been driven by competition for diamonds, oil and other precious minerals.

Until recently, few recognized climate change as one of the causal factors of human conflict. The authors of a study published in 2007 pointed out that "scientific research on the social effects of climate change has tended to focus on the economic costs of current and future climate change and has neglected the study of how societies have historically reacted to long-term climate change",[2] but in the past decade there has been a marked increase in research investigating the relationship between climate change and violent conflict. Their study, using high-resolution paleo-climatic data, explored the effects of climate change at a macroscale on the outbreak of war and population decline in the preindustrial era, demonstrating that long-term fluctuations of war frequency and population changes were significantly correlated with the cycles of temperature change. Dramatic changes in temperature, they show, affect the carrying capacity of the land and hence the per capita food supply. Another study, focused not on the preindustrial past but on modern day sub-Saharan Africa, found a similar correlation.[3] Looking at the period between 1981 and 2002, the authors provide evidence linking past internal armed conflict incidence to variations in temperature, finding substantial increases in conflict during warmer years. They use this relationship and climate model projections of future temperature trends to build projections of the potential effect of climate change on future conflict risk in Africa, warning that unless African governments are reformed and foreign aid donors' policies are changed we could see a 54 per cent increase in armed conflict by 2030.

The Intergovernmental Panel on Climate Change (IPCC) has warned that temperatures above 2°C could result in "severe, widespread, possibly abrupt and irreversible impacts that civilizations would be unable to cope with".[4] While characteristically cautious in concluding that there is a straight link between climate change and violent conflict, the Report nevertheless states that there is "robust evidence" and "high agreement" that "human security will be progressively threatened as the climate changes".[5] This is especially true for populations "that are already socially marginalized, resource dependent, and have limited capital assets".[6] One example is the rise of Boko Haram in Nigeria, which has been traced to severe drought in the region causing the disruption of normal patterns of life and leaving people "in a chaotic state of absolute poverty and social dislocation".[7] The so-called pastoralist corridor straddling the borderlands of Kenya, Uganda, Sudan, Ethiopia and Somalia has seen increasingly extreme weather patterns – drought and flash flooding – which has resulted in a toxic convergence of poverty, climate change, failed or failing states and consequent inter-tribal or religiously infused violence.[8] The Rwandan genocide of 1994 was preceded by years of declining agricultural production.[9]

Central and South Asia are other areas that are already experiencing the inter-mixture of extreme climate events with political violence. "Read the history of the war in Afghanistan closely", Christian Parenti says in his book, *Tropic of Chaos*, "and a climate angle emerges".[10] The area has been suffering droughts and floods "that fit the pattern of anthropogenic global warming". The region saw increasing violence even before the American presence, starting with a drought in 1969 that resulted in a mass famine and consequent political chaos. More recently, the Taliban have taken advantage of the dissolution of traditional bonds of solidarity and loss of livelihoods that have resulted from such weather events. For example, drought has wiped out Afghanistan's wheat crop, leaving thousands of farmers destitute. The Taliban reintroduced poppy cultivation (which they once prohibited) as a means of providing these farmers with a living as poppies require only a third of the water needed by wheat. Profits from heroin sales, in turn, finance their movement.

The modern era's first climate change refugees were some 500,000 Bangladeshis, left homeless by the flooding of their island in 2005. Bangladesh has seen frequent outbreaks of public unrest and terrorism as a result of such upheaval. Due to rising sea levels, an estimated 22 million people in Bangladesh will become climate refugees by 2050. Anticipating this event, India is already completing a militarized border fence along its 2,500-mile frontier with Bangladesh.[11] But India, too, is suffering from internal conflict related to climate change. The often violent guerrilla movement known as the Naxalite movement is situated along the fault lines of a pattern of drought.[12] An estimated 1.3 billion people on the Indian sub-continent depend on runoff from the Himalayan Glaciers.[13] Erratic monsoon patterns resulting in both massive floods and droughts have contributed to numerous intercommunal conflicts and mass suicides across the region. If the glaciers disappear, one can only imagine the chaos that will ensue. Add to this volatile mix the violence that wracks Pakistan. The religious fanaticism of the area is also linked to declining water resources.[14]

Although political violence in the Middle East is riven by interreligious con-flict, beneath these conflicts also lurks the reality of climate change. The so-called Arab Spring began as a protest against the rising cost of food – the result of droughts and floods in the wheat producing countries that supply Egypt with wheat – and then morphed into a movement for democratization.[15] The Syrian war was ignited when the government failed to respond to the most severe drought in the Fertile Crescent's history which occurred between 2006 and 2011, driving dispossessed farmers in Syria's breadbasket into the cities where they became recruits for the Syrian resistance movement.[16] The UN reported that since the start of the Syrian war, the supply of safe water has dropped by two-thirds and that drought also threatens to put new pressures on Jordan, Lebanon and Iraq, exacerbating conflict and refugee flows in an area that is already roiling with them.[17] The Israeli-Palestinian conflict also has scarce water as one of its drivers. Israel controls the greater part of the Jordan River basin and the aquifers that run under the West Bank – the major source of water for this arid region, depriving Palestinian communities of access to fresh water.

While maldistribution is currently one of the sources of this conflict, as climate change worsens, there will only be more conflict over this vital source of life.

Historian, Timothy Snyder has recently argued that Hitler's search for *lebensraum* and hence his desire to conquer the world was due to his fear that Germany was running out of resources to feed its people and maintain their standard of living.[18] Snyder warns that such genocidal instincts are not beyond the realm of possibility today even in the more industrialized parts of the world:

> The anxieties of our own era could once again give rise to scapegoats and imagined enemies, while contemporary environmental stresses could encourage new variations on Hitler's ideas, especially in countries anxious about feeding their growing populations or maintaining a rising standard of living.[19]

As refugees from the Middle East and North Africa flood into Europe today we are seeing ominous signs of Snyder's warning in the rise of neofascist groups, especially in countries that are struggling with stagnant or slipping economies, and even in the United States a tide of anger and resentment against immigrants is on the rise. In a world faced with dwindling resources – of potable water, arable land, fisheries, easily extracted fossil fuels – and growing consumption due to population increase and a rapacious economic system, resource-related conflicts are becoming more widespread and violent. But the very conflicts generated by such resource competition end up despoiling the environment, thus creating more conflict. It is a vicious circle.

Even the US Pentagon now acknowledges the connection between climate change and conflict. A 2003 Pentagon study raised the possibility that should GHG emissions rise above a certain level there could be an abrupt change in weather patterns causing a decrease in net global agricultural production that would destabilize the geopolitical environment and create a greater threat to world stability than terrorism.[20] This was a worst case scenario when the report was written but one that scientists fear is becoming more and more possible unless we move quickly to drastically reduce global emissions. The last three Pentagon *Quadrennial Defense Reviews* also characterize climate change as a "threat multiplier".[21] Yet the Pentagon's response is not to reduce its commitment to war-making (which is, after-all its *raison d'être*), but to prepare for what it sees as the inevitable increase in global conflict as a result of climate change.

The impact of militarism on climate change

As a response to political conflict militarism seems to be the world's modus operandi. Yet militarism itself is a major, but overlooked contributor to climate change. While the majority of the world's most polluting nations recognize the need to reduce greenhouse gases, none of the world's major powers has been willing to renounce militarism or to curb the trade in weapons.[22] Between 2001 and 2013, global military spending increased by an estimated 90 per cent.[23] Although world military spending decreased by 0.4 per cent between

2013 and 2014, the world continued to spend $1.8 trillion on weapons.[24] Despite a series of bilateral arms control agreements which drastically reduced the nuclear arsenals of the US and Russia, these two states still possess a total of 15,000 nuclear warheads, any one of which would cause catastrophic human suffering and ecological disaster if detonated, and they are modernizing their delivery systems.[25] In recent years, US military expenditures had begun to fall as a result of the recession, the alleged ending of the wars in Iraq and Afghanistan and congressional budget tightening, yet current US military spending is still 45 per cent higher than in 2001, just before the 11 September terrorist attacks. In November 2014, the US announced that it would be pursuing an ambitious program to identify and develop *new* weapons systems in a bid to maintain its global dominance.[26] In October 2015 President Obama announced that, contrary to his pledge to end the war in Afghanistan, thousands of troops would continue there indefinitely, thus prolonging a war that has lasted for fourteen years.[27]

The US Pentagon is the world's largest industrial consumer of fossil fuels. Fighter jets, destroyers, tanks and other weapons systems emit highly toxic, carbon-intensive emissions, not to mention the greenhouse gases (GHG) that are released from the detonation of bombs. To understand the impact of our commitment to militarism, one has to take into account not only the carbon dioxide equivalent[28] of the fuels that are directly used by the military (for transportation and in the building of military infrastructure, weapons, etc.), but those related to the US military's role in protecting global maritime petroleum distribution, as well as the wars in the Middle East related to securing global petroleum reserves. Despite the Pentagon's decision to include more biofuels in its energy mix, the GHG emissions from the production of biofuels as well as their indirect effects also need to be measured. A study published in 2010 estimated that total emissions from US conventional military fuel use and acquisitions totalled about 172 million metric tons (MMt) of CO_2 equivalent per year. Of that, an estimated 16 MMt of CO_2 equivalent per year may be attributed to the military's role in securing global oil supplies.[29] An additional 43.3 MMt is attributed to the Iraq war. To get some idea of this military contribution to global warming consider that 172 million MMt is roughly equivalent to driving an average passenger vehicle four and one-half billion miles a year, or burning almost 185 billion pounds of coal.[30]

Unlike most warfare before the twentieth century where weapons were used to kill by piercing and crushing human bodies, twentieth and twenty-first century warfare is designed to kill by destroying and poisoning the enemy's environment, making such warfare, even setting aside its contribution to GHG emissions, qualitatively more environmentally destructive.[31] Not only are modern weapons environmentally destructive when used, but their extraction, production, transportation, storage and testing leaves its own toxic trail. Think of the history of uranium production and testing or the 900 toxic military production sites that dot the American landscape turning thousands of acres into uninhabitable "sacrifice zones". More than 12,000 US military sites where weapons testing takes

Table 13.1 US military's contribution to climate change

Included in UN greenhouse gas inventories	Excluded from UN gas inventories
Domestic fossil fuel use (transportation, heating and cooling, building of infrastructure, weapons construction)	Aviation outside country (troop and equipment transportation, security for global petroleum trade)
	Fuel use related to the building and maintenance of between 700 and 800 military bases around the world
	Direct and indirect effects of biofuel production and use
	Extraction, production, transporting, storing and testing of war materials
	Bomb detonation

place have been found to release perchlorate, a toxic chemical that affects thyroid and respiratory function, into the groundwater.[32]

The United States and its allies have spent trillions financing anti-terrorist military operations in Iraq, Afghanistan and Pakistan and now they are moving into Africa. The US alone spent $2.6 trillion between 2001 and 2014 but has taken obligations to spend as much as $4.4 trillion.[33] As the largest arms supplier (and importer) in the world, the US fuels conflicts directly and indirectly. Many of these weapons fall into the hands of non-state actors like ISIS, al-Qaeda and Boko Haram thus multiplying the arenas of military conflict. While the terrible social, cultural and economic costs are publicly discussed, little is said about the environmental costs of such wars and little research has been conducted into these costs, leaving the public in ignorance and without the knowledge necessary to question and challenge the military's role.[34] To make matters worse, the military sector – with the exception of the military's *domestic* fuel use, that is, the emissions related to the military's use of fossil fuels within the United States but not abroad – is excluded from UN inventories of national greenhouse gas emissions thanks to intensive lobbying by the United States at the Kyoto Protocol negotiations (Table 13.1).[35] The exclusion of much of the military sector from national greenhouse gas inventories thus makes a mockery of the UN climate negotiations process.

The toxic legacy of past wars

To get some idea of the environmental costs of our commitment to militarism, one only has to look at the lingering effects of past wars. During the Vietnam War the US army sprayed 80 million litres (21 million gallons) of the defoliant, Agent Orange, containing the highly toxic chemical, dioxin, as well as other herbicides over more than 24 per cent of the land area of South Vietnam, leaving nearly 5 million acres of denuded or heavily defoliated upland and coastal forests – about 36 per cent of the total mangrove forest, an ecosystem rich in biodiversity – and damaging some 500,000 acres of rice and other crops. Although the US government has finally begun to clean up the mess it left, hundreds of toxic sites still remain and it will take centuries to reproduce the ecologically balanced mix of flora and fauna that once thrived there.[36] Between 2.1 to 4.8 million Vietnamese were directly exposed to Agent Orange and other herbicides during the War as were thousands of US soldiers. As late as 2014, both Vietnamese and American veterans were dying from cancers likely caused by the poison and children were still being born with birth defects.[37] During the first Gulf War, Saddam Hussein set an estimated 600 Kuwaiti oil wells ablaze. The thick plumes of smoke that rose as high as six kilometres into the atmosphere produced an estimated 3,400 metric tons of soot per day. An estimated 11 million gallons of oil, more than 20 times the amount of oil that was spilled in the Exxon Valdez disaster in Alaska, flowed into the Persian Gulf and an estimated 80 ships containing oil and munitions were sunk, causing irreparable harm to the Gulf's ecological integrity.[38] Having learned nothing from these disasters we now have the spectacle of the US bombing oil refineries in Syria in an attempt to cripple the oil revenue stream to ISIS. In numerous abandoned battlefields across the world land that could be used for growing food is dotted with land mines and innocents, many of them children, continue to be blown up by these hidden weapons long after the wars have ended. The terrorist attack on New York City's World Trade Center in 2011 is still demonstrating the long term effects on the air and human health of the bombing of modern infrastructures with their load of toxins, yet this aspect of war's ravages is little noticed in the coverage of the wars in the Middle East and Africa.

Challenging militarism's role in climate change

While those most responsible for militarism (the world's political leaders) have failed to acknowledge, let alone act on the connections between war/militarism and environmental destruction, few others have made this connection central to their campaigns for a sustainable environment. For the most part, the environmental movement has not made the case that climate change and war are Siamese twins. Nor has a less visible peace movement linked war with climate change, although this is beginning to change.[39] Although trade unions in other parts of the world have taken up the fight against climate change, the mainstream US labour movement, with some exceptions,[40] has been slow to get involved and, in fact, is riven by internal conflict over the issue.[41] To his credit, in his encyclical, *Laudato Si'*, Pope Francis devoted a small section to the connection between militarism and environmental destruction. Section 57 states:

> War always does grave harm to the environment and to the cultural riches of peoples, risks which are magnified when one considers nuclear arms and biological weapons. Despite the international agreements which prohibit chemical, bacteriological and biological warfare, the fact is that laboratory research continues to develop new offensive weapons capable of altering the balance of nature. Politics must pay greater attention to foreseeing new conflicts and addressing the causes which can lead to them.[42]

In his speech to the United Nations in September 2015 the Pope called for a complete prohibition of nuclear weapons.[43] Yet the Catholic Church is still mired in a pre-climate change mindset in which "just wars" can be waged. Thus the militarism/environment connection did not rate the kind of centrality it deserves. While research on climate change as a causal factor in violent conflict has begun to proliferate, only a handful of researchers have looked at militarism's contribution to climate change.[44] They are the exceedingly rare voices making the case that there can be no climate change mitigation without peace and no peace without moving swiftly to provide the poorer parts of the world with the resources needed to adapt to climate change and build resilient economies.

In 2014 world military expenditures amounted to $1.776 trillion.[45] Such yearly outlays rob national treasuries of the resources needed to provide funds for climate mitigation and adaptation. At the United Nations summit on climate change in Lima, Peru in November 2014, the world's nations pledged $10.2 billion to finance the Green Climate Fund – a pool of money dedicated to helping developing countries reduce their greenhouse gases and adapt to inevitable climate changes. That is a mere fraction of what they spend in just one year on militarism. If the world's leaders were serious about climate change, they would be placing an emphasis on diplomacy and development over their knee jerk military responses to conflict, drastically reducing their military arsenals, abolishing nuclear weapons altogether, and putting much more money into helping poorer nations adapt to climate change and leapfrog into green economies. This is the only way of ensuring international security. Just think what even a quarter of that $1.776 trillion spent on weapons of war could do to reduce international tensions and build sustainable communities.

In 1990, the Executive Secretary of the United Nations Framework Convention on Climate Change (UNFCCC), Christiana Figueres, in a speech to the Congress of Deputies of Spain, argued that "the very scale of the security problem in a world that begins to panic over the advanced impacts of climate change could overwhelm any single country's ability to defend against it, let alone pay the cost to do so". She then went on to ask:

> What will be better? To continue to support a traditional global military budget that has risen 50 percent in real terms from 2000 to 2009 and continues to increase? Or to increase a preventive military budget investing into adaptation and low-carbon growth and avoid climate chaos that would demand a defence response that makes even today's spending burden look light?[46]

Today, that question remains to be answered. It is time to connect the dots.

Notes

1 Michael D. Shear, Helene Cooper and Eric Schmitt, "Obama Administration Ends Effort to Train Syrians to Combat ISIS", *New York Times*, 9 October 2015, available at

www.nytimes.com/2015/10/10/world/middleeast/pentagon-program-islamic-state-syria.html (accessed 14 October 2015); Mark Mazzetti, "CIA Study of Covert Aid Fueled Skepticism About Helping Syrian Rebels", *New York Times*, 14 October 2015, available at www.nytimes.com/2014/10/15/us/politics/cia-study-says-arming-rebels-seldom-works.html?_r=0 (accessed 14 October 2015).

2　David D. Zhang, Peter Brecke, Harry F. Lee, Yuan-Qing He and Jane Zhang, "Global Climate Change, War, and Population Decline in Recent Human History", *PNAS*, vol. 104, 49 (4 December 2007): 19,214–19,219.

3　Marshall B. Burke, Edward Miguel, Shanker Satyanath, John A. Dykema, and David B. Lobell, "Warming Increases the Risk of Civil War in Africa", *PNAS*, vol. 106, 49 (8 December 2009): 20,670–20,674.

4　The Intergovernmental Panel on Climate Change has warned that to keep average global temperatures below 2°C relative to pre-industrial levels GHG emissions would have to be reduced between 40–70 per cent by 2050 compared to 2010 and to zero by the end of the century. IPCC Fifth Assessment Synthesis Report, *Climate Change 2014: Impacts, Adaptation and Vulnerability*, Summary for Policymakers, available at www.ipcc.ch/report/ar5/syr (accessed 21 December 2014).

5　IPCC, Fifth Assessment Synthesis Report, Executive Summary, 758.

6　Ibid., Chapter 12, 762.

7　Nafeez Ahmed, "Behind the Rise of Boko Haram – Ecological Disaster, Oil Crisis, Spy Games", *The Guardian*, available at www.theguardian.com/environment/earth-insight/2014/may/09/behind-rise-nigeria-boko-haram-climate-disaster-peak-oil-depletion (accessed 10 May 2014).

8　Christian Parenti, *Tropic of Chaos: Climate Change and the New Geography of Violence* (New York: Nation Books, 2011), 46–49. For more on this thesis, see *Journey to Planet Earth: Extreme Realities*, available at http://video.pbs.org/video/2365380402/.

9　Timothy Snyder, "The Next Genocide", *New York Times*, 12 September 2015, available at www.nytimes.com/2015/09/13/opinion/sunday/the-next-genocide.html?_r=0 (accessed 12 September 2015).

10　Parenti, *Tropic of Chaos*, 99.

11　Ibid., 7.

12　Ibid., 135.

13　"Measuring Glacial Change in the Himalayas", United Nations Environmental Program, September 2012, available at http://na.unep.net/geas/getUNEPPageWith ArticleIDScript.php?article_id=91 (accessed 5 January 2015); "Water and Climate in the Himalayas", World View of Global Warming, available at www.worldview ofglobalwarming.org/himalaya_1/index.php (accessed 5 January 2015).

14　Parenti, *Tropic of Chaos*, 129.

15　David Biello, "Are High Food Prices Fueling Revolution in Egypt?" *Scientific American*, 1 February 2011, available at http://blogs.scientificamerican.com/observations/2011/02/01/are-high-food-prices-fueling-revolution-in-egypt (accessed 21 December 2014).

16　Colin P. Kelley, Shahrzad Mohtadi, Mark A. Cane, Richard Seager, and Yochanan Kushnir, "Climate Change in The Fertile Crescent and Implications of the Recent Syrian Drought", *PNAS*, vol. 112, 11 (17 March 2015): 3241–3246; Thomas L. Friedman, "Without Water, Revolution", *New York Times*, 18 May 2013, available at www.nytimes.com/2013/05/19/opinion/sunday/friedman-without-water-revolution.html?pagewanted=all&_r=2& (accessed 19 May 2013).

17　UNICEF, "Lowest Rainfall in Over 50 Years is Latest Threat to Children in Syria and Region", 6 June 2014, available at http://childrenofsyria.info/2014/06/06/lowest-rainfall-in-over-50-years-is-latest-threat-to-children-in-syria-and-region (accessed 18 September 2014).

18　Timothy Snyder, *Black Earth: The Holocaust as History and Warning* (New York: Tim Duggan Books/Random House, 2015).

19 Snyder, "The Next Genocide"; See also, Timothy Snyder, "Hitler's World", *New York Review of Books*, 24 September 2015, available at www.nybooks.com/articles/archives/2015/sep/24/hitlers-world (accessed 16 October 2015).

20 Peter Schwartz and Doug Randall, "An Abrupt Climate Change Scenario and its Implications for United States National Security", October 2003, available at www.epa.gov/cleanenergy/energy-resources/calculator.html#results (accessed 18 October 2014).

21 *Quadrennial Defense Review*, available at www.defense.gov/home/features/2014/0314_sdr/qdr.aspx.

22 Since the end of WWII, Japan was the only major industrial power constitutionally forbidden to have an offensive military capacity, but that appears to be changing as Prime Minister Shinzo Abe works to reconfigure Japan as a military power, a move that appears to have the blessing of the American administration. Debito Arudou, "US Greenlight's Japan's March Back to Militarism", *Japan Times*, 21 May 2015; "Shock Doctrine in Japan: Shinzo Abe's Rightward Shift to Militarism, Secrecy in Fukushima's Wake", transcript of interview with Koichi Nakano, *DemocracyNow*, 15 January 2014, available at www.democracynow.org/2014/1/15/shock_doctrine_in_japan_shinzo_abes (accessed 18 October 2015); Jonathan Soble, "Japan Moves to Allow Military Combat for First Time in 70 Years", *New York Times*, 16 July 2015, available at www.nytimes.com/2015/07/17/world/asia/japans-lower-house-passes-bills-giving-military-freer-hand-to-fight.html (accessed 18 October 2015).

23 World military expenditures totalled $1.75 trillion in 2013, a drop of 1.9 per cent in real terms from 2012 though still far too high. Stockholm International Peace Research Institute, available at www.sipri.org/media/pressreleases/2014/Milex_April_2014 (accessed 4 December 2014).

24 Stockholm International Peace Research Institute, Press Release, 13 April 2015: "US Military Spending Falls, Increases in Eastern Europe, Middle East, Africa and Asia says SIPRI", available at www.sipri.org/media/pressreleases/2015/milex-april-2015 (accessed 16 October 2015).

25 The world's nuclear-armed nations altogether possess a total of 16,000 nuclear weapons. Arms Control Association, "Nuclear Weapons: Who Has What at a Glance", available at www.armscontrol.org/factsheets/Nuclearweaponswhohaswhat (accessed 18 October 2015).

26 "White House Announces Push for Next Generation of High-Tech Weapons", *The Guardian*, 16 November 2014, available at www.theguardian.com/us-news/2014/nov/16/white-house-announces-push-for-next-generation-of-hi-tech-weapons?CMP=ema_565 (accessed 16 November 2014).

27 Matthew Rosenberg and Michael D. Shear, "In Reversal, Obama Says US Soldiers Will Stay in Afghanistan to 2017", *New York Times*, 15 October 2015, available at www.nytimes.com/2015/10/16/world/asia/obama-troop-withdrawal-afghanistan.html (accessed 15 October 2015).

28 A metric ton is an international unit of measurement equivalent to approximately 2,200 pounds. CO_2 equivalents (CO_2-e) offer a universal standard measurement that allows for the comparison of different greenhouse gases based on their ability to trap heat in the atmosphere. Joe Abraham, "What is One Million Ton of Carbon Dioxide Equivalent?" Southwest Climate Change Network, available at www.southwestclimatechange.org/solutions/reducing-emissions/mmtco2-e (accessed 21 December 2014).

29 Adam J. Liska and Richard K. Perrin, "Securing Foreign Oil: A Case for Including Military Operations in the Climate Change Impact of Fuels", *Environment Magazine* (July–August 2010), available at www.environmentmagazine.org/Archives/Back%20Issues/July-August%202010/securing-foreign-oil-full.html (accessed 17 December 2014).

30 Environmental Protection Agency, greenhouse gases equivalency calculator, available at www.epa.gov/cleanenergy/energy-resources/calculator.html#results.

31 Gregory Hooks and Chad L. Smith, "Treadmills of Production and Destruction: Threats to the Environment Posed by Militarism", *Organization and Environment* 18, no. 1, 19–37.

32 Patricia Hynes, "Military Hazardous Waste Sickens Land and People", *Truthout*, 4 August 2011, available at www.truth-out.org/news/item/2377:military-hazardous-waste-sickens-land-and-people (accessed 17 December 2014).

33 Neta Crawford, "US Costs of War Through 2014: $4.4 Trillion and Counting", available at http://costsofwar.org/article/economic-cost-summary (accessed 1 December 2014).

34 The United Nations Environment Program acknowledged that there has been little oversight or research on the military's impact on the natural environment or on climate change. Tamara Lorincz, *Demilitarization for Deep Decarbonization: Reducing Militarism and Military Expenditures to Invest in the UN Green Climate Fund and to Create Low-Carbon Economies and Resilient Communities*, Draft Working Paper for the International Peace Bureau, available at www.ipb.org/web/index.php?mostra=content &menu=Resources&submenu=Books#, 18.

35 Lorincz, Ibid., 22–25.

36 "Environmental Impact of Agent Orange/Dioxin", available at http://makeagent orangehistory.org/agent-orange-resources/background/environmental-impact-of-agent-orange-dioxin (accessed 18 December 2014). Phung Tuu Boi, *Vietnam: War's Lasting Legacy*, PowerPoint, available at www.agentorangerecord.com/impact_on_vietnam/environment/defoliation (accessed 17 December 2014).

37 Clyde Haberman, "Agent Orange's Long Legacy, for Vietnam and Veterans", *New York Times*, 11 May 2014, available at www.nytimes.com/2014/05/12/us/agent-oranges-long-legacy-for-vietnam-and-veterans.html (accessed 18 October 2015).

38 Melissa Krupa, "Environmental and Economic Repercussions of the Gulf War on Kuwait, "Case Number 9, May 1997, Trade and Environment Database, American University, available at www1.american.edu/ted/ice/kuwait.htm (accessed 19 October 2015).

39 The massive climate march in the fall of 2014 began to make these connections; and for the first time in this author's memory, a large international gathering of peace activists in New York City in late April 2015 entitled "Peace and Planet" brought the two issues together.

40 The independent Labor Network for Sustainability, the Global Labor Institute at Cornell and the Blue-Green Alliance have all spearheaded labour involvement with the issue of climate change and several unions participated in the mass climate march in NYC in September 2014; but the AFL-CIO has an "all of the above" energy policy and several unions whose members have jobs in the fossil fuel industry are vehemently opposed to eliminating oil and gas from the energy mix.

41 Remarks by Joe Uhlein to the Joint Seminar on Full Employment, Social Welfare and Equity and Globalization, Labor and Popular Struggles, Columbia University, 17 November 2014.

42 Pope Francis, "Encyclical Letter *Laudato Si'* of the Holy Father Francis on Care for Our Common Home", 24 May 2015, available at http://w2.vatican.va/content/ francesco/en/encyclicals/documents/papa-francesco_20150524_enciclica-laudato-si. html.

43 "Full Text of Pope Francis's Speech to the United Nations", PBS Newshour, available at www.pbs.org/newshour/rundown/full-text-pope-francis-speech-united-nations (accessed 18 October 2015).

44 Among these researchers are: Michael Klare, *Resource Wars: The New Landscape of Global Conflict* (New York: Henry Holt and Company, 2002); Parenti, *Tropic of Chaos*; H. Patricia Hynes, "The Military Assault on Global Climate", *Truthout*, 8 September 2011, available at www.truth-out.org/news/item/3181:the-military-assault-on-global-climate; Hynes, "Military Hazardous Waste Sickens Land and People"; researchers at

the International Peace Bureau, available at www.ipb.org/web (accessed 18 October 2014); Hooks and Smith, "Treadmills of Production and Destruction".

45 Swedish International Peace Institute, SIPRI Fact Sheet, "Trends in Military Expenditure", April 2015, available at http://books.sipri.org/product_info?c_product_id=496, available at www.sipri.org/yearbook/2014/04 (accessed 19 October 2015).

46 Address by Christiana Figueres, Executive Secretary United Nations Framework Convention on Climate Change, to the Congress of Deputies of Spain at the Centro Superior de Estudios de la Defensa Nacional in Madrid, 15 February 2011, available at http://unfccc.int/files/press/statements/application/pdf/speech_segurida (accessed 9 October 2014).

14 The Canadian government's anti-democratic attack on asylum seekers

Anne Venton

Introduction

The publisher's synopsis of the 2014 book, *Tragedy in the Commons*, by Samara Canada authors Allison Loat and Michael MacMillan states that "Canada's democracy has lost its way, its purpose and the support of the public it is meant to serve. How did one of the world's most functional democracies go so very wrong?" This essay provides one answer to the question with a case study of how the Canadian Government's attack on asylum seekers was undemocratic. The attack began with a 2012 order-in-council by the Government of Canada led by Prime Minster Steven Harper (hereafter "Harper government") that modified and reduced the funding of the Interim Federal Health Program (IFHP) for refugees. The Harper government's order-in-council violated the principles of democracy but the attempt was foiled by the Federal Court that ruled it unconstitutional.

Constitutions and the rule of law are features of democracy that provide for the equality of justice. They complement the other central feature of democracy: the equality of all to participate in the legislative process. Constitutions were introduced in the Athenian democracy in the fourth century BC to address the problem of demagogues in the fifth century BC. "Demagogues preyed on the desires of the multitude and urged them to reject the established norms" of democracy (Neill 2011: 26). The strategies and practices that the Harper government used in its attempt to persuade the Federal Court of the efficacy of cutting funding for health support for refugee applicants are the twenty-first century analogue of the strategies and tactics of the demagogues in ancient Athens.

Canada's Immigration and Refugee Protection Act

The Immigration and Refugee Protection Act (IRPA) is an Act of Parliament passed in 2002 which replaced the Immigration Act, 1976 as the primary federal legislation regulating immigration to Canada. The Immigration and Refugee Protection Regulations (IRPR) contain the laws created to fit within the IRPA in order to specify how the IRPA is to be applied.

The Act grants protection to an individual who is a United Nations convention refugee or a person in need of protection. A convention refugee is an individual that has a well-founded fear of persecution based on: race, religion, nationality, political opinion or membership in a particular social group.

The Act is administered by the Immigration and Refugee Board (IRB), an independent tribunal established by the Parliament of Canada in 1989. The mission of the IRB, on behalf of Canadians, is to resolve immigration and refugee cases efficiently, fairly and in accordance with the law.

The legislation is derived from, or based upon, Canada's obligations to the international community as incorporated in the following United Nations conventions to which Canada is a signatory. These include:

- 1951: Convention Relating to the Status of Refugees.
- 1967: Protocol Relating to the Status of Refugees.
- 1984: Convention Against Torture and Other Cruel, Inhuman and Degrading Treatment or Punishment.

The objective of security is also found in the 1982 Canadian Charter of Rights and Freedoms, especially in its sections that relate to the "security of the person".[1] Security of the person includes both physical and economic security. The latter element is of special importance because refugee claimants in Canada are often vulnerable in the sense that many have insufficient economic resources to provide for their basic necessities of food, shelter, clothing and health while their application for refugee status is under consideration by the IRB tribunal. Therefore, the Canadian government needs to provide them with the same support that it provides to similarly situated Canadian citizens through various social welfare and assistance programs such as welfare for the unemployed or disabled and universal health care benefits under the Canada Health Act. One of the specific programs for refugee applicants is the Federal Interim Health Program.

Democracy and the Immigration and Refugee Protection Act

The rationale for democracy is the "common good" which comprises several elements that are the aims of society. These aims are multiple, diverse and shared by all citizens. The notion of the common good was enunciated in the writings of Aristotle who defined democratic regimes as being governed by all citizens who share in deliberations about the common good. "The Aristotelian conception of the polis (city state) was that of a self-sufficient cooperative arrangement that is centred on a particular understanding of the good" (Neill 2011: 40). Further "The good life can only be actualized in the public deliberations of the citizenry" (Newell 2011: 9).

One of the fundamental elements of the common good is personal security and hence an objective of governance in a democratic country. The importance of the ideals of security that shaped the concepts of refugee protection in Canada dates back 800 years to the Magna Carta circa 1215 when the Barons in England sought predictability and fairness in their relationships with the king and the state. Without predictability in their role with the king and the state, citizens would be subject to arbitrary judgements. Fair procedural guarantees for the barons *vis-à-vis* the king and the state was the goal (comment recorded by the author at a panel discussion on the Magna Carta at the Munk School of Global Affairs, Toronto, 4 May 2015).

In the first part of *Leviathan*, Thomas Hobbes, a seventeenth-century British political philosopher who founded the social contract tradition, echoed the Magna Carta when he described life in the state of nature as "solitary, poor, nasty, brutish and short" (Hobbes 2001). The social contract is a fundamental feature of democracy which represents a contract in which citizens give up some individual liberty in exchange for collective benefits. As Aristotle put it, freedom and equality especially characterize democracy and one element of freedom is "for all to rule and for all to be ruled in turn" (Cooper 2011: 197, 200). The second element of freedom is to live as one likes since not to live as one likes is the life of a slave (ibid.: 200). As well, these sentiments are echoed in the United Nations resolution 217 A (III) of 10 December 1948 wherein the United Nations General Assembly adopted the Universal Declaration of Human Rights.

The sentiments about personal security are also found in the 1982 Canadian Charter of Rights and Freedoms, particularly in sections relating to "security of the person". The Charter is a Bill of Rights entrenched in the Constitution of Canada and forms the first part of The Constitution Act, 1982. The Charter guarantees certain political rights to Canadian citizens and civil rights to everyone in Canada regarding policies and actions of all areas and levels of government. It was designed to unify Canadians around a set of principles that embody those rights. The Charter was signed into law by Queen Elizabeth II of Canada on 17 April 1982 in Ottawa, Canada, along with the rest of the Constitution Act.

Principles of democracy

One of the principles of democracy is that of equal treatment. This principle provides equality for all citizens and groups of citizens before the law, whether they are members of religious, cultural, ethnic or racial groups. In granting refugees asylum in Canada, it is implicit that they would be accorded the same protections to ensure their personal security as provided to all Canadian citizens and residents. Thus the democratic principle of equal treatment would apply to refugees.

Another principle of democracy is John Rawls's principle of social justice: any increases in the inequality of the distribution of wealth and authority are just if they result in compensating benefits for everyone, and in particular the least-advantaged members of society (Kloppenberg 2011: 90–92).

A third principle of democracy is equality of justice which requires the freedom of speech and the rule of law. As previously noted, Magna Carta still forms an important symbol of liberty today. Aristotle's idea of freedom "for all to rule and for all to be ruled in turn" implies equality for all in political participation in the legislative process and the selection of the (elected) officials. But it also implies "equality of justice". Justice for all means that justice of the multitude must, by necessity, be sovereign and whatever is decided by the majority must be final and constitute justice. More specifically this means that judges are chosen by lot and in this sense are representatives of the multitude (Cooper 2011: 200). Sovereign implies that the judges' interpretations of the laws are superior in authority to the legislature and to the executive branches of government.

The achievement of the objective of equality of justice requires that the democratic state has a constitution and is governed by the rule of law (ibid.: 197). By the rule of law Aristotle means a manner of governance in which the political leaders are constrained beforehand by principles (in the constitution) which are known to the citizen body. A political regime that defers to the rule of law was, in Aristotle's opinion, superior to a regime in which the present generation alone is dominant (Neill 2011: 26). Aristotle's view reflects his measures for safeguarding democratic constitutions – namely to protect the better classes and limit the scope of the popular courts (Cooper 2011: 199). The problem with the popular courts in Athens in the fifth century BC was that they tended to be influenced by social demagogues and, as a consequence, acted in ways inconsistent with the principles of democracy that are incorporated in a constitution. In effect "The importance of demagogues stemmed from the fact that the people are sovereign over all things while demagogues are sovereign over the opinion of the people" (ibid.: 198).

In ancient Athens the language of demagogues was the antithesis of the language of free speech because freedom of speech entailed not only equal opportunity to speak but also speech that was frank and open. It was speech that was egalitarian, rejected hierarchy and revealed and uncovered the truth as one sees it.[2] This definition of free speech is in sharp contrast to speeches expressed by the spokespersons of modern governments which tend to be scripted in order to produce a pleasing sound bite that keeps the government's message on track. Such speech is not meant to be open and frank; it limits debate and it rarely reveals the truth. For Athenians this kind of speech characterizes life under tyranny and oligarchy where there is no rule of law but only fear (ibid.: 204). In summary, prevalent throughout the Athenian notion of the administration of equal justice and its corollary, the rule of law, is unencumbered free speech (ibid.: 201).

The Harper government's denial of basic health care to refugee applicants

Canada is a country of immigrants. Having a safe place to grow roots attracts people from around the world while Canada's inclusiveness helps them build better lives once they reach the country. In 2012, the Harper government issued an order-in-council to make changes to the Interim Federal Health Program (IFHP) for refugees – a service meant to address refugees' basic needs. A number of civil society groups including the Canadian Association of Refugee Lawyers (CARL) challenged the order-in-council and took the government to the Federal Court.

The Harper government gave the following arguments to the Federal Court for the changes in its order-in-council:

- The costs of the IFHP were increasing.
- The program was not fair because refugees were getting more support from this program than working Canadians.
- The cuts would leave basic health of refugees protected.
- The cuts were necessary to protect the integrity of the system that the government claimed was being corrupted by bogus refugees.
- The cuts were needed to help balance the budget.

The hearings of the Federal Court and its decision reveal that these rationales were undemocratic in three respects. First, the argument that the Government needed to cut spending on the program to balance the budget ran counter to the democratic principle of social justice: that any increase in the inequality of the distribution of wealth and authority is just only if it results in compensating benefits for everyone, and in particular the least-advantaged members of society. The Harper government's proposed reduction in spending on the IFHP for refugees clearly targeted refugee claimants who were generally vulnerable and who consequently could be considered to be among the least advantaged members of society.

In general the Harper government's argument that it needed to balance the budget was a self-imposed policy objective necessitated by the major cuts in taxes (General Sales Tax and Corporate Income Taxes) that disproportionately bene-fited high income earners. These were undemocratic in the sense that they effectively transferred wealth from poor households to rich households contrary to the preferences of a majority of Canadians. A Broadbent Institute survey in 2012 revealed that 77 per cent of the respondents believe that income inequality is a serious issue; 83 per cent support higher income taxes for the affluent and nearly 75 per cent said they wanted corporations to pay higher tax rates than 2008 levels (Moran 2012).

The Harper government's austerity program that began after 2010 was neces-sitated by two of its objectives. The first objective was to reduce taxes; the rates of General Sales Tax (GST) were reduced from 7 to 5 per cent over a period from July 2006 to January of 2008 and the rate of Corporate Income Tax was reduced over a period from 2008 to 2012 from 22.12 to 15.0 per cent. The second objec-tive was to balance the federal budget in time for the general election in 2015. For the fiscal year 2011–2012, the annual loss in tax revenues was estimated to be $26 billion. Table 14.1 shows the net impacts on each of the three household income groups on the hypothetical assumption that the $26 billion in reduced taxes in 2011–2012 was financed by across-the-board cuts in federal expenditure totalling $26 billion in that year.

Note that the estimates in the table are for one fiscal year only; the cumulative loss in GST and corporate income tax revenues from 2006–2007 to 2014–2015 are estimated to be $156 billion – 6 times the $26 billion estimated for

Table 14.1 Net benefits from the government's tax reductions and hypothetical matching
expenditure reductions, 2011–2012

Household income group	Share of total income[a]	Share of GST and corporate income taxes[a]	Share of federal expenditures[b]	Net benefit per household[c]
Highest 20%	48.0%	44.8%	19.5%	+$2,469
Middle 60%	48.1%	49.2%	64.0%	–$482
Lowest 20%	3.8%	6.0%	16.6%	–$1,049
Totals	100%	100.0%	100.0%	0

a. Assumes that the shares of total income and taxes in 2011 were the same as they were in 2005. See the Appendix for details.
b. Assumes that the shares of federal expenditures in 2011–2012 were the same as they were in 2006. See the Appendix for details.
c. There were 13.321 million households in Canada in 2011. Net benefits equal the benefit of tax reductions minus the loss in benefits from expenditure reductions.

2011–2012. Much of the expenditure cuts to match these revenue losses will not be realized for many years because most of them were financed by an increase in the federal debt of $130.8 billion over the period from March of 2006 to March of 2015.

Second, the intent of the IFHP was to provide health benefits to refugee claimants that were the same as benefits provided to Canadian citizens and residents under *The Canada Health Act* and administered by provincial governments. The program's intent was consistent with the democratic principle of equal treatment of all citizens before the law. However, the Harper government's argument was that the program was not fair because refugees were getting more support from this program than working Canadians. The implication was that it was necessary to cut program benefits to restore equality between Canadian citizens and refugee applicants. However, the Federal Court found that there was no evidence to support the government's argument that refugees were getting better treatment than Canadians (*Doctors v. Canada*, para. 947).[3] Specifically, the Harper government's argument was based on an incorrect comparison of low income refugees with high income Canadians. Low income Canadians were getting the same support from the program as low income refugees.

Third, the Federal Court found that there was evidence of program discrimination based on the nationality of the refugees. Specifically the Court ruled that section 15 of the Charter was violated because a lesser level of health insurance coverage was being provided to refugee claimants from the government's DCO (designated country of origin) countries in comparison with that provided from non-DCO countries (*Doctors v. Canada*, para. 1081). Such discrimination was clearly contrary to the democratic principle of equal treatment.

Finally, the Federal Court found that the Government had in several instances not tried to assemble evidence to prove its assertions. Insofar as the assertions were pleasing to large segments of the public and insofar as the government did not seek to uncover the truth of its assertions it did not meet the Athenian definition of free speech. During the hearings of the Federal Court, lawyers for the Court cross examined several civil servants in the government and found several instances where there was no evidence to support the government's arguments for the changes to the program. As Ontario Bar Association expert on refugee law Chantal Desloges reported, the Court determined that the government had made no effort to determine the validity of its argument that the changes in the order-in-council would protect the integrity of the system by deterring bogus refugees (recorded at a presentation on refugee health care to a Why Should I Care forum, Toronto, 20 April 2015). Nor had the government done any study to determine the extent to which, if any, refugees were coming to Canada solely to get free health services. The Court also found that the government had done no study on the impact of the cuts on the children of refugees. The absence of any study called into question whether there was evidence to support the Government argument that the cuts in spending would leave basic health of refugees (and their families) protected. Finally the court found that the Government had done no study to support its assertion that there would be cost savings. Justice Anne McTavish concluded that there was no reliable evidence that the 2012 changes to the IFHP would, on their own, result in cost savings at the federal level and that some of the cost of medical services that were previously covered under the IFHP had simply been downloaded to the provinces (*Doctors v. Canada*, para. 1012).

Postscript on the federal court ruling

In November 2014, the Harper government temporarily restored health care coverage to refugee claimants pending the outcome of its appeal of the Federal Court's ruling in 2012. The temporary plan was described as punitive and selective by the Canadian Association of Refugee Lawyers (CARL) which was one of the groups that took the government to court over the Government's 2012 order-in-council to cut benefits of the program. In the opinion of Peter Showler, co-chair of CARL, the government's temporary plan does not comply with the Federal Court ruling in 2014. He stated that "The government is still being punitive and is being selective when the Federal Court told them to reinstate all benefits" (Mas 2014).

The Harper government's publicity strategies and practices

The Harper government's violation of democratic free speech with respect to its arguments to the Federal Court provides an example of strategies and practices that are common in modern democratic states. They fit Cooper's conclusion that the Athenian definition of free speech is in "Sharp contrast to speeches expressed by the handlers of modern governments which tend to be scripted in order to produce a pleasing sound bite that keep the government's message on track. Such speech is not meant to be open and frank; it limits debate and it rarely reveals the truth. For Athenians this kind of speech characterizes life under tyranny and oligarchy where there is no rule of law but only fear" (Cooper 2011: 204).

Refugee protection is an example of an issue that the Harper government attempted to manipulate through its communications and advertising in the commercial media. The Harper government communications leading up to the elections of 2011 and 2015 used refugees as scapegoats by suggesting that domestic security issues were linked to crime, immigration and refugees. Print and non-print media stressed that Canada was not a safe haven and has porous borders. Therefore, security and defence were needed so that Canada will not be hijacked by criminals and fake refugees who jump the queue. Press releases from the Minister of Public Safety stressed security and defence issues as a government priority (Kozolanka et al. 2015). Traditionally, scapegoating uses propaganda and disinformation to attack minorities and terrorists. It is used as a means to divert the people's attention from other problems, to shift blame for failures, and to channel frustration in controlled directions.

In the early twentieth century, Edward Bernays (1891–1995), the Austrian–American pioneer in the field of public relations and propaganda referred to as the "father of public relations" anticipated the influence of publicity strategies and practices in the world of politics. In his 1928 book, *Propaganda,* he stated:

> Good government can be sold to a community just as any other commodity can be sold. I often wonder whether the politicians of the future, who are responsible for maintaining the prestige and effectiveness of their party, will not endeavour to train politicians who are at the same time propagandists.
>
> (Cited in Kozolanka 2014: 3)

Creating a climate of fear

Another tactic is to create a climate of fear among the general public to induce a certain response to achieve the politician's aims. In the general federal election of 2011 the Conservative party led by Prime Minister Harper focused on a theme of "Who are Canada's enemies?". The campaign described Canada's enemies as including violent criminals, refugee claimants who jump the queue and enter the country illegally, those who are soft on terror and anyone who is against the government. In the lead-up to the 2015 election, the Harper government continued to create a climate of fear by associating refugees with terrorists as possible suspects. The real or imagined fear of terrorists who attacked members of Parliament and Canadian citizens in early 2015 has been given much publicity and the evidence that the shooter was not a terrorist has been downplayed or simply ignored. This is the same tactic used by corporate advertising and public relations activities that downplay, minimize, omit or simply deny any negative aspects that the public may perceive about their products or services.

Harper continues to promote a culture of fear of terrorists among Canadians by accentuating the probability of terrorist threats on Canadian soil. He is on record as saying, "Make no mistake: by fighting this enemy here you are protecting

Canadians at home. Because this evil knows no borders, and left uncontained, it will spread like a plague" (*Toronto Star* 2015). Meanwhile journalists report that the Islamic State of Iraq and Syria (ISIS) remains focused on a war within Islam.

Critics of Bill C-51, the Harper government's anti-terrorism Act passed in early 2015 say that the Bill's drastic measures are an unjustified infringement of Canadians' rights. The Bill's most troubling provisions give unprecedented powers to government departments, to police and to the Canadian Security Intelligence Service (CSIS), the country's spy agency.

The rise of consumerism in politics today and the advent of "consumer" democracy

In her book, *Shopping for Votes*, Susan Delacourt argues that a culture of "consumer democracy" has developed among most citizens. In a consumer democracy, citizens want only answers to questions about what tangible specific things the government or party has done for them lately. In a consumer democracy citizens have different attitudes than democratic citizens. Examples are provided in Table 14.2 (Delacourt 2013: 15, 16).

Note that the attitudes of consumer citizens are almost the polar opposite to democratic citizens as it was defined by Aristotle some 2,500 years ago. In Aristotelian terms, democratic regimes are governed by all citizens who share in deliberations about the common good. Further, the good life can only be actualized in the public deliberations of the citizenry (Newell 2011: 9). In a democratic regime the virtues of citizens include moderation in the pursuit of wealth. Moderation for households means that each household (family) has no more income than what is required to meet its basic needs for food, shelter and clothing plus an amount of wealth that will enable the householders to have sufficient leisure for the study of politics and philosophy and for deliberations in public about the common good. Thus, man's two chief virtues are participation in politics and learning philosophy (ibid.: 17). In ancient Greece the term philosophy included the modern disciplines of philosophy, ethics, economics, political science, human psychology, and sociology – bodies of knowledge that are relevant for a modern democratic government's management of the economy and institutions for serving the elements of the common good.

Table 14.2 Consumer citizen versus democratic citizen

Issue	Consumer citizen	Democratic citizen
Sacrifice for the common good	Unwilling to make some sacrifice	Willing to make some sacrifice
Taxes	Taxes are a cost	Taxes are payments for benefits
Civics education	Not entertaining	A prerequisite for political participation
Satisfaction	It's all about wants	It's all about needs

Conclusions

The Harper government's 2012 order-in-council to cut health benefits for refugee applicants to Canada was clearly undemocratic in three respects. First, it violated the principle of equal treatment. Secondly it violated the principle of social justice. And thirdly it violated freedom of speech as it is defined in democracy because it did not seek to uncover the truth but rather used statements of public prejudice.

On 4 July 2014 Justice Anne McTavish for the Federal Court declared the order-in-council to be unconstitutional. She concluded:

> I have found that the affected individuals are being subjected to "treatment" as contemplated by section 12 of the Charter, and that this treatment is indeed "cruel and unusual". The 2012 modification to the IFHP potentially jeopardizes the health and indeed the very lives of these innocent and vulnerable children in a manner that shocks the conscience and outrages our standards of decency. They violate section 12 of the Charter.
>
> (*Doctors v. Canada*, para. 1080)

The culture and views of citizens in a consumer society have spilled over into political life and have affected how political leaders and parties treat their citizenry. Politicians and political parties resort to publicity strategies and practices to persuade the public to accept the initiative using deceptive tactics more commonly known as spin. These strategies were used by the Harper government in an attempt to persuade the Federal Court. However, the Federal Court saw through the duplicity of the arguments and quashed the order-in-council declaring it to be unconstitutional.

The presence of a democratic constitution that included Canada's Charter of Rights and Freedoms backed up with the rule of law triumphed over the undemocratic initiatives of the Harper government. The sacrosanct separation of powers between the judiciary and the executive branches enabled the judiciary to save the day for asylum seekers to Canada. Without a constitution that is sovereign over the executive branch of government, the Harper government's modification of the IFHP would have meted out "cruel and unusual punishment" to refugees.

Appendix: net benefits from the Harper government's tax reductions and hypothetical expenditure reductions, 2011–2012

Economist Marc Lee's study for the Canadian Centre for Policy Alternatives estimated the shares of federal GST and corporate income taxes directly or indirectly paid by Canadian families in ten income groups (deciles) in 2005 (Lee 2007: 16, 17, 31, 32). The data from Lee's study are aggregated for three broad household income groups and presented in Table 14.3.

Economists Hugh Mackenzie and Richard Shillington's study for the Canadian Centre for Policy Alternatives used a series of formulae to allocate different kinds

Table 14.3 Shares of federal taxes among household income groups, 2005

Household income group	Total income	GST	Corporate income tax	Both taxes
Highest 20%	48.0%	37.2%	53.7%	44.8%
Middle 60%	48.1%	55.1%	42.2%	49.2%
Lowest 20%	3.8%	7.7%	4.1%	6.0%
All groups	100.0%	100.0	100.00	100.0%

Source: Derived from Lee (2007: 16, 17, 31, 32).

Table 14.4 Shares of federal government expenditure benefits among household income groups, 2006

Household income group	Share of federal expenditures
Highest 20%	19.5%
Middle 60%	64.0%
Lowest 20%	16.6%
Totals	100.0%

Source: derived from Mackenzie and Shillington (2009: 34–38).

of federal government expenditures to 17 different household income groups for the year 2006 (Mackenzie and Shillington 2009: 34–38). The results of their study are aggregated and presented in Table 14.4.

Notes

1   Canadian Charter of Rights and Freedoms, 18 April 1982, available at publications. gc.ca/collections/Collection/CH37-4-3-2002E.pdf.
2   The "truth as one sees" presumably means the perspective with which reality is viewed – as distinct from absolute truth.
3   *Canadian Doctors for Refugee Care v Canada* (Attorney General) 2014 FC651 (4 July).

References

Cooper, C. (2011) "Oligarchy and the Rule of Law". In D. Tabachnick and T. Koivukoski (eds), *On Oligarchy. Ancient Lessons for Global Politics*, University of Toronto Press, Toronto.

Delacourt, S. (2013) *Shopping for Votes: How Politicians Choose Us and We Choose Them*, Douglas and McIntyre, Madeira Park, BC.

Hobbes, T. (2001) *Of Man, Being the First Part of Leviathan*, available at http://www.gutenberg.org/ebooks/3207.

Kloppenberg, J. (2011) *Reading Obama: Dreams, Hope and the American Political Tradition*, Princeton University Press, Princeton, NJ.

Kozolanka, Kirsten (2014) "Communicating for Hegemony: The Making of the Publicity State in Canada". In Kirsten Kozolanka (ed.), *Publicity and the Canadian State: Critical Communications Perspectives*, University of Toronto Press, Toronto.

Kozolanka, Kirsten et al. (2015) Roundtable discussion at the Union for Democratic Communications: Circuits of Struggle, University of Toronto, 1–3 May.

Lee, M. (2007) *Eroding Tax Fairness: Tax Incidence in Canada, 1990 to 2005*. November, Canadian Centre for Policy Alternatives, Ottawa.

Loat, A. and MacMillan, M. (2014) *Tragedy in the Commons*. Toronto: Penguin Random House Canada.

Mackenzie, M. and Shillington, R. (2009) *Canada's Quiet Bargain: The Benefits of Public Spending*, April, Canadian Centre for Policy Alternatives, Ottawa.

Mas, S. (2014) "Refugee Health Care Temporarily Restored in Most Categories". CBC *News* (4 November).

Moran, A. (2012) "Ed Broadbent Calls for More Taxes to Battle Income Gaps in Canada". Available at www.broadbentinstitute.ca/blog/ed-broadbent-calls-more-taxes-battle-income-gaps-canada.

Neill, J. S. (2011) "Aristotle and American Oligarchy: A Study in Political Influence". In D. Tabachnick and T. Koivukoski (eds), *On Oligarchy: Ancient Lessons for Global Politics*, University of Toronto Press, Toronto.

Newell, W. R. (2011) "Oligarchy and Oikonomia: Aristotle's Ambivalent Assessment of Private Property". In D. Tabachnick and T. Koivukoski (eds), *On Oligarchy: Ancient Lessons for Global Politics*, University of Toronto Press, Toronto.

Toronto Star (2015) "Editorial". *Toronto Star* (9 May): IN 1.

15 Climate-induced migration

What legal protection for climate migrants at the international level?

Francesca Mussi

During the 1980s and 1990s, climate change was predominantly conceived as a scientific, economic and environmental issue. Then, since 1990, the focus of the climate change debate has progressively shifted to the potential impacts on human migration,[1] thus receiving considerable scholarly attention.[2]

In recent years, the speed with which climate change is magnifying the risk of extreme weather events and, consequently, the increased number of people thought to be susceptible to it,[3] have renewed attention to the need to address the issue of people moving in response to the effects of climate change, especially those forced to flee across international borders.

At the international level, there are no specific legal frameworks that can provide protection and assistance for people displaced by climate change,[4] possibly as a result of the ambiguity over what is meant by "climate migrants" or "climate displaced persons".[5]

Contemporary international law, for its part, recognizes only two categories of people who are forced to move, typically due to an armed conflict or a severe weather event: refugees and internally displaced persons (hereinafter: IDPs). However, existing legal frameworks concerning refugee protection and internal displacement are not sufficiently well equipped to comprehensively deal with cross-border climate migration: the Convention Relating to the Status of Refugees (Geneva, 28 July 1951) (hereinafter: 1951 Refugee Convention)[6] does not cover climate displaced people, as they do not fulfil the legal conditions to be treated as "refugees"; the Guiding Principles on Internal Displacement developed by the UN Commission on Human Rights in 1998 only apply to those who have not crossed an international border.[7]

Furthermore, because those displaced by climate change are not specifically recognized as needing protection and assistance, no international agency is specifically mandated to assist them, thus undermining the preparedness of the responses.

In light of what has been called a wide "protection gap",[8] the present chapter aims at analysing what efforts are being taken at the international level to address the issue of people displaced across borders as a consequence of climate change. To this end, the discussion is organized in two sections. The first section looks at the inapplicability of existing international legal frameworks concerning refugee protection and internal displacement. The second part considers current responses aiming at closing the normative gaps that exist with regard to climate migrants who cross international borders: after the assessment of the UN approach, attention will focus on the Nansen Initiative on Disaster-Induced Cross-Border Displacement, an intergovernmental process which today appears as one of the most promising initiatives to address climate cross-border migration.

Inapplicability of existing international legal frameworks to climate migrants

In the past, group-based mechanisms have been established to protect the rights of specific categories of individuals, such as women and children, racial, national or religious minorities, refugees and stateless persons, migrant workers and members of their families, and people with disabilities.[9] As Roger Zetter pointed out "(. . .) [I]t is therefore surprising that a similar framework to protect the rights of people forced to move because of climate-induced environmental change does not exist".[10]

In light of the absence of a specific international legal instrument applicable to those displaced by climate change, this section explores the limits of existing legal regimes applicable to refugees and IDPs in responding to the need of protection for climate migrants.

Inapplicability of refugee law

Even though a glance at some scientific literature shows the recurrence of the expressions "climate refugees"[11] or – more generally – "environmental refugees"[12] or "ecological refugees",[13] the choice of the term "refugee" is highly controversial,[14] as it implies several conditions that do not exist from the perspective of climate migration.

Art. 1, para. A (2) of the 1951 Refugee Convention, as modified by the 1967 Protocol, defines a refugee as any person who

> owing to well-founded fear of being persecuted for reasons of race, religion, nationality, membership of a particular social group or political opinion, is outside the country of his nationality and is unable or, owing to such fear, is unwilling to avail himself of the protection of that country; or who, not having a nationality and being outside the country of his former habitual residence as a result of such events, is unable or, owing to such fear, is unwilling to return to it.

The requirement to show "a well-founded fear of being persecuted" is at the heart of the refugee definition. However, the term "persecution" is not defined in normative terms in international refugee law.[15] Thus, it can be difficult to know precisely what persecution is.

Some guidance on this matter is nonetheless offered by Art. 33, para. 1 of the 1951 Refugee Convention, according to which, at least, but not only, threats to life or freedom on account of race, religion, nationality, political opinion or membership of a particular social group constitute persecution.[16] Outside of this core meaning, the notion of persecution has been interpreted in a way to include other serious violations of human rights for the same reasons.[17]

In light of the above, while some have sought to argue the case,[18] there are difficulties in characterizing climate change consequences as "persecution": storms, earthquakes and floods may be harmful, but they do not meet the threshold of seriousness to constitute serious violations of human rights.[19]

The 1951 Refugee Convention poses an additional hurdle for environmental and climate migrants: namely, that persecution is on account of the individual's race, religion, nationality, political opinion, or membership of a particular social group. It is true that the 1951 Refugee Convention is characterized by a universalist concept of refugees, as it may apply to all people fleeing persecution, whatever their country of origin.[20] Such a universalist approach, however, is balanced by the necessity to individualize the fear. As far as climate migrants are concerned, environmental and climatic events do not choose and do not target a specific person or a specific group.[21] As a consequence, movement caused by environment or climate change is inevitably indiscriminate – at least with respect to the five conditions of the 1951 Refugee Convention – and an argument that environmental migrants might together constitute a "particular social group" would be difficult to establish, for the reason that people must be connected by a fundamental, immutable characteristic other than the risk of persecution itself.[22]

Such an approach has been unsurprisingly confirmed by judgements delivered by Australian and New Zealand's courts, which have denied refugee protection when it was invoked on the ground of environmental factors. For instance, the Supreme Court of New Zealand considered inadmissible a request for refugee protection of individuals who highlighted environmental circumstances in their country of origin, Kiribati. The Court noted growing scientific evidence on the impacts of climate change on this small island developing State, but it rejected the application as

> [a] person who becomes a refugee because of an earthquake or growing aridity of agricultural land cannot possibly argue, for that reason alone, that he or she is being persecuted for reasons of religion, nationality, political opinion, or membership of a particular social group.[23]

The second impediment to the eligibility for refugee status and international protection concerns the "source-of-persecution" requirement that the cause of the harm be either the government or a person or group of persons that the government is unwilling or unable to prevent from continuing the persecution. During an environmental or climatic crisis, most people will be helped and protected by domestic authorities. If these authorities may be unable to help them, nevertheless they are not persecuting them either.[24] Furthermore, in such a context, it is not possible to assign a single State the responsibility for a certain extreme weather event.

The last obstacle that makes it very difficult to argue that people displaced by the impacts of climate change are refugees within the meaning of the 1951 Refugee Convention is the fact that the Convention definition only applies to people who have already crossed an international border. Unlike persecution, environmental and climate factors mostly generate internal migration (that is to say, migration within national borders), and thus do not meet this requirement. This cross-border displacement is indeed an indispensable requirement of being a "refugee" and triggers the need for international protection.

Paradoxically, while climate displaced persons who cross international borders face a particularly concerning "legal and operational limbo",[25] international law offers some protection to those who prefer to relocate to safe places within, rather than leaving, their own country.

The limits of applicability of internal displacement regime

As long as the victims of climate change do not cross a border, they benefit from the Guiding Principles on Internal Displacement. This set of principles, developed in 1998 by the former UN Secretary-General Representative on IDPs, Francis M. Deng, and afterwards adopted by the UN, aims at guiding States on the protection of IDPs against, during and after displacement.

The Guiding Principles do not seek to create new law, but rather to codify and gather existing norms from different branches of international law that had already been agreed upon by States – namely, international humanitarian rights law, human rights law and international criminal law.[26] They apply to any person or group of persons

> who have been forced or obliged to flee or to leave their homes or places of habitual residence, in particular as a result of or in order to avoid the effects of ... natural or human-made disasters, and who have not crossed an internationally recognized State border.[27]

The inclusion of natural disaster as a cause of internal displacement indicates that the Guiding Principles play a role in climate migration, provided that degradation caused by climate change amounts to a disaster.[28] In this case, IDPs have the right to receive protection and humanitarian assistance from their State's authorities (Principle 3), and a State shall not arbitrarily refuse international humanitarian assistance. Once persons have been displaced, they retain a broad range of economic, social, cultural, civil and political rights, including the right to basic humanitarian assistance (such as food, medicine, shelter) (Principle 18), the right to be protected from physical violence, the right to education (Principle 23), freedom of movement and residence (Principle 14), political rights such as the right to participate in public affairs and the right to participate in economic activities (Principle 22). Displaced persons also have the right to assistance from competent authorities in voluntary, dignified, and safe return, resettlement, or local integration, including help in recovering lost property and possessions

(Principle 28). When restitution is not possible, the Guiding Principles call for compensation or just reparation (Principle 29).[29]

Questions arise as to what extent the Guiding Principles, as a non-legally binding instrument, are enforced by States in practice. Several States have incorporated the Guiding Principles or adapted versions into their national legislation and policy.[30]

The UNHCR has so far interpreted its mandate on IDPs protection as limited to those displaced by conflict.[31] This is much narrower than the notion provided by the Guiding Principles on Internal Displacement. Furthermore, due to its limited resources, the UNHCR has constantly maintained that it does not have a general competence for IDPs and its intervention is far from automatic.[32] Thus, this result excludes most of the climate IDPs from the UNHCR's mandate.

Addressing the international legal gaps in climate induced migration: current initiatives

The above analysis has shown that, in the absence of a specific regime applicable to climate migrants, at least in certain cases, existing international law offers little protection only to people displaced by climate change inside their own country. Thus, as Walter Kälin, the former UN Special Representative for the human rights of IDPs and current Special Envoy of the Chairmanship of the Nansen Initiative, rightly pointed out, "[t]he main challenge is to clarify or even develop the normative framework applicable to persons *crossing internationally recognized state borders*" (emphasis added).[33]

This section will therefore provide an overview of current attempts by the international community to address the normative gaps that exist with regard to climate migrants who cross international borders.

The UN approach to the issue of climate migration

Bearing in mind that no authoritative international institution is responsible for assisting climate migrants, it is noteworthy that a number of UN institutions have acknowledged the impact of climate change consequences on mass movement of people in a set of relatively recent instruments. However, thus far, little attention has been paid to the specific issue of climate migration.

In 2004, the Commission on Human Rights invited the Sub-Commission on the Promotion and Protection of Human Rights to prepare a report on the legal implications of the disappearance of States for environmental reasons.[34] The Sub-Commission appointed Françoise Hampson as the Special Rapporteur on the issue to draft a comprehensive study on the same topic and to send a questionnaire to affected States.[35] Her appointment, however, required the endorsement of the Commission on Human Rights and this did not occur before it was disbanded in 2006. When it was superseded the same year by the Human Rights Council, this working topic was not continued.[36]

The Human Rights Council adopted several resolutions from 2008 to late 2014,[37] in which it recognized the effects of climate change on the population and its human rights, remarked upon the need for effective international cooperation, and supported rather long-term political solutions at the international and national policy-making levels. However, thus far, the Human Rights Council has never explicitly mentioned climate change induced displacement in a resolution.

Meanwhile, the Security Council first considered the implications of climate change on international peace and security regarding forced or voluntary population movements during a debate in April 2007.[38] Later, in 2011, it again debated about the impacts of climate change on international peace and security,[39] but has not passed any resolution yet, possibly in light of the fact that there is considerable uncertainty about the extent to which climate change can be said to cause conflict (especially cross-border conflict).[40]

As far as the General Assembly is concerned, the most significant initiative about climate migration dates back to 2009, with the submission of a comprehensive report on "Climate Change and its Possible Security Implications" to the UN Secretary-General.[41] On that occasion, the General Assembly underlined the inapplicability of the 1951 Refugees Convention and the Guiding Principles on Internal Displacement and the need to develop a new and climate-focused legal framework to protect persons displaced by climate change.[42]

Among the documents adopted at UN climate summits, the humanitarian consequences of climate change-related population movements are explicitly mentioned only at Paragraph 14(f) of the Cancún Adaptation Framework, adopted in December 2010,[43] which was the result of three years of negotiations on adaptation by States parties to the United Nations Framework Convention on Climate Change (New York, 9 May 1992) (hereinafter: UNFCCC).[44] The provision invited all States parties to enhance action on adaptation by undertaking, inter alia, "[m]easures to enhance understanding, coordination and cooperation with regard to climate change induced displacement, migration and planned relocation, where appropriate at the national, regional and international levels."

From a legal point of view this provision is very weak, as it is couched within a non-binding "decision" of the States parties to the UNFCCC and imposes no formal obligations on them, instead simply "inviting" them to undertake measures that assist "understanding, coordination and cooperation" on climate change-related mobility.[45] However, it provided an important reference point and the impetus for UNHCR's leading role in securing States' agreement to develop a global guiding framework on protection in the context of climate change and displacement.

Within the UN, the UNHCR is the institution which has played the most significant role in the debate on climate change and displacement – not least because of the early (mis)framing of the issue as being about "climate refugees",[46] – even though its legal mandate does not cover *per se* the issue.[47]

Besides drafting a number of important submissions outlining the relationship between climate change and displacement,[48] in 2011 the UNHCR took the initiative to organize a closed Expert Meeting on Climate Change and Displacement to discuss options for addressing climate-related displacement, internal as well as across borders. Although the expert group could not reach consensus on all issues, there was sufficient agreement on the "need to develop a global guiding framework or instrument to apply to situations of external displacement other than those covered by the 1951 Convention, especially displacement resulting from sudden-onset disasters".[49] On that occasion, given the magnitude of the issues involved, also highlighted was UNHCR's expertise on the protection dimensions of displacement, which makes it a particularly valuable actor.[50]

Although UNHCR was not formally a co-sponsor of the event, it was a key player in the Nansen Conference on Climate Change and Displacement in the 21st Century, convened by the government of Norway in June 2011 to explore responses to the challenges posed by climate change and forced migration. The conference concluded with the adoption of the Nansen Principles, a series of ten recommendations whose main goal is "to guide responses to some of the urgent and complex challenges raised by displacement in the context of climate change and other environmental hazards".[51]

The relatively successful action of UNHCR was abruptly interrupted when States met at the Standing Committee of UNHCR's Executive Committee (hereinafter: ExCom) in June 2011. There, UNHCR put forward a proposal by the Inter-Agency Standing Committee (hereinafter: IASC) – the primary mechanism for the coordination of humanitarian assistance between relevant international UN and non-UN agencies – for UNHCR to become the lead agency for coordinating protection responses in situations of natural disaster.[52]

The proposal aimed at formalizing a role that UNHCR had already assumed in seven natural disasters between 2004 and 2011,[53] and at that point was to be only a one-year pilot scheme. Thus, in operational terms at least, it did not signify a radical shift in practice. However, it was controversial, not least because it dealt with IDPs – an already politically sensitive issue – and generated concerns among States about UNHCR exceeding its mandate and operational capacity. Many delegates had reservations about the desirability of permitting UNHCR to assume additional responsibilities and activities, thus becoming a "forced migration" agency, rather than a "refugee" agency. Consequently, "[t]here was a clear call for postponement of any designation of responsibility as lead agency for protection in situations of natural disaster until outstanding questions were answered".[54]

The issue of displacement on account of climate and environmental disasters was raised again during the UNHCR Ministerial Meeting in December 2011. Even though in his opening address the High Commissioner highlighted the lack of a coherent international framework for protecting the rights of persons who are displaced across borders owing to forces other than persecution, serious human rights violations and ongoing conflict,[55] during the discussion only five States (Norway, Switzerland, Costa Rica, Germany and Mexico) pledged "to cooperate with interested States, UNHCR and other relevant actors to obtain a better understanding of such cross-borders at relevant regional and sub-regional levels, identifying best practices and developing consensus on how best to assist and protect the affected people".[56]

At the end of the meeting, a Ministerial Communiqué was issued. However, it did not even mention the topic of natural disasters or climate change. As pointed out by Walter Kälin in his capacity as Special Envoy of the Chairmanship of the Initiative, this was no accident, but rather the expression of a lack of willingness by a majority of governments, whether for reasons of sovereignty, competing priorities or the lead role of UNHCR in the process.[57]

So far, the UN approach to climate migration has proven to be unsuccessful. However, a step forward could potentially be the provisional adoption of the Draft Articles on the Protection of Persons in the Event of Disasters, approved on first reading by the International Law Commission (hereinafter: ILC) in 2014.[58]

Art. 3 of the Draft Articles on the Protection of Persons in the Event of Disasters defines a "disaster" as "a calamitous event resulting in . . . environmental damage". Thus, the provision explicitly covers natural disasters. However, the Draft Articles do not aim, specifically, at addressing the impact of natural disasters on migration, but rather at protecting people affected by a disaster, whether forced to move or not. Thus, they cannot be considered as a ready-made tool for addressing environmental and climate change displacement. However, as Jane McAdam correctly points out "[e]ven though the draft articles do not directly bear on the displacement of people as a result of natural disasters, their normative underpinning is very important and would be equally applicable in the displacement context."[59]

Future responses to climate-induced displacement: the Nansen Initiative on Disaster-Induced Cross-Border Displacement

When it comes to legal protection in the context of climate change and migration, the most significant initiative is thus far represented by the Nansen Initiative on Disaster-Induced Cross-Border Displacement, which is not a UN initiative as such, but "a state-led, bottom-up consultative process intended to identify effective practices and build consensus on key principles and elements to address the protection and assistance needs of persons displaced across borders in the context of disasters, including the adverse effects of climate change."[60]

The origins of the Nansen Initiative may be traced back to the aftermath of the UNHCR Ministerial Conference in mid-2011, when the international community of States refused to extend the UNHCR mandate to address climate migration at the UN level. Thus, to overcome this impasse, the government of Norway teamed up with the government of Switzerland to launch the Nansen Initiative in October 2012.

In order to gather more information about the varying challenges faced in different parts of the world as well as good practices adopted, the Nansen Initiative has held a series of five Regional Consultations in the Pacific, the Americas, Africa, Asia and Europe between 2013 to 2015,[61] and has been formally supported by a steering group made up of Australia, Bangladesh, Costa Rica, Germany,

Kenya, Norway, Mexico, the Philippines and Switzerland. It has been also supported by a Consultative Committee made up of academics and practitioners, a small Secretariat, and has been headed by a Special Envoy of the Chairmanship of the Initiative.[62]

On 12–13 October 2015, States met in Geneva to adopt the final draft of the "Agenda for the Protection of Cross-Border Displaced Persons in the Context of Disasters and Climate Change",[63] a non-binding instrument addressing the needs of people displaced across international borders as a consequence of natural disasters, including the effects of climate change.

The Agenda identifies three priority areas for action: collecting data and enhancing knowledge; enhancing the use of humanitarian protection measures for cross-border disaster-displaced persons; and strengthening the management of disaster displacement risk in the country of origin. The latter may entail, as relevant, integrating human mobility within disaster risk reduction and climate change adaptation strategies; facilitating migration with dignity as a potentially positive way to cope with natural hazards and adverse effects of climate change; improving the use of planned relocation; and ensuring that the needs of internally displaced persons in disaster situations are specifically addressed.

The Protection Agenda does not suggest creating new international law (either in the form of a treaty or as a soft law instrument) but rather includes a set of common understandings of the issue, its dimensions and the challenges faced by relevant stakeholders. It identifies and reiterates key principles in the areas of protection and international and regional cooperation, and provides examples of existing practices and tools to prevent, prepare for and respond to internal and, in particular, cross-border displacement in disaster contexts. Finally, it includes recommendations on the way ahead for follow-up when the Nansen Initiative ends in December 2015.

Since the "Protection Agenda" has been recently adopted, it would be premature to assess its impact on the improvement of legal protection for people forced by climate change to flee their countries. However, some preliminary evaluations can already be made.

From a procedural perspective, even though the Nansen Initiative's approach was at first considered too tentative, at the moment this seems to be the only feasible strategy. As correctly pointed out by Walter Kälin, an absence of sufficient knowledge about "what happens on the ground" would make it difficult to draft a set of guiding principles adequately reflecting realities, and there is of course the practical obstacle as well, namely "the difficulty of getting governments on board with the idea of producing a normative framework right away".[64]

From a substantial point of view, the Draft Agenda is of particular interest as it specifically deals with cross-border displacement caused in the context of disasters and climate change. On the one side, the advantage of such a narrow focus is that it keeps the problem specification clear; on the other side, however, the disadvantage is that the number of people who cross borders may be relatively low.[65] Furthermore, a precise definition of those whose protection rights are to be improved is far more difficult than it might at first appear. Migration, after all, is rarely attributable solely to climatic factors. Social, economic or political factors almost always play an important part in migration decisions. Similarly, it is difficult to make a clear distinction between voluntary migration and forced migration.

Final remarks

In light of the lack of a specific international legal instrument for recognizing, protecting, and resettling cross-border climate migrants, the analysis above has shown that they do not even fit neatly into the category of either "refugee" or "IDP". As a consequence, the existing legal lacuna with regard to people displaced across international borders by climate change is currently the subject of much debate.

Thus far, the UN approach has proven to be more theoretical than practical. Moreover, it has shown that States want to retain control over developments about climate migration and they are reluctant to assume formal obligations or to "delegate" responsibility to international organizations. A step forward could be potentially represented by the ILC Draft Articles on the Protection of Persons in the Event of Disasters, which, however, do not directly bear on the displacement of people as a result of climate disasters, but whose normative underpinning is very important and would be equally applicable in the displacement context.

To date, the most promising initiative to address climate cross-border migration is represented by the Nansen Initiative, which is a State-owned consultative process, outside the UN, aiming at addressing the challenges of cross-border displacement in the context of disasters and the effects of climate change.

In the aftermath of the adoption of the Agenda for the Protection of Cross-Border Displaced Persons in the Context of Disasters and Climate Change, it is difficult to evaluate its impact on the improvement of legal protection for people forced by natural and climate disasters to flee their countries. However, the hope is that the international community will build on the groundwork of the Nansen Initiative, taking this instrument as a first step towards developing the "more coherent and consistent approach at the international level . . . to meet the protection needs of people displaced"[66] across borders in the context of natural disasters.

Notes

1 In 1990, the first evaluation report of the Intergovernmental Panel on Climate Change (IPCC) stated that "[m]igration and resettlement may be the most threatening short-term effects of climate change on human settlements": see IPCC, *First Assessment Report*, Intergovernmental Panel on Climate Change and World Meteorological Organization, 1990, p. 103.
2 On the link between climate change and forced migration see, among others, Piguet, Pécoud and de Guchteneire (eds), *Migration and Climate Change*, Cambridge, 2011; Stojanov and Novosà (eds), *Migration, Development and Environment: Migration Processes from the Perspective of Environmental Change and Development Approach at the*

Beginning of the 21st Century, Newcastle upon Tyne, 2008; Perch-Nielsen, Bättig and Imboden, "Exploring the Link between Climate Change and Migration", in *Climatic Change*, 2008, p. 375 ff.

3 According to scientific estimations, the number of those likely to relocate due to climatic reasons ranges between 50 and 350 million by 2050: see UNHCR, *Forced Migration in the Context of Climate Change: Challenges for States under International Law*, 2009, p. 3. However, debates about numbers remain highly contentious: see McAdam, *Climate Change, Forced Migration, and International Law*, Oxford, 2012, pp. 24–30.

4 Several authors have strongly advocated the need for a new treaty or a new protection regime: see, among others, Docherty and Giannini, "Confronting a Rising Tide: A Proposal for a Convention on Climate Change Refugees", in *Harvard Environmental Law Review*, 2009, p. 349 ff.; Williams, "Turning the Tide: Recognizing Climate Change Refugees in International Law", *Law and Policy*, 2008, p. 502 ff.; Biermann and Boas, *Preparing for a Warmer World: Towards a Global Governance System to Protect Climate Refugees*, Global Governance Working Paper, 2007. One of the most complete proposals to date is represented by the Draft Convention on the International Status of Environmentally-Displaced Persons, drawn up in 2010 by specialists from the University of Limoges (text available at www.cidce.org/pdf/Projet%20de%20 convention%20relative%20au%20statut%20international%20des%20d%C3%A9 plac%C3%A9s%20environnementaux%20%28deuxi%C3%A8me%20version%29. pdf). For the opposite opinion see McAdam, "Swimming against the Tide: Why a Climate Change Displacement Treaty is Not the Answer", *International Journal of Refugee Law*, 2011, p. 2 ff.

5 There is no uniform terminology used to describe people who move in response to the impacts of climate change, because, in most cases, climate change is only one of a number of reasons why people decide to move. Moreover, it is difficult to establish a causal link between climate phenomena and migration. For the analysis of the various terms used by scholars to describe climate-induced migration, see Atapattu, *Human Rights Approaches to Climate Change: Challenges and Opportunities*, London, 2015, pp. 162–164; Vlassopoulos, "Defining Environmental Migration in the Climate Change Era: Problem, Consequence or Solution?", in Faist and Schade (eds), *Disentangling Migration and Climate Change. Methodologies, Political Discourses and Human Rights*, The Hague, 2013, pp. 145–163. As a consequence, it has been highlighted that in this field of study there is currently no consensus on definitions and "the resulting variety of terms is not just confusing but unhelpful": see Dun and Gemenne, "Defining 'Environmental Migration'", *Forced Migration Review*, 2008, p. 10.

6 The 1951 Refugee Convention entered into force on 22 April 1954.

7 UN Commission on Human Rights, Guiding Principles on Internal Displacement, Doc. E/CN.4/1998/53/Add.2, 11 February 1998, available at http://daccess-dds-ny. un.org/doc/UNDOC/GEN/G98/104/93/PDF/G9810493.pdf?OpenElement.

8 Zetter and Morrisey, "Environmental Stress, Displacement and the Challenge of Rights Protection", in Martin, Weerasinghe and Taylor (eds), *Humanitarian Crises and Migration: Causes, Consequences and Responses*, New York, 2014, p. 179. Similarly, Atapattu, *Human Rights Approaches to Climate Change: Challenges and Opportunities*, p. 160.

9 These instances have been considered to represent the broader trend described as "pluralization of human rights", described as "the phenomenon whereby human rights, as law and ideology, has increasingly recognized the needs of specific groups or categories within humanity as worthy of a specific human rights protection": see Mégret, "The Disabilities Convention: Human Rights of Persons with Disabilities or Disability Rights?", *Human Rights Quarterly*, 2008, p. 495. Similarly, Mayer and Cournil, "Climate Change, Migration and Human Rights: Toward Group Specific Protection?", in Quirico and Boumghar (eds), *Climate Change and Human Rights: An International and Comparative Law Perspective*, London, 2015, p. 174.

10 Zetter, "Legal and Normative Frameworks", *Forced Migration Review*, 2008, p. 62.
11 See, in particular, the notable exception of Gemenne, "One Good Reason to Speak of 'Climate Refugees'", *Forced Migration Review*, 2015, pp. 70–71: the author, though admitting that the concept of "climate refugee" has been progressively abandoned as it has no legal basis, argues that such a notion recognizes that climate migrations are first and foremost the result of a persecution inflicted on the most vulnerable.
12 The notion of "environmental refugee" first appeared in 1948, but the extended concept appeared in the 1980s through the work of El Hinnawi, *Environmental Refugees* (United Nations Environmental Programme), Nairobi, 1985. Similarly, Black, *Environmental Refugees: Myth or Reality?*, UNHCR New Issues in Refugee Research, Working Paper 34, 2001; Myers, "Environmental Refugees in a Globally Warm World", *Bioscience*, 1993, p. 752 ff.
13 See, e.g., Westra, *Environmental Justice and the Rights of Ecological Refugees*, London, 2009.
14 In 2008, António Guterres, the UN High Commissioner for Refugees, stated that he had "serious reservations with respect to the terminology and notion of environmental refugees or climate refugees" and indicated that "these terms have no basis in international refugee law" as the majority of those had not crossed an international border. See UNHCR, "Climate Change, Natural Disasters and Human Displacement: A UNHCR Perspective", 23 October 2008, p. 8, available at www.unhcr.org/4901 e81a4.html. It is noteworthy that the expressions "environmental refugees" or "climate refugees" are not even mentioned in the main handbooks on refugee law as they are not considered legal concepts: see, among others, Zimmermann and Mahler, "Article 1 A, para. 2 1951 Convention", in Zimmermann (ed.), *The 1951 Convention Relating to the Status of Refugees and Its 1967 Protocol: A Commentary*, Oxford, 2011, p. 439; Hathaway, *The Rights of Refugees under International Law*, Cambridge, 2005, p. 1285.
15 UNHCR, *Handbook on Procedures and Criteria for Determining Refugee Status Under the 1951 Convention and the 1967 Protocol*, Doc. HCR/IP/4/Eng/REV.3, December 2011, para. 51: "There is no universally accepted definition of 'persecution' and various attempts to formulate such a definition have met with little success". The text of the Handbook is available at www.unhcr.org/3d58e13b4.html.
16 Ibid.
17 The "human rights approach" is endorsed by the UNHCR: see, respectively, Guidelines on International Protection no. 6: Religion-Based Refugee Claims under Article 1A(2) of the 1951 Convention and/or its 1967 Protocol relating to the Status of Refugees, Doc. HCR/GIP/04/06, 28 April 2004, paras. 2, 11, 15–16, available at www.unhcr.org/ 40d8427a4.html; Guidelines on International Protection no. 1: Gender-Related Persecution Within the Context of Article 1A(2) of the 1951 Convention and/or its 1967 Protocol relating to the Status of Refugees, Doc. HCR/GIP/02/01, 7 May 2002, paras. 5 and 9, available at www.unhcr.org/3d58ddef4.html. Such an approach to the notion of persecution is also overwhelmingly approved by scholars: see, among others, Hathaway and Foster (eds), *The Law of Refugee Status*, Oxford, 2014, pp. 193–208; Storey, "Persecution: Towards a Working Definition", in Chetail and Bauloz (eds), *Research Handbook on Migration and International Law*, Cheltenham, 2014, pp. 469–478; Goodwin-Gill and McAdam, *The Refugee in International Law*, Oxford, 2007, pp. 131–133.
18 See, e.g., Kozoll, "Poisoning the Well: Persecution, the Environment, and Refugee Status", *Colorado Journal of International Environmental Law and Policy*, 2004, p. 271; Cooper, "Environmental Refugees: Meeting the Requirements of the Refugee Definition", *New York University Environmental Law Journal*, 1998, p. 480.
19 McAdam, *Climate Change, Forced Migration, and International Law*, pp. 43–45. See also Goodwin-Gill and McAdam, *The Refugee in International Law*, p. 92.
20 Chassin, "Dealing with International Vulnerability: European Law and Climate-Induced Migrants", in Ippolito and Iglesias Sánchez (eds), *Protecting Vulnerable Groups: The European Human Rights Framework*, Oxford, 2015, p. 274.

21 McAdam, *Climate Change, Forced Migration, and International Law*, p. 46.
22 Goodwin-Gill and McAdam, *The Refugee in International Law*, pp. 79–80. The UNHCR has, however, noted that situations might arise where victims of a natural disaster flee their country because their Government has consciously withheld assistance to punish or marginalize them on one of the five persecution grounds set out in the refugee definition: see UNHCR, *Climate Change, Natural Disasters and Human Displacement: A UNHCR Perspective*, 14 August 2009, p. 7, available at www.unhcr.org/refworld/docid/4a8e4f8b2.html.
23 *Ioane Teitiota v. the Chief Executive of the Ministry of Business, Innovation and Employment*, CIV-2013-404-3528 [2013] NZHC 3125, 26 November 2013, para. 11. For further details on this case, see, among others, Scott, "Refuge from Climate Change-Related Harm: Evaluating the Scope of International Protection within the Common European Asylum System", in Bauloz, Ineli-Ciger, Singer and Stoyanova (eds), *Seeking Asylum in the European Union: Selected Protection Issues Raised by the Second Phase of the Common European Asylum System*, Leiden, 2015, pp. 199–296; McAdam, "The Emerging New Zealand Jurisprudence on Climate Change, Disasters and Displacement", *Migration Studies*, 2015, pp. 131–142; Ni, "A Nation Going Under: Legal Protection for 'Climate Change Refugees'", *Boston College International and Comparative Law Review*, 2015, pp. 336–351.
24 Chassin, *Dealing with International Vulnerability*, p. 275.
25 Kälin, *Displacement Caused by the Effects of Climate Change: Who Will Be Affected and What Are the Gaps in the Normative Framework for Their Protection?*, Background Paper for the IASC Informal Working Group on Climate Change and Migration/displacement held on 15 September 2008, available at www.brookings.edu/research/papers/2008/10/16-climate-change-kalin.
26 Introductory Note by the Representative of the Secretary-General on Internally Displaced Persons, Mr. Francis M. Deng, available at http://daccess-dds-ny.un.org/doc/UNDOC/GEN/G98/104/93/PDF/G9810493.pdf?OpenElement.
27 Guiding Principles, Introduction – Scope and Purpose, para. 2.
28 As pointed out by Atapattu, *Human Rights Approaches to Climate Change: Challenges and Opportunities*, p. 166, it is questionable whether people leaving their homes because of incidents caused by creeping degradation or pollution are covered by the notion of IDPs.
29 For a general analysis on how the protection of IDPs could inform the current debates on the protection of climate migrants, see Gemenne and Brücker, "From the Guiding Principles on Internal Displacement to the Nansen Initiative: What the Governance of Environmental Migration Can Learn from the Governance of Internal Displacement", *International Journal of Refugee Law*, 2015, p. 245ff.
30 Such as Angola, Colombia, Peru and the U.S. other governments (for example, Burundi, the Philippines, Sri Lanka and Uganda) have developed national policies based on the principles: see Kälin and Williams, "Introduction", in Kälin, Williams, Koser and Solomon (eds), *Incorporating the Guiding Principles on Internal Displacement into Domestic Law: Issues and Challenges*, Washington, 2010, p. 3.
31 See, e.g., UNHCR, "Information Note: UNHCR's Role with Internally Displaced Persons", 20 November 1998, available at www.refworld.org/docid/3ae6b31b87.html.
32 Docherty and Giannini, "Confronting a Rising Tide", p. 359.
33 Kälin, "Climate Change Induced Displacement: A Challenge for International Law", Distinguished Lecture Series – 3, 2011, p. 23, available at www.mcrg.ac.in/DL3.pdf.
34 Decision 2004/122, UNCHR, 60th Session, Doc E/CN.4/2004/L.49, 4 December 2004, available at http://ap.ohchr.org/documents/E/CHR/decisions/E-CN_4-DEC-20041_22.doc.
35 See UN Commission on Human Rights, Doc. E/CN.4/Sub.2/AC.4/2006/CRP.2, 30 June 2006, available at www.ohchr.org/Documents/Issues/IPeoples/WG/E-CN4-Sub2-AC4-2006-CRP2.doc.

36 See UN Commission on Human Rights, Resolution 2006/16, 24 August 2006, available at www2.ohchr.org/english/bodies/subcom/docs/58/A.HRC.Sub.1.58.L.11. pdf.

37 UN Human Rights Council, Resolution 25/21, 28 March 2014, available at http:// daccess-dds-ny.un.org/doc/UNDOC/GEN/G14/136/17/PDF/G1413617. pdf?OpenElement; Resolution 19/10, 22 March 2012, available at http://daccess-ods. un.org/TMP/763163.566589355.html; Resolution 18/22, 30 September 2011, available at www.ohchr.org/Documents/Issues/ClimateChange/A.HRC.RES.18.22. pdf; Resolution 10/4, 25 March 2009, available at http://ap.ohchr.org/documents/E/ HRC/resolutions/A_HRC_RES_10_4.pdf; Resolution 7/23, 28 March 2008, available at http://ap.ohchr.org/documents/E/HRC/resolutions/A_HRC_RES_7_23.pdf.

38 UN Security Council, Doc. S/PV.5663, 5663rd Meeting, 17 April 2007, available at www.securitycouncilreport.org/atf/cf/%7B65BFCF9B-6D27-4E9C-8CD3 CF6E4FF96FF9%7D/Ener%20SPV%205663.pdf. For an analysis of the debate, see Sindico, "Climate Change: A Security (Council) Issue?", *Climate Change Law Review*, 2007, p. 29ff.

39 UN Security Council, Statement by the President of the Security Council, Doc. S/PRST/2011/15, 20 July 2011, available at www.un.org/en/ga/search/view_doc. asp?symbol=S/PRST/2011/15.

40 See, e.g., Hartmann, "Rethinking Climate Refugees and Climate Conflict: Rhetoric, Reality and the Politics of Policy Discourse", *Journal of International Development*, 2010, p. 233 ff; Barnett and Adger, "Climate Change, Human Security and Violent Conflict", *Political Geography*, 2007, p. 639 ff.

41 UN General Assembly, Report of the Secretary-General: Climate Change and its Possible Security Implications, Doc. A/64/350, 11 September 2009, available at www. un.org/ga/search/view_doc.asp?symbol=A/64/350.

42 Ibid, para. 59.

43 UNFCCC, Report of the Conference of the Parties on its sixteenth session, held in Cancun from 29 November to 10 December 2010, The Cancun Agreements: Outcome of the Work of the Ad Hoc Working Group on Long-term Cooperative Action under the Convention, Doc FCCC/CP/2010/7/Add.1, 15 March 2011, available at http:// unfccc.int/resource/docs/2010/cop16/eng/07a01.pdf#page=4.

44 The UNFCCC entered into force 21 March 1994.

45 McAdam, "Creating New Norms on Climate Change, Natural Disasters and Dis-placement: International Developments 2010–2013", *Refuge*, 2013, p. 23: according to the author, this was in line with context, which was not a suitable forum in which to examine the complexity of these issues in a structured or comprehensive way.

46 Ibid., p. 13. The UNHCR engagement with the issue has been strongly encouraged by António Guterres, the UN High Commissioner for Refugees, who has described climate change as "the defining challenge of our time" and has called on States to address contemporary circumstances in which there are "more and more people forced to move because of extreme deprivation, environmental degradation and climate change": see Guterres, "Maintenance of International Peace and Security: New Challenges to International Peace and Security and Conflict Prevention", Statement by UN High Commissioner for Refugees, United Nations Security Council Briefing, 23 November 2011, available at www.unhcr.org/4ee21edc9.html.

47 This remains a key obstacle to UNHCR's formal involvement.

48 See, among others, UNHCR's "Background Paper to the High Commissioner's Dialogue on Protection Challenges, Protection Gaps and Responses", 30 November 2010, available at www.unhcr.org/4cebeeee9.html: for the first time, after identifying a lack of international guidance on meeting the protection needs of people forcibly displaced as a result of climate change, natural disasters and other circumstances that may fall outside the scope of the 1951 Refugee Convention, the paper directed attention towards possible normative responses.

49 UNHCR, "Summary of Deliberations on Climate Change and Displacement", 22–25 February 2011, p. 1, available at www.unhcr.org/4da2b5e19.pdf.
50 Ibid., p. 2.
51 Nansen Conference on Climate Change and Displacement in the 21st Century, Chairperson's Summary, 6–7 June 2011, available at www.regjeringen.no/globalassets/upload/ud/vedlegg/hum/nansen_summary.pdf.
52 UNHCR ExCom, Report of the 51st Meeting of the Standing Committee (21–23 June 2011), Doc. EC/62/SC/CRP.25 16 September 2011, paras. 30–36, available at www.unhcr.org/4e79b1f59.pdf.
53 UNHCR ExCom, "UNHCR's Role in Support of an Enhanced Humanitarian Response", Doc. EC/58/SC/CRP.18, 4 June 2007, para. 9, available at www.unhcr.org/46641fff2.html.
54 UNHCR ExCom, Report of the 51st Meeting of the Standing Committee, para. 32.
55 Guterres, Address, Statement by UN High Commissioner for Refugees, delivered at the Intergovernmental Meeting at Ministerial Level to mark the 60th Anniversary of the1951 Convention relating to the Status of Refugees and the 50th Anniversary of the 1961 Convention on the Reduction of Statelessness, Geneva, 7 December 2011, in UNHCR, Pledges 2011: Ministerial Intergovernmental Event on Refugees and Stateless Persons, Geneva, 7–8 December 2011, para. 141, available at www.unhcr.org/commemorations/Pledges2011-preview-compilation-analysis.pdf.
56 Ibid., paras. 38 (Costa Rica), 76 (Germany), 96 (Mexico), 102 (Norway), 119 (Switzerland).
57 Ministerial Communiqué, Intergovernmental Event at the Ministerial Level of Member States of the United Nations on the Occasion of the 60th Anniversary of the 1951 Convention relating to the Status of Refugees and the 50th Anniversary of the 1961 Convention on the Reduction of Statelessness, Doc. HCR/MINCOMMS/2011/6, 7–8 December 2011, available at www.unhcr.org/4ee210d89.pdf.
58 International Law Commission, Protection of Persons in the Event of Disasters: Text and Titles of the Draft Articles on the Protection of Persons in the Event of Disasters, Doc. A/CN.4/L.831, 15 May 2014, available at www.ifrc.org/PageFiles/126924/A-CN.4-L.831.pdf.
59 McAdam, *Climate Change Displacement and International Law: Complementary Protection Standards*, UNHCR Legal and Protection Policy Research Series, PPLA/2011/03, 2011, p. 49.
60 The Nansen Initiative, "The Response", available at www.nanseninitiative.org/.
61 The Nansen Initiative, "Concept Note", available at www.nanseninitiative.org/global-consultations/.
62 Kälin, "From the Nansen Principles to the Nansen Initiative", *Forced Migration Review*, 2012, p. 49.
63 Text available at https://www.nanseninitiative.org/global-consultations/.
64 McAdam, *Creating New Norms on Climate Change, Natural Disasters and Displacement*, p. 18.
65 On this aspect, see Betts, "Governance Questions for the International Community", *Forced Migration Review*, 2015, p. 72.
66 Nansen Principle no. 9.

16 Democratic capitalism for realizing the Earth Charter vision

Peter Venton

Introduction to democratic capitalism

The Earth Charter is an international declaration of fundamental values and principles for building a just, sustainable, and peaceful global society in the twenty-first century. "It is unique in the articulation of an holistic ethical vision" (Taylor 2014: 12). It is argued that the realization of the Earth Charter vision can best be achieved through "democratic capitalism" which is an ideology comprised of three interrelated systems: a political regime of democracy, a capitalist economic system, and a moral-cultural system (Novak 1982: 14).

The democratic regime

In Aristotle's writing, democratic regimes are governed by all citizens who share in deliberations about the common good: "The Aristotelian conception of the polis (city state) is that of a self-sufficient cooperative arrangement that is centered on a particular understanding of the good" (Neill 2011: 40), and "the good life ... can only be actualized in public deliberations of the citizenry" (ibid.: 9). The common good is a collection of society's ends that are numerous, diverse and shared. They include elements such as peace, physical security, health, ecological integrity, environmental sustainability, work-life balance, and democratic engagement. They also include the absence of "bads" such as crime, family breakdown and pollution of air, water and land. Most of these elements could be viewed as common human needs. In addition the common good includes two elements of "social cohesion": economic security as indicated by full employment and shared prosperity as indicated by limits on inequality in the distribution of wealth and income that reflect the principles of social justice.

"Freedom and equality especially characterize democracy" (Cooper 2011: 197). There are two elements of freedom: positive and negative. Positive freedom is "for all to rule and for all to be ruled in turn" (ibid.: 200). In other words positive freedom is the freedom for all to band together for collective benefits and to be bound by the legislation of the officials (i.e. rulers). Positive freedom implies equality for all in political participation in the legislative process and the selection of officials. To this end, one of Aristotle's features of a democratic form of government is the "election of officials from all, to rule and be ruled (by the) use of the lot" (ibid.: 200). For Aristotle "all" refers to a majority of citizens "by which Aristotle means the free, poor, and underprivileged classes [who] are dominant over the political process" (Neill 2011: 25). "A second element of freedom is to live as one likes since, not to live as one likes, is the life of a slave" (Cooper 2011: 200). It's negative in the sense that it's freedom from responsibility for all fellow citizens.

A feature of democracy is the equality of justice which requires that the democracy has a constitution and is governed by the rule of law (ibid.: 197). By rule of law Aristotle meant a manner of governance in which the political leaders are constrained beforehand by principles (in the constitution) which are known to the citizen body. A political regime that defers to the rule of law is in his opinion superior to a regime in which the present generation alone is dominant (Neill 2011: 26). The objective of equality of justice requires freedom of speech which, in turn, entails not only equal opportunity to speak but speech that is frank and open and without fear of recrimination. It is speech that is egalitarian, rejects hierarchy and reveals and uncovers the truth as one sees it (Cooper 2011: 204). The rule of law was important because it protected the interests of the people against the demagogues who influenced public opinion in ways that were not in the interests of the people (ibid.: 198).

Capitalist economic systems

In the first chapter of his book *An Introduction to Actually Existing Capitalism*, political economist Geoffrey Mann defines capitalism as an economic system for organizing society that is "sustained by a set of institutions, techniques and ideas about human affairs and social goals" (Mann 2013: 5). More specifically it is a mode of production of goods and services and their distribution. In Mann's view, Geoffrey Ingham, in his 2008 book *Capitalism*, "has most effectively conceptualized the essentials of capitalism as private enterprise for producing commodities, market exchange, a monetary system based on bank credit and a distinctive role for the state in relation to these features" (Mann 2013: 13). With respect to the distinctive role of the state:

> Democratic government is obviously the overarching mechanism for resolving social priorities in the face of market failures ... and the public–private boundary (between private firms and government) reflects voter preferences, the specifics of history, changes in technology and the availability of resources.
> (Coyle 2011: 251, 252)

To sum up the role of the government in a democracy is to coordinate private firms, markets and the monetary system (which includes financial markets) for the achievement of the common good.

Democratic capitalism's moral-cultural system

The moral-cultural system comprises the principles of democracy, the virtues of citizens in democratic participation, and a set of individual virtues. These are listed in the chart below. Some of them (e.g. sentiments of feeling, common sympathy and benevolence) were espoused by Adam Smith, a pioneer in political economy who was famous for his 1776 book, *The Wealth of Nations*. However, most of them are found in Aristotle's writing. Foremost among Aristotle's views was moderation in the pursuit of wealth:

> Aristotle argues that attachment to property (i.e. wealth) actually lowers the mind by directing it to 'bodily gratifications' rather than to the nobler goods associated with the higher activities of the soul. An education in noble virtue was therefore required to free citizens from the lure of private

Table 16.1 Democratic capitalism's moral-cultural system

Principles of democracy	Virtues of citizens in democratic discourse	Individual virtues and vices
Equal treatment. All groups, whether religious, cultural, or ethnic, are treated equally. Equal treatment includes the idea of fraternity which implies treating common people and elites as equals in pursuing common interests and ideals. In reality humans are roughly equal with respect to the amount of time available for work as will be explained later.	Learning philosophy and politics. Participating actively in shared deliberations about the common good. Sacrificing some freedom for the common good of the community. Cooperating in communal tasks for reaching communal goals (Novak 1982: 134, 138). Having moral sentiments of "feeling" (emotion of empathy), common sympathy and benevolence (Novak 1982: 145, 142).	*Virtues* Moderation in the pursuit of wealth. Moderation in consumption (for needs only). Hard work (effort). Self-restraint on impulses or emotions. Discipline (self-control or orderly conduct). *Vices* Unlimited pursuit of wealth (greed). Consumption to satisfy all wants. Easy work (effortless). Lack of restraint on impulses or emotions. Lack of discipline.
Social justice. John Rawls's theory requires that any increases in the inequality of the distribution of wealth and authority is just only if it results in compensating benefits for everyone and, in particular, the least-advantaged members of society (Kloppenberg 2011: 90–92).		
Equal opportunity to participate in the economic, political and social life of society. The purpose of economic systems should be to give citizen-households sufficient income to fund basic needs and leisure sufficient for engaging in shared deliberations about politics and philosophy (Newell 2011: 5, 17).		

wealth, with its ensnarement of the mind in merely necessary and self-regarding goods.

(Sikkenga 2011: 52)

Specifically, education was intended to be about politics and philosophy. Indeed "The primary purpose of household management (i.e. economics) was to give citizen-householders the leisure for pursuing man's two chief ends, politics and philosophy" (Newell: 2011: 17). In this regard, "Ameliorating tensions between the haves and have-nots was a precondition for establishing an environment of minimal stability and civil order, only on the basis of which could citizens find the safety and leisure to cultivate higher forms of virtuous character through shared deliberation" (ibid.: 5).

In ancient Athens philosophy included considerations of virtues (i.e. human morals, character and behaviour) and the principles and laws in the fields of economics, human psychology, sociology and science (Tabachnick and Koivukoski 2011: 3, 24, 47, 90, 110). Knowledge in all these fields was required in the process of democratic discourse for the purpose of defining the common good. The knowledge base represents a huge curriculum today even for 12 to 16 years of formal education of the young. But for Aristotle it was the subject of life-long learning.

Integration of the political, economic and moral-cultural system

The capitalist economic system is subordinate to the political regime in the sense that it was instrumental for achieving the regime purpose which was the common good. Specifically this hierarchy implies that government intervene in the activities of private firms, in markets and in monetary systems to ensure the "social cohesion" elements of the common good, namely full employment and limited inequality in the distribution of wealth and income in accordance with the principle of social justice. It also implies that governments, on behalf of the ruled, have to decide on "trade-offs" between the efficiency of capitalist firms and markets on the one hand and their effectiveness in achieving the common good on the other hand.

On behalf of the ruled the government plays all the following roles related to the moral-cultural system of virtues of individual citizens and their participation in democratic discourse:

- fosters democratic discourse and supports the education of the citizenry in politics and philosophy;
- maintains equality in the legislative process through the machinery of government and equality of justice by upholding the constitution with the rule of law and the maintenance of free speech;
- maintains a balance between positive and negative freedoms consistent with the expressed wishes of the majority of citizens; and
- promotes individual citizen virtues with incentives in the capitalist economic system and political institutions that encourage virtuous behaviour and discourage non-virtuous behaviour.

The rest of this chapter

With reference to Canada and its major trading partner, the United States, this chapter compares how government policies about wages, profit rates for private firms, taxation of income and wealth, control over land and natural resources and trade and investment would be determined under the ideal of democratic capitalism compared with how they are actually determined in modern capitalism. It concludes with a broad outline of the implications for Canada if it were to fully adopt democratic capitalism in terms of economic inequality, economic growth and realizing the Earth Charter vision for addressing the threat of climate change. But first the next section outlines the nature of capitalism.

Modern capitalism

As mentioned earlier the four institutions of capitalism are private firms, market exchange, a monetary system based on bank credit and a distinctive role for government in the coordination of the first three institutions. Each of the institutions has technical features and ideas about individual and social goals that have underlying values and objectives that are often not apparent to most citizens.

Private firms

Private firms are organized for producing goods and services and distributing and selling them in exchange for money with the purpose of making a profit. Firms use human labour in combination with real capital (e.g. tools, machinery, equipment, factories) that is either made by the firm or purchased from other firms to produce the product. It is the firm's use of capital in combination with labour that gives rise to the name, capitalism.

The return to the owners on the cost of their investment in productive capital assets (e.g. tools, equipment, buildings) of the firm will be realized over a period of years. More specifically their annual returns will be the cash flows that are equivalent to the annual amortization expense of the capital. Under generally accepted accounting principles the cost of the capital assets is amortized to estimate the portion of it that is used up in the process of production of goods for that year. This rather technical feature is important because it distinguishes the return of the shareholder's capital investment from the firm's profit.

The conventional profit rate for a private firm is the income from sales of its products less the cost of its operations including the annual amortization expense. Profits do not represent a return for hard work or a return of the shareholder's capital investment. The returns for hard work in the firm are the salaries and benefits for the managers and the wages and benefits for the workers. This is significant when it is recalled that one of the virtues of democratic capitalism is hard work. Profits are not a virtue in capitalism in the sense that they are realized for no effort on the part of the shareholder. Instead profits are a reward for the risk that the shareholders of the firm take when they make their investment in the firm's capital. The risk is that, in circumstances where there is genuine uncertainty about whether the firm will succeed or fail, the investor/owner may lose his or her entire investment. So, without the incentive of profit, no one would make an investment in a firm's capital and no goods would ever be produced. Thus one of the aspects of profits in capitalism is that they are a necessary evil.

The mission of the private firm is to earn ever increasing profits for the share-holders/owners. In this regard size (in terms of sales and production) matters very much for success. As sales and production increase, some costs are fixed with the result that sales increase faster than costs and profits increase even faster. As production increases there are more opportunities to specialize in certain areas of production – specialization that increases productivity that reduces cost per unit of output and increases profits. As sales and production increase the firm is able to crowd out competing firms and dominate the industry or sector in which the firm operates. As a result the firm will have greater market power. This "mono-polistic" power enables it to set higher prices for its product without worrying that competitor firms will respond by reducing their prices to take sales away. Its "monopolistic" power also enables it to determine its rate of profit through a combination of setting prices and restricting its investment in productive capital. In summary the mission of private firms is for unlimited expansion and its ulti-mate success is the domination of its industry. Its success is limited only by the size of the market in its industry.

Market exchange

Markets are places where commodities are exchanged for money prices. Prices in markets send signals to both supplier firms and potential purchasers. So the setting of prices is related to the amount of the supply and the amount of demand from prospective buyers and the purpose of setting prices is to "clear" the market. An implication of this process is that prices are substantially affected by relative scarcity. If many people demand the product but there is limited supply, there is a scarcity and prices will increase. Conversely, if few demand the product, there is an excess and firms will reduce prices to clear the market.

Another significant feature of markets is that large majorities of prospective buyers would be willing to pay more for the product than the market price. This is a feature of typically downward sloping demand curves for products. The sig-nificance of this observation is that, for many prospective buyers, market prices are not related to value that they place on the commodity or service. It also means that prospective buyers who have a relatively low preference for the product might be induced to buy it if the price were lower. These observations are reflected in the common statement that "Some people know the price of everything and the value of nothing". In summary, for many prospective buyers, market prices are not realistic measures of the intrinsic value that they place on the product owing to the reality that there is a large degree of heterogeneity among prospective buyers in terms of wealth, in terms of preferences and in terms of other circumstances among prospective buyers.

The extension of markets to the services of labour in the capitalist economy means that the price of labour is determined by the scarcity of labour relative to the capitalists' need/demand for it. Wage rates consequently will differ for different kinds of labour service. The implication of this fact will be discussed later.

A monetary system based on bank-credit money

Governments control the money supply, interest rates and rates of inflation through the policies of central banks that intervene in money and capital markets. Controls over interest rates and the rate of inflation are a big benefit for private firms because they greatly reduce the risk of losing returns on their investments due to major changes in inflation or the cost of credit between the time when costs of production are incurred and when products are sold.

Privately owned banks collectively create money because they are allowed to lend most of their deposits and keep only a fraction of their deposits (e.g. 10%) in reserve (Lipsey et al. 1988: 722). One bank's loan to an individual ends up as a deposit in another bank which then lends 90 per cent of that deposit to a third person. This process continues for many iterations with the result that bank credit increases to a level many times the amount of money (notes and coins). This is how the system of privately owned banks increases the money supply.

The amount of money created by the banking system depends on the degree of leverage allowed as manifested in the fractional reserve requirement. In times when the process of money creation is reversed, many customers decide to withdraw their deposits to repay their loans at the same time, the banks run out of sufficient reserves to pay for the withdrawals and they become bankrupt. The lower the fractional reserve requirement the greater the risk of failure in the banking system from this reverse process. In the 1930s the banks ran out of reserves and a Great Depression ensued. In the 2008 financial crisis the US central bank (the federal reserve) intervened so that the banks did not run out of reserves and a great depression was averted.

The mission of the private banks is to make ever increasing loans from which they derive profits. This mission may be at variance with the virtues of moderation in consumption and self-restraint in the moral-cultural system of democratic capitalism.

Government coordination of the institutions of capitalist economic systems

Governments perform the coordination role by assessing the strengths and weaknesses of these institutions. On the basis of their assessments they make decisions to intervene where these institutions fail to serve the common good. Governments may intervene by taxation, by regulating private firm activities and operations, by owning private firms, or by public relations to advertise ideas to shape public opinion. There are many variations of capitalist economic systems depending on the nature of how, and degree to which, governments intervene in the activities of private firms, the operations of markets and the monetary system that governs the operations of financial markets and activities of financial institutions.

All modern capitalist economies are socialist

A central feature of modern capitalist economies is the symbiotic relationship between governments and private firms. Governments are in partnership with private firms in the monopolistic sector of the economy. The monopolistic sector is comprised of oligopolistic enterprises characterized by a few large private firms supplying a given market. Governments are also in partnership with private firms in the competitive sector of the economy comprised of many small private firms supplying a given market. Governments depend on the taxation of high profits and high employee wages in the monopolistic sector and also on the taxation of lower profits and relatively low wages paid by private firms in the competitive sector. In return for taxes, the government supports the ever increasing accumulation of capital of private firms in the monopolistic sector by making "social investments" that enhance their productivity, by incurring "social expenses" that reduce their labour costs and by Keynesian stabilization policies that periodically indirectly increase aggregate demand for their products. "Social investments" include expenditure for public research, roads, industrial parks, and human capital that benefit private firms. "Social expenses" include expenditures for public schools, day care provision, social security, and unemployment insurance. This role of governments was labelled by US sociologist James O'Connor as "accumulation" (Lichty 2005: 1–5).

Ever increasing productivity of labour in the monopolistic sector reduces jobs in that sector and increases the supply of labour in the competitive sector with the result that wage rates are depressed and some workers are unemployed. However, to maintain harmony in a democratic society, governments compensate groups that are negatively affected by the firms in the monopolistic sector in order to legitimize its support for them. To this end governments typically charge lower rates of corporate income taxes and/or provide subsidies to small businesses to enable them to survive and pay at least subsistence wages to their employees. Furthermore, governments support the unemployed, the working poor, those on welfare and the disabled. O'Connor labelled this second role as one of "legitimation" (ibid.: 3, 4).

The fact of a symbiotic relationship between governments and private firms means that all capitalist economies in democratic states (e.g. US, Canada) are in fact socialist. The older definition of socialism relies on government ownership of the means of production but the more modern European vision of socialism believes in social justice through social reform and wealth redistribution via state welfare and taxation. It believes that "there is a capitalist market in which the government must operate" (Carlson 2011). Thus interventions, whether to support these institutions or restrict their activities to meet the goals of democracy, represent socialism. The significant differences among countries are with respect to the degrees of corporate welfare and the degree of social welfare. Corporate welfare arises when the costs of government support in terms of social investment, social expenses and stabilization costs exceed the tax revenues on corporate income. Social welfare arises when the benefits to the poor, disabled and disadvantaged exceed the (indirect) taxes on them.

Democratic capitalism

Joseph Stiglitz, winner of the Nobel Prize in Economics, states that "Capitalist market economies come in many variations" (Stiglitz 2010: xii). Democratic capitalism is a variation of a capitalist market economic system that is based upon the principles of democracy, particularly the principles of equality of opportunity and social justice; the latter principle is about the degree of inequality that is fair. The production and distribution of products and services by private firms, though profitable, must also have a positive impact on one or more of the elements of the common good. The following sections suggest policies for changing the way wages are determined, the ways that private firms and individuals are taxed, the ways that land and natural resources are controlled or owned, the way private firms share productivity gains with labour, and the strategies for international trade that would be necessary to realize a regime of democratic capitalism in Canada.

Wages are based on Marx's labour theory of value in use

> The democrats according to Aristotle are correct when they claim that all men are equal (simply by virtue of existing) but not in all respects.
>
> (Newell 2011: 6)

> The task of statesmanship is to sift out the respects in which citizens are equal and unequal, both vis-à-vis one another and within the parts of their own souls and this requires nothing less than political philosophy the first time that the study of politiea is explicitly raised to philosophical rank in the *Politics*.
>
> (Newell 2011: 5)

From the standpoint of society as a whole, one important respect in which humans are realistically roughly equal is the time available for them for work. In using the term roughly I mean that the variations in time that people have to work are relatively small given needs for sleep, household activities, leisure and particularly institutional norms of standard work weeks that typically range from 35 to 40 hours in modern societies like Canada.

In this context, the respect in which all are equal in a democratic society is taken from Marx's labour theory of value in use. That is the idea that all labour, regardless of the type, is "akin to average labour" and should be valued at some average hourly rate (Wikipedia 2015: §4.4). This idea provides a measure of achievement of the democratic principle of equal treatment of all groups regardless of their differences – in this respect differences in the particular characteristics of labour. The idea of a common hourly rate for different types of labour can also be used for policies to realize the other democratic principle of equal opportunity. Specifically a standard hourly wage could be set at an amount that is lower than the average such that it would provide sufficient income for any citizen to enable him or her to provide for his or her basic needs plus sufficient safety and leisure to participate in shared deliberations about politics and philosophy and the common good. Admittedly the idea of a common hourly wage for all kinds of different labour is controversial because it appears to foreclose on the other issue of statecraft, namely how to address individual differences such as labour productivity that would merit unequal treatment and different wages for different workers. This issue will be discussed later.

By contrast, modern capitalist economic systems value labour time on the basis of relative scarcity. This design fault leads to the incidence of poverty in circumstances where the aggregate need for labour by firms is less than the aggregate supply due to changes in population. Over the last hundred years this design default has been implicitly recognized but only partly corrected by government interventions in labour markets. These interventions have included policies of minimum wages, legislation supporting the practice of collective bargaining by unions to secure fairer wages, compensatory social welfare assistance and experiments in the establishment of a guaranteed annual income. For decades governments in Canada have refused to adopt, or fully adopt, such policies.

For other kinds of labour, the laws of supply and demand are suspended. For workers in professions, in government, in hospitals, in educational institutions, wages are "administered prices" based on historical, social norms or union bargaining power – regardless of whether or not there is an excess supply of workers in these groups. Administered prices for human labour are extended to private firms. The most dramatic effects are found with chief executives in large business firms among the so called one percent. Here compensation is based on historical, social and cultural norms and the relative bargaining power of managers *vis-à-vis* the representatives of shareholders.

Profits related to labour productivity are shared with labour

Firms can increase profits by reducing the amount of labour in the production process either by changing production processes or substituting capital (equipment, tools and facilities) for labour. The result is that output per unit of the remaining labour in the firm increases. However, none of the increased profits that result from the higher productivity of the remaining labour gets shared with labour; all of it is retained for the shareholders. Meanwhile the cost of the capital that has substituted for labour is largely recovered through the annual amortization expenses related to it. The fact that none of the profits related to the productivity increase are shared with labour runs contrary to the principle of social justice: that any increase in wealth inequality is just only if there are compensating benefits for everyone.

The failure to share profits from productivity is reflected in the empirical law recently discovered by Thomas Piketty: the rate of return on capital always exceeds the rate of growth in the national income in capitalist economic systems. This phenomenon leads to an ever increasing concentration of wealth in the hands of a few in the society (Piketty 2014). Furthermore the wealthy heavily influence the policies of elected representatives so that nominally democratic states tend to become *de facto* oligarchies. In a democratic capitalist regime, the government intervenes with taxation policies to reduce the after-tax rate of return to a level at or below the growth rate in national income.

The government recoups excessive profits of "monopolistic" firms

Democratic capitalism calls for taxation policies that recoup excessive rates of return on capital (i.e. the ratio of profit to the cost of the capital) to private firms in the "monopolistic sector" of the economy. Owing to their strong market power, these firms are able to charge higher prices and earn higher rates of profit than they would if markets were perfectly competitive as Adam Smith assumed was more or less the case in 1776. To this end they can restrict the amount that they invest in capital in order to achieve a rate of return to which they have become accustomed – or indeed to increase it.

An important implication of stronger market power is that it dramatically reduces the risk that their owners/investors will lose their investment in capital as a result of the firm becoming bankrupt. Indeed, if some firms become too big to fail, they will not fail due to government bailouts. This fact is significant because it relates to modern portfolio theory which posits that capital markets are efficient if the rate of return is positively correlated with the degree of genuine risk. Conversely, the theory suggests that, when the risk of bankruptcy of a firm is reduced, the rate of return on its capital should be reduced.

In reality the modern portfolio theory has not been born out as Thomas Piketty explains:

> For the last 200 years or more a great debate has been conducted about the distribution of wealth and income notably between Marxists, neo-classical economists, Keynesians and more recently neoconservatives. The debate has been conducted with an abundance of prejudice and a paucity of fact.
>
> (Piketty 2014: 2)

With systematic data in hand Piketty concludes that the increasing accumulation of capital results in a shrinking share of labour income in the national income which, in turn, results in insufficient consumption to meet the output of the productive capital. Consequently there is an under-utilization of capital and a falling rate of profit. In these circumstances:

> Capitalists dig their own grave; either they tear each other apart in a desperate attempt to combat the falling rate of profit . . . or they force

labour to accept a smaller and smaller share of national income which ultimately leads to a proletarian revolution and general expropriation.

(Piketty 2014: 228–229)

For capitalists the way out of their dilemma is to induce the government to reduce taxes on corporate income in order to maintain high after-tax profit rates. Another way is to seek freer trade and investment liberalization to expand demand for their products and reduce costs through off shoring production in low wage countries. A third way is to encourage labourers to go further into debt to enable them to purchase their products and services. Indeed this is exactly what happened in the decades leading up to the global financial crisis of 2008. As Thomas Piketty explains:

> There is absolutely no doubt that the increase of inequality in the United States contributed to the nation's financial instability. The reason is simple: one consequence of increasing inequality was virtual stagnation of the purchasing power of the lower and middle classes in the United States which inevitably made it more likely that modest households would take on debt especially since unscrupulous banks and financial intermediaries, freed from regulation and eager to earn good yields on the enormous savings injected into the system by the well-to-do, offered credit on increasingly generous terms.
>
> (Piketty 2014: 297)

Higher prices for consumers and excessive profit rates of firms are unnecessary costs in the economy and for the majority of the public. In democratic capitalist regimes higher rates of tax on corporate profits are initiated to recover some of those costs and return them to the public. In so doing they avoid the effect of increasing inequality on periodic instability in financial markets. Also governments in democratic capitalist regimes are much smarter at regulating transactions in the financial sector to discourage speculative trading in existing financial assets.

Since most of the transactions for speculation have no material utility in terms of the common good they can be taxed with a "Tobin" type small tax rate on all the transaction. In retrospect the enormous increase in the size of the financial sector in the US by four percentage points of GDP over the last few decades was a hugely uneconomic waste of resources.

An alternative to higher rates of taxation of private firms in the monopolistic sector is for government to purchase them and retain the excess profits for the treasury. Economist John Kenneth Galbraith made such a proposal in his 1973 book, *Economics and the Public Purpose*. Specifically he proposed to have the treasury buy the shares of the largest several hundred corporate giants and place them under public ownership. Rich stockholders would be fully compensated, but over time, their fortunes would be lessened and this would decrease wealth and income inequality (Parker 2006: 514).

Galbraith also suggested that, under government ownership, senior managers' salaries would be reduced. Thomas Piketty's recent analysis suggests that the reduction in CEO salaries would not have a negative effect on their productivity. A study of 3,000 firms in fourteen countries, which Piketty co-authored, found little correlation between skyrocketing executive pay for US senior managers and their hypothesized managerial productivity. Rather he argued that CEO pay was fairly well explained by the model of CEO bargaining with shareholder compensation committees and their self-serving rules of thumb (Piketty 2014: 512, 639 n46).

Galbraith also argued that government ownership would put government in a position to control the "technostructure" of giant firms (Parker 2006: 514) and thus prevent the negative instances of the "caprice of technology". For example, government-owned natural gas firms could have prevented the fracking of natural gas for increased corporate profits, reduction in jobs and degradation of ecological systems and water quality.

Governments control land on behalf of all citizens

In democratic capitalism, land should either be owned or controlled by government on behalf of all citizens. This policy is predicated on the assumption that land is inherently a public good; therefore the benefits of land that is privately owned or occupied should be shared with all citizens in the community. Specific policies include municipal ownership of all urban lands used for public housing and public buildings, and ownership of land occupied by private firms that is leased to those firms. An alternative to ownership is the taxation of real capital gains on the land component of properties such as principal residences. Such taxation recoups the capital gain that has been generated by population growth in the community, local government expenditure on infrastructure, or lack of it, and local government land use/zoning policies.

Tax policies encourage virtues and discourage vices

Two of the virtues in democratic capitalism are hard work and moderation in the pursuit of wealth; the corresponding vices are lack of effort and greed. Policies of Canadian governments, for example, feature lower tax rates on income from capital, which is earned with no effort, than on employment income, which is earned from hard work. Income from capital includes capital gains on stocks and real estate and dividends paid to shareholders. These preferential tax policies represent incentives for the opposite vice of easy work for money. Another example of this vice is the Canadian government's elimination of estate taxation on intergeneration transfer of wealth in the 1970s, a policy that provides wealth to recipients for no effort on their part. Furthermore, these policies exacerbate the negative effect of intergenerational transfers of wealth on increasing the inequality in the distribution of wealth in the US and Europe that Thomas Picketty has observed. So in democratic capitalist regimes taxation policies include the full taxation of real capital gains on stocks and real estate and dividends from corporations, taxation of inheritances and taxes on capital (net worth) of the wealthiest.

"Economic rent" on natural resource extraction is eliminated

In simple terms, economic rent on natural resources represents the excess profits that extractive firms earn from their investment in capital equipment for the activities of extraction, processing and distribution to customers. The excess profits arise from the government's sale or grant of the rights to extraction too cheaply. Economic theory regards economic rent as uneconomic in the sense that it is a misallocation of resources. Furthermore, economic rent represents easy money for the owners of the extracting firms – as opposed to the hard work of their employees. One of the collateral effects of free rent to the extractive firms is the lost opportunity of increased government revenues that could have been used to fund social welfare for the poor or to reduce taxes on the middle class. Another collateral effect is far more serious. Extraordinary profits of the extracting firms have encouraged an expansion in the exploration of oil reserves and increased production of oil and gas that have contributed to world climate change and the prospect of a "terror of an unliveable future" (Klein 2014: 28).

In a democratic capitalist regime, the government charges extractive companies for large portions, if not all, the economic rent. An example of this policy is the Norwegian government's 50 per cent tax on oil companies to capture economic rent on top of its ordinary 28 per cent tax rate (Campbell 2012).

No more free trade for Canada – only fair trade

In Canada the free Trade and Investment Liberalization (TAIL) regime began in 1994 with the implementation of the North American Free Trade Agreement (NAFTA) of 1988:

> The public promise and theoretical predictions of TAIL in the mid-1980s have gone unfulfilled. The TAIL era has not delivered broad-based prosperity for Canadians, but instead has brought differential income gain for the few and income stagnation or outright decline for the many.
>
> (Brennan 2011: 28)

In other words the TAIL regime has run counter to the democratic principle of social justice – a principle dubbed as "justice as fairness" (Kloppenberg 2011: 91). More specifically the TAIL regime "Opened the door to the two largest mergers and acquisitions waves in Canadian history which led to increasing inequality. The linkages between corporate concentration and personal income inequality run as follows: amalgamation increases concentration; increased concentration translates into less competition; less competition translates into enlarged earnings margins, greater profits and increased cash flow; the resulting increase in cash flow has the potential to translate into highest executive salaries and dividends; and it is the very high executive salaries that are playing a key role in driving Canadian income inequality (Brennan 2015: 7–9)."

In 2011 Governor of the Bank of Canada, Mark Carney, warned about the risks to Canada of further free trade: "Major advanced economies are attempting to increase demand for their overproduction by increasing their exports while, at the same time, emerging markets are refusing to let their exchange rates materially adjust. Both sides are doubling down on losing strategies" (Carney 2011: 8). In this context, the risks of failing to capitalize on the potential for free trade are even more significant for Canada because, as Carney explains, "Canadian firms should realize that they are not as productive as they could be and that it is imperative to invest in improving their productivity" (ibid.: 11).

The idea that both countries can win from free trade is based on the theory of comparative advantage that was developed by economist David Ricardo in 1817 (Lipsey et al. 1988: 394–400). It does not work well today because comparative advantage depends on differences in education and innovation between countries rather than differences in natural resource endowments. For example:

> It was argued that, under NAFTA, labour intensive manufacturing in Canada would shift to Mexico while Canadian manufacturing would restructure based on Canada's alleged comparative advantage in technological sophistication, innovation and skills. But it turns out that low wage countries have created a competitive edge in sophisticated industries.
>
> (Jackson 2015: 1)

Foreign trade is often resorted to when governments fail to create employment through fiscal stimulus and/or fail to limit growing inequality in income which reduces aggregate domestic demand and increases unemployment (Watkins 2014: xvi).

Some trade agreements reduce Canada's sovereignty to make public policy. For example Canadian negotiators propose to include in the Canada Europe Trade Agreement (CETA) NAFTA's investor state dispute resolution mechanisms which enable foreign investors to directly sue the national government for potential losses caused by government policies and deregulations. CETA negotiators are intent on prohibiting "buy Ontario" polices and other supports for local producers (Jacobs 2012: 16, 17). By way of comparison, see Gray on the issue of investor state dispute resolution mechanisms in the Australian environmental context (Gray 2015: 139–144).

Democratic capitalism requires strategic trade and investment strategies. The most important objectives for these strategies are job creation and the reduction in economic inequality. The other objectives are limited encumbrances on Canada's capacity to pursue independent monetary, fiscal, and social development policies and neutralizing impacts of international investment flows on the exchange rate that would significantly alter the trade balance and the real economy.

Democratic capitalism needs less economic growth which will reduce environmental degradation

In modern capitalist economies there is a tendency for total income in the economy to exceed total expenditure with the result that there is insufficient demand in the economy to employ all available labour; the result is some unemployment. For clarification total income is the sum of wages of workers, rental income of landlords and profits of the capitalists. Total expenditure is the sum of expenditures by workers, by landlords and by capitalists for personal consumption and investment in capital. The main reason for the gap between income and expenditure is due to the fact that profits are too high so that not all of them are reinvested in capital; in other words there is a leakage out of the real economy in the form of the part of profits that are unspent. To compensate for this leakage, governments borrow money to spend on infrastructure to generate more economic output and more employment so that unemployment is eliminated. Government investment in infrastructure has the effect of increasing economic output (i.e. GDP) through "multiplier effects".

In a democratic capitalist regime, rates of return on capital are lower than the rate of growth in national income. The lower profits result in a smaller leakage and less need for government expenditure to offset the leakage with the result that there is less GDP.

Economic capitalism would correct for a major mistake in the 1960s

A major opportunity to limit the rate of economic growth in the United States arose in the late 1950s when John Kenneth Galbraith published his 1958 book, *The Affluent Society*. Galbraith claimed that American households had reached a state of affluence that met the Aristotelian ideal: having enough income to meet the basic needs of all plus sufficient safety and leisure for politics and philosophy. His advice to the American government was to spend more on social security, to reduce social tensions, and enhance the quality of life and less on private consumption. In other words his advice called for increasing the balance of positive freedom relative to negative freedom. However, Galbraith observed that the American government's priority was for "military and strategic goals" and short term Keynesian macro management policies. "Meanwhile, the domestic economy was left to advertisers and marketers in an endless celebration of private consumption . . . that increasingly met not needs so much as wants carefully manipulated by private-goods producers. Pressing public needs were not filled" (Parker 2006: 302).

In his 13 October 1954 speech at Wellesley College Massachusetts on "Economic Freedom", Galbraith had argued that "Market capitalism promoted the coarse ideological claim that ever increasing consumption is freedom." To Galbraith this was a "false ideal". He argued, or hoped, that, given more affluence, more and more people would come to realize that "leisure, free time and intellectual achievement are the real thing" (Parker 2006: 303, 724 n89).

Galbraith's advice fell on deaf ears and the results have been tragic for the majority of American citizens. While US real GDP per capita increased by 105 per cent over the 34 year period from 1970 to 2004, the General Progress Indicator for the country, which is arguably an index of improvement in general welfare or the common good, increased by only 5 per cent (Victor 2008: 129). In a regime of democratic capitalism more progress could have been made with less economic growth and less environmental degradation.

Conclusions

In the US, and to a lesser degree in Canada, citizens live in a system of oligarchic capitalism which is presently infused with a moral-cultural system of consumerism, entertainment, spectacle and the acquisition of wealth reminiscent of the culture of bread and circuses in Imperial Rome. For Canada a transformation to democratic capitalism would entail a significant reduction in the size of the private sector in the economy and a corresponding increase in the size of the government sector. Poverty would be eliminated and the inequality in the distribution of wealth and income would be reduced. As a result the consumption of uneconomic luxury and status goods would be reduced while the consumption of life-giving basic goods and services would be increased. The rate of economic growth would slow but the reduced rate of return on capital would avert future increases in economic inequality. The elimination of poverty and a reduction in inequality in the distribution of wealth and income would significantly reduce the need for economic growth and thus address climate change.

In his recent Encyclical on the Environment, *Laudato Si'*, Pope Francis "Urgently appeals for a new dialogue that includes everyone about how we are shaping the future of our planet" (Pope Francis 2015: para. 14). He observed that "The financial crisis of 2007–2008 provided an opportunity to develop a new economy more attentive to ethical principles and new ways of regulating speculative financial practices and virtual wealth" (ibid.: para. 189). Democratic capitalism represents such an economy.

References

Brennan, Jordan (2011) "Has Free Trade Fulfilled its Promise in Canada?: Contesting a 'Sacred Tenet' of Globalization Theory" Progressive Economics Forum. June, available at www.progressive-economics.ca/wp-content/uploads/2011/06/Brennan.pdf.

Brennan, Jordan (2015) "NAFTA Corporate Power and Growing Income Inequality". *CCPA Monitor* 21(8) (February): 7–9.

Campbell, Bruce (2012) "Managing Oil Wealth: The Alberta/Canada Model vs. the Norwegian Model." Ottawa: *The Hill Times On Line*, available at www.hilltimes.com/opinion-piece/2012/04/30/managing-oil-wealth-alberta/Canada_model_vs_the_Norwegian_model.

Carlson, K. (2011) "What Are NDP Talking About When They Talk about Socialism?" *National Post*, 25 August.

Carney, Mark (2011) "Growth in the Age of Deleveraging." Remarks to the Empire Club of Canada/and Canadian Club of Toronto, 12 December.

Cooper, C. (2011) "Oligarchy and the Rule of Law". In D. Tabachnick and T. Koivukoski (eds), *On Oligarchy: Ancient Lessons for Global Politics*, University of Toronto Press, Toronto, pp. 196–216.

Coyle, Diane (2011) *The Economics of Enough: How to Run the Economy as if the Future Matters*, Princeton University Press, Princeton, NJ.

Galbraith, John Kenneth (1958) *The Affluent Society*, Houghton Mifflin, Boston, MA.

Galbraith, John Kenneth (1973) *Economics and the Public Purpose*, Houghton Mifflin, Boston, MA.

Gray, J. (2015) "Unconventional Gas Mining – What a Fracking Story! Policy, Regulation, Law and Trade Agreements". In L. Westra, J. Gray and V. Karageorgou (eds), *Ecological Systems Integrity: Governance, Law and Human Rights*, Routledge, Abingdon, pp. 133–148.

Ingham, Geoffrey (2008) *Capitalism*, Polity, London.

Jackson, Andrew (2015) "Canadian Economy Suffers from Myth of Comparative Advantage". Available at www.broadbentinstitute.ca/en/blog/canadian_economy_suffers_from_myth_of_comparative_ advantage.

Jacobs, J. (2012) "Trade Deal with Europe Threatens Many Costly Ill-Effects such as Job Losses, Higher Drug Prices and 'Buy Local' Curbs". *CCPA Monitor* 19(5) (October): 16–17.

Klein, Naomi (2014) *This Changes Everything: Capitalism Versus the Climate*, Alfred A. Knopf Canada, Toronto.

Kloppenberg. J. (2011) *Reading Obama: Dreams, Hope and the American Political Tradition*, Princeton University Press, Princeton, NJ.

Lichty, W. R. (2005) "Lecture on James O'Connor's *Fiscal Crisis of the State*, 1970." 4 April, available at www.d.umn.eduw-rlichty/Radical%20Lectures/radles10.pdf (accessed 18 February 2013).

Lipsey, R., Purvis, D. and Steiner, P. (1988) *Economics Sixth Edition*, Harper & Row, New York.

Mann, Geoff (2013) *Disassembly Required: A Field Guide to Actually Existing Capitalism*, AK Press, Oakland, CA.

Neill, J. (2011) "Aristotle and American Oligarchy: A Study in Political Influence". In D. Tabachnick and T. Koivukoski (eds), *On Oligarchy: Ancient Lessons for Global Politics*, University of Toronto Press, Toronto, pp. 24–46.

Newell, W. R. (2011) "Oligarchy and *Oikonomia*: Aristotle's Ambivalent Assessment of Private Property". In D. Tabachnick and T. Koivukoski (eds), *On Oligarchy: Ancient Lessons for Global Politics*, University of Toronto Press, Toronto, pp. 3–23.

Novak, Michael (1982) *The Spirit of Democratic Capitalism*, Simon & Schuster, New York.

Parker, Richard (2006) *John Kenneth Galbraith: His Life, His Politics, His Economics*, Harper Collins, Harper Perennial Edition, Toronto.

Piketty, Thomas (2014) *Capital in the Twenty-First Century*, Belknap Press of Harvard University Press, Cambridge, MA.

Pope Francis (2015) *Praise Be to You Laudato Si': On Care for Our Common Home*, Ignatius, San Francisco.

Sikkenga, J. (2011) "Overcoming Oligarchy: Republicanism and the Right of Property in *The Federalist*". In D. Tabachnick and T. Koivukoski (eds), *On Oligarchy: Ancient Lessons for Global Politics*, University of Toronto Press, Toronto, pp. 47–69.

Stiglitz, Joseph (2010) *Freefall: America, Free Markets and the Sinking of the World Economy*, W. W. Norton & Co., New York.

Tabachnick, D. and Koivukoski, T. (eds) (2011) *On Oligarchy: Ancient Lessons for Global Politics*, University of Toronto Press, Toronto.

Taylor, Prue (2014) "The Earth Charter, the Commons and the Common Heritage of Mankind Principle". In L. Westra and M. Vilela (eds), *The Earth Charter, Ecological Integrity and Social Movements*, Routledge, Abingdon, pp. 12–23.

Victor, Peter (2008) *Managing Without Growth: Slower by Design, Not Disaster*, Edward Elgar Publishing, Cheltenham.

Watkins, Mel (2014) "Keynes Can Teach Us as Much Today as He Did in the 1930s". *CPRA Monitor* (April): xvi.

Wikipedia (2015) "Labor Theory of Value". Available at http://wikipedia.org/wiki/Labor_theory_of_value (accessed 20 November 2015).

17 Constitution and future generations

A new challenge for law's theory

Antonio D'Aloia

Constitution language and the future: new words for new problems

Talking about future generations as a new category of juridical reasoning: it means to place in law the theme of the time (and of the future), from an original point of view. If it is true that "the law use the time as a tool that sorts human actions" (Casavola 2004) in this new perspective, the time is imposed on the law, forces it to take on meanings and objectives, distorts (or enriches) some of its ordinary categories (such as responsibility).

Actually the intergenerational issue is now on the table in an overwhelming way, and concerns several material fields. We went from the "if" to the "how" of the legal protection of future generations. The problem is to find resources by which the law can and indeed must translate this "intellectual leitmotif", almost a new narrative in the legal dimension.

The change of paradigm was in the discovery of our ability to change also in a negative sense, in the discovery of the ability to alter profoundly the sequence apparently linear between phases of time. As E. Brown Weiss wrote, "what is new is that now we have the power to change our global environment irreversibly, with profoundly damaging effects on the robustness and integrity of the planet and the heritage that we pass on to future generations" (Weiss 1990: 198).

According to Al Gore, "We became a geologic strength . . . the principal evolutionary agent of the world, and the fact that we know that, the fact that we are aware now of this impact, it increases of course our moral responsibility" (Gore 2013: 218). So, the future could really no longer be what it once was. We are facing a radical change of the order of problems which the law should face.

The law is forced to face the duty to think of the future, to make it possible, or at least not to download on it and on who will be called to live it the irreversibly negative effects of the current choices.

Exemplary is the parallel that Zagrebelsky (2011: 2), recalling Jared Diamond, explains in the history of Easter Island:

> Every generation has behaved as if it was the last dealing with the resources it disposes as they were their unique properties to use and abuse . . . constitutionalism has not yet had to deal with bullying intergenerational reasons . . .

but today we see the separation in time of benefits – early – about costs – postponed – happiness, the wellness, the strength of current generations in exchange of unhappiness, discomfort, impotence or even extinction of the impossibility of being in the world of the future ones. The break of the contextuality of time marks a turning point that cannot allow the indifference of morality and law.

(Zagrebelsky 2011: 2; my translation)

An unprecedented perspective, which leaves the law disoriented as it has to look for new tools even new words, words that redirect behaviours and decision models – like sustainability, sustainable development, precaution – containing this intertemporal opening, and the goal to preserve not to compromise and to keep for the future.

The fields of intergenerational justice: the weakness of future generations as common fact

Actually, the theme is not very new: the first elaborations of a risk for future generations connected with actions in the present time date back to some documents just further to the end of World War II, because at that time the danger of nuclear war emerged.

I refer to the San Francisco UN Charter, of 1945, which explicitly links in its Preamble the need to prevent the scourge of the war with the will of the people to preserve the future generations. The war especially with weapons of mass destruction endangers human survival; the value (also) Constitutional of the peace becomes one of the first contents of the law "oriented" to the future.

The risk came from a clearly unacceptable behaviour, which would have hit in a serious and irreversible way even the aggressor, if it had been possible to identify it; as well as the present generations.

From here also a kind of abnormal guarantee: nuclear war was and is a huge mistake regardless of who should win it; indeed to take the metaphor from a famous movie of the 1980s (*Wargames*, dir. J. Badham), "it is a game with no winner and it does not suit anyone".

At the same time, an explicit reference to the protection of future generations also appears in an international legal document such as the Convention for the Regulation of Whaling (1946), which is the first showing of the intergenerational instance in the field of environment preservation and natural resources (in this case, "whale stocks"). In the following decades, many international acts will show an increasing awareness of the intergenerational importance of goods and interests related to the ecological and natural system: from the Convention on the Conservation of Migratory Species of Wild Animals in 1979, to the Agreement on the conservation and management of fishing in high seas in 1995 to the Code of Conduct for a Responsible Fishing of 1995 to the Convention on Biological Diversity in Nairobi of 1992 (for these and other documents see Bifulco 2008: 90).

The real breakthrough of this new theoretical and cultural perspective is recorded in the 1970s of the last century hooking into the development of a deeper environmental awareness and the need to respect the "natural" limits of human activities:

- the debate opened by the reflections of the Club of Rome in 1972 (Meadows et al. 2006);
- the bioethics of the survival is oriented to "granting biosphere" of Van Rensselaer Potter (1971);
- the theoretical elaborations of Stone (1972), Jonas (1979) and Passmore (1974);
- the campaign of Jacques Cousteau on the Declaration of the Rights of Future Generations (1979);
- the progressive construction of a theological thought driven by strong ecological and custodial inspirations, starting from Jurgen Moltmann's works (Moltmann 1971), which can find today a strong emphasis in the recent encyclical of Pope Francis (*Laudato si'* – On the care of our common home); and
- Rawls's justice theory (1971).

Those are the cornerstones of an extraordinary and diverse scholarly work, which raises the intergenerational issue between the basic aspects of ethical reflection, as well as the legal and regulatory implications that follow.

The environment and the maintenance of ecological and "natural" balances in time, are at the heart of this "narrative", its main field, and continued to be so throughout the journey of improvement and consolidation of "intergenerational justice". In fact, today the theme "monster" of the debate on the protection of future generations is climate change (Posner and Weisbach 2010: 144; Mulgan 2006), whose link with human activity is now established incontrovertibly.

The environmental topic was then joined by other sides and "goods": the cultural heritage, the consequences of genetic engineering and of the bioethics' development, and especially the welfare and the economic dynamics around two "subjects" inherently transgenerational such as public debt or pensions.

The common element between the different hypotheses is given by the weakness or vulnerability of future generations' interests. Politics, democracy and economy, are overwhelmingly crushed by the needs of the present. Hence all efforts to deal with the problem of the future: Hans Jonas points out that "What is not existing doesn't possess any lobby and the unborn are powerless" (Jonas 1979: 30). The rights of future generations fail precisely because the process of bargaining between self-interested individuals who are able to harm each other and that is one of preconditions of the contractual vision of utilitarian moral obligations. Future generations are likely to appear in the same way as "ghosts", in front of choices (legislative, administrative, political) and behaviours in which dominates a logic of "here" and "now" (Tremmel 2006: 187).

So it is clear that the intergenerational issue forces to rethink the mechanisms of democratic politics and of the legislative majorities. The legislator must be "the man of the future" as a subject who is able to make informed decisions of future consequences and respectful of the need of preservation of resources and necessary opportunities for the decent survival of the human race. The survival reprieves decision-making procedures and forms of consultation and impact assessments which may not follow the traditional models, too exposed to the temptation of the social discounting, of the underestimation of the future effects of current choices. But we must build a framework of intertemporal co-operation governed by the same conception of justice that regulates the cooperation between contemporaries.

Who are future generations? Parfit's paradox and the arguments to overcome it

The perspective of future generations is very intriguing, but controversial and difficult.

The first problem is the following: Who are the future generations? Is there a temporal boundary beyond which it doesn't make any sense to commit in a discourse about protection and acknowledgement of duties and rights? If we look at the entire field of intergenerational issues the category of future generations takes very different forms.

The future is a dynamic and indeterminate dimension. At the same time, in the words of Jonas the future is never completely detached from the present: every time different generations coexist, those are subjects "belonging to all ages of life". First, it is very difficult to try to divide clearly present and future, current generations and future generations, holders of moral obligations of responsibility or beneficiaries for these.

Second, the impact on the future of current choices is not the same in the various fields in which the intergenerational issue appears. The sequence of choices and effects can be realized in a time far or very far as some consequences of climate change or in the event of nuclear waste ; it can give immediate and not reversible results (so apt to continue in the future even more remote), such as the possibility of genetic interventions that may affect the germ line.

It may finally determine the situation of intergenerationality "contextual" in which present and future generations share different stages of the same problem (the examples of public debt and the intergenerational operation of pension systems are in this sense emblematic).

According to a common theory (Passmore 1974: 102), the point of reference of this new "perspectival responsibility" should be essentially (and only) the generations immediately following to us, or even "already born". Towards the generations more distant in time would be at most configurable humanitarian duties, but not obligations of intergenerational justice.

The motivations of the thesis are mainly practical: the farther we get over time, the more difficult it becomes to analyse the relationship between the current benefits and future costs of certain decisions, and vice versa. Moreover, and this seems a reasonable point of view, the willingness to give up something today can find a more intense motivation in the possibility to avoid certain harmful consequences at least for their direct descendants, with whom "we have a much stronger and direct emotional connection . . . [and] we can imagine, with a good deal of precision, preferences, desires and needs, while this exercise is more difficult towards generations distant in time" (Barry 1999: 99).

More generally there is no doubt that the familiar/parenting background can be an important laboratory to learn and configure the responsibility towards those who follow us in the time line.

The apparent limitation of the view to the closer generations, in the opinions expressed would however find an element of gradual correction in the continuing chain of responsibility of progressively following generations, so that would be able to produce a kind of "generational continuum" capable to transmit to our successors that same concern and solidarity that we were able to have and demonstrate towards them.

Nevertheless the normative version of intergenerational issue seems to look to the whole humanity without distinctions related to greater or lesser temporal proximity. For international and constitutional law, future generations are "simply" and "generally" the entire set of human beings following the current generation (Bifulco 2008: 75).

Moreover, the intergenerational issue is also a shape of the principle of reasonableness: to think to the future, to the consequences of their own behaviours is a way to use reasonably rights resources and available goods and it is therefore positive even currently, and as an element of sustainability and "intergenerational" responsibility.

In the same manner, it does not seem so crucial to face the problem of what future generations, and until what future, it is necessary to protect. The intergenerational issues show an objective imprint: future generations rather than people who have to be identified are projections of goods and principles that can be seen also as the inherent limitations of the human experience.

The theories about responsibility towards future generations or about future generations' rights, crash with a theoretical paradox elaborated by English philosopher Derek Parfit. On the basis of the "non-identity problem" of Parfit, can we say we damaged someone making a choice without which he wouldn't exist? And if we have damaged no one, how can our choice be morally wrong?

Parfit starts from the time-dependence claim. According to him, if a specific person has not been conceived when in fact he has been conceived, it would be true that this person would never exist. In other words, the choice of a policy of resources depletion, above all for a farer future (Parfit speaks about a future after three centuries) doesn't damage anyone in the future, for the simple and paradoxical reason that if we had adopted a different conduct, no one would have been conceived in the same moment in which he was effectively conceived, and therefore no one would have existed. According to Parfit, our choice to deplete the resources, for them isn't a worsening, but even a benefit because if we had chosen protection of resources, they would have never existed (Parfit 1984: 447, 464). This argument is certainly strong, but not invincible. First of all, Parfit himself tries to suggest a way to bypass the non-identity obstacle. We need to try to define the principle of beneficence in a way more "impersonal", where it is not important what is good or bad for the people who are affected by our behaviour.

The depletion of non-renewable resources is an absolute evil, which doesn't simply harm an individual right, but a fundamental right (or better a fundamental good), for these reasons independently from specific individuals. In other words, the non-identity problem is solved, distinguishing the concepts of harm and tort. In summary, I can suffer an injustice even if I did not have any damage as a result of this action; what matters is that someone has acted deliberately failing to respect my interests.

The damage does not have to be a comparison between how I was before and how I am now: according to Kumar, what is important to have a legitimate claim to a material repair is that "those who have suffered a tort is faced with serious disadvantages, or is worse in ways systematically related to the action of those who committed the wrong" (Kumar 2007: 67).

The configuration of duties (moral and legal) towards future generations does not need to be perfectly identified who can receive damage from current choices and behaviours, or that this/these person(s) will not exist, just "because" actions "wrong" committed in the past.

The intergenerational responsibility has a collective structure, where it is not necessary (or even possible) to define a perfectly symmetrical relationship between those who produce the damage (and are therefore obliged to repair it or to prevent it) and those who suffer it, also because generations (present and future) ensue one to another in movements of overlapping and coexistence, uninterrupted and partial. To use a metaphor, the passage of generations is like turning the pages of a book that remains (the previous ones and subsequent ones) simultaneously open.

What appears sufficient to justify a moral responsibility, and to give legitimacy to the law to imagine and structure models of protection, is the evidence of a causal link between today's choices and the effects that these choices can have on key elements of the quality of life in the future.

Posterity to the dawn of modern constitutionalism

In terms of constitutional theory, the apparent novelty of the intergenerational issue is in very truth a kind of "back to the past". Some of the first manifestations of modern constitutionalism emphasize, in different ways, the problem of the relationship between current generations (and political choices), and posterity.

The first section of the Bill of Rights of the Constitution, Virginia, in 1776, considers the enjoyment of life and liberty, the possibility of acquiring and possessing property, and the pursuit of happiness and security, as fundamental rights of which no one may be deprived or stripped, even future generations.

Article 28 of the French Constitution of 1793 stipulated that "a people always has the right to review, reform and change its Constitution. One generation cannot subject to its laws future generations." Apparently, this provision seems a statement of (mutual) freedom of each generation respect than intertemporal constraints, an exaltation of political and popular sovereignty in each historical moment. However, the assertion of a condition of freedom of each generation compared to the previous one presupposes the idea of the reverse constitutional constraint (i.e. of each generation compared to those following), precisely because this freedom needs to be able to be exercised as part of a legal-material context, that, in the passage of generations, should not be altered or suffer irreversible reductions.

Moreover, the theme of the duties to posterity belongs to the dawn of the American constitutional debate, with the famous letter of Thomas Jefferson to Madison in 1789, when the future President of the United States said his doctrine that "the earth belongs to the living in usufruct", a usufruct to which generations to come are also entitled, so if each generation in turn present abused this opportunity to occupy the lands, or obliged the future people to pay the debts that it has bargained, then "the earth would belong to the dead".

Constitutional law is naturally planned to go beyond present time and beyond the territorial space of each State. The key feature of Constitutions is the pretension of stability, the ambition to last without limits of time. All fundamental categories of constitutional language, its principles, the goods that Constitutions want to protect, have a value that exceeds the contingency. In the constitutional field, above all the rights have a transgenerational feature. The rights are inviolable, and therefore they are "prior, but also following after the specific time of the existing political regime".

In other words, "the will to bring the future" that animates the Constitution implies, at the same time, to maintain the possibility of the future. Then, speaking of future generations and trying to outline a theory of rights (or of responsibility towards) future generations, it means proposing actually a complete theory of the Constitution and constitutional law. Not only rights, but also the structures of democracy, the state, political representation, the work and the competence of the legislator are normally projected to take account of the interests of time and future humanity.

The same State is not a provisional figure, but a subject by nature, by definition, "permanent", its action can only point to the maintenance (through the succession of generations) of the living conditions of the community, working at the same time for the collective interest in the continuity in the time of its fractions.

However, as already pointed out, the gap between the theory and practice of constitutional democracy and the modern representative state, has been and continues to be particularly strong. Decision-making procedures look to short time, chasing the consensus today, even at the expense of the objective interest (of the present moment as well as) of the long time. The need to meet the demands of the electorate who vote and decide governments and majorities produces an inevitable bias in favour of present time.

The limit of modern democracy is this absolute "presentism", and in this context it is difficult to find space for the interests and expectations of future generations. The challenge, of democracy and constitutionalism, is instead to imagine institutions, procedures, tools, rules that give substance to this principle of intergenerational responsibility (or solidarity, equity).

Protect future generations: a commitment between international and constitutional law

As I formerly wrote, the irruption of the intergenerational issue on legal grounds took place mainly through international law, with many documents, perhaps starting with the Stockholm Declaration on the Human Environment in 1972.[1]

With these documents begins a journey in which the idea of responsibility towards future generations assumes more and more defined contours and branches out into a plurality of "fields" linked to an open concept of "human environment": not only the environmental perspective in the strict sense (natural resources, environmental goods), but the biosphere, human interactions with the environment, economic development, scientific and technological progress and its influence on genetic heritage.

International law accommodates a variety of notions incorporating a transgenerational dimension, such as common concern, the common heritage of mankind, the precautionary principle and the sustainable development principle. The progression of the intergenerational issue and its projections moves from international to national law, when gradually the two dimensions become highly interrelated.

Since World War II and especially in the last 40 years, international law and constitutional law have begun to move toward each other in a two-way process of mutual influence and contamination especially in the field of rights. At the same time has occurred a process of constitutional law's internationalization and of international law's constitutionalization.

Coming to more recent constitutional documents the constitutional "intergenerational" clauses can be of different types and their location varies from the Preambles to the rules on the various "objects" that have a intertemporal consistency (such as environment, nature, cultural heritage, education, welfare, financial stability).

In particular Constitution and environment seem mutually indispensable terms: the constitutional state of the twenty-first century as an environmental state, according to Häberle (2006: 224). In some recent constituent choices for example in Bolivia and Ecuador (beyond the particular cultural and ideological matrices of these Constitutions), Nature is designed in the same way as a "total" entity, legally relevant which is everywhere, anywhere and at any time. Remarkable and important is also article 20a of the German Constitution, according to which "Mindful also of its responsibility toward future generations, the state shall protect the natural bases of life by legislation and, in accordance with law and justice, by executive and judicial action, all within the framework of the constitutional order."

The accentuation of a perspective focused on duties is common to these trends of thought or constitutional experience, on the responsibilities of each to everyone and to nature a responsibility that looks forward, that finds legitimacy also in the ability to show awareness and in taking care of people yet unborn and that requires an active and conscious participation in the decisions regarding the "own living environment".

Bringing in the Constitution, thus, the protection of the interests of posterity, also simply re-reading in an intergenerational key, already existing clauses and naturally suited to this perspective (such as the rules on cultural heritage, on the environment and on natural resources, and others), is certainly a strengthening factor for policies and measures taken in the legislative and administrative field, helps to stabilize them, and to put them on top of the current political direction.

Moreover, it gives to the constitutional courts and to the common judges, a (direct or indirect) parameter of review and invalidation of choices and (even) legislative omissions. Constitutional norms also play a cultural function (of orientation and communication), triggering factors that can "react" on the effectiveness on the concrete relevance and definitively on the substantial quality of the rule. Nevertheless, we must be aware of the fact that the constitutional norms are at the starting point of the political discourse, which needs to be completed and accompanied by laws, administrative measures, *ad hoc* institutions.

Future people: rights, duties, responsibilities

The most important and intricate theoretical knot when we try to give a legal framework to the category of future generations is the question of what situations are configurable to them: Do future generations have rights? Or again, do present generations have obligations towards the future ones? And what kind of obligations are these: simply moral obligations or real demands of justice? How can these rights or these obligations be claimed/protected/translated in action?

The subjective (individual) right in its traditional version, presupposes a present holder while future generations just as such have no right to praise toward previous generations.

It says that future generations are a "fiction" (Haverkate 1992: 250). It may be true. But fictions are an important resource of the law, a way that the law (legislative or judicial) has to overcome the obstacles and aporias that the natural and factual structures pose to its need for consistency and adaptation to real situations.

Of course the difficulty in qualifying the subjective situations that move around the figure of future generations is reflected in the same regulatory pathway (both international and national) that has gradually built (or consolidated) this new category of law. The language of the duties (so the responsibility "for" or "toward" or "in the interest of" the future generations) seems to be prevalent on a reconstruction in terms of the rights of future generations; even when this representation is used it is not easy to understand whether it is a rhetorical element, or rather a conscious choice of legal policy assisted by a number of consequences and operational possibilities.

According to my point of view the real theme is "upstream": whatever they are, if rights of future generations or obligations of the present generations towards them we still talk about situations involving a bond, a bond with legal consequences, that can be controlled in its results. My argument is that they are at least duties of justice; so that, the intergenerational issue can (and in some cases must) justify a series of current consequences of the law on behaviours and choices of the present time that can assume many shapes progressively binding however attached to the more or less explicit meanings of the constitutional principles. In short the love of the distant (in this case over time) is not only a spontaneous feeling moved by ethical reasons or religious beliefs, but can be somehow imposed or at least driven by a right of constitutional significance aware that the time, not less than the space is the stage on which it performs its function.

So, what are the obstacles speaking of future generations' rights? The first drastic objection to the use of the subject future generations' rights is the "non-existence": according to Beckerman "future generations cannot have any rights" simply because "they do not exist". From this premise comes a tight syllogism, on the basis of which, given that "any coherent theory of justice implies conferring rights on people", the conclusion is that "the interests of future generations cannot be protected or promoted within the framework of any theory of justice" (Beckerman 2006: 53–54).

This argument is normally accompanied by the problem of who and how it can defend the rights/interests of future generations. In summary the underlying reality to the concept of future generations' rights would see rights without persons in a position to exercise them and even before to claim them fielding a series of instruments for the protection and the promotion. Steiner for example reasons that a future person cannot prevent or impose anything to a present person, or react to his failure for the simple fact that the two parties lack the requirement of "modernity" (Steiner and Madden 2008).

This asymmetry creates a seemingly insurmountable obstacle which makes impractical even a contractual perspective: future generations cannot negotiate the contents of an alleged "intergenerational pact"; they have nothing to offer in return or to use as a deterrent or incentive argument. In summary, then, if rights need someone who can ask and even before want them, and that it is able to claim a certain content or to react against a violation when it is perpetrated or simply attempted, talking about future generations' rights appears a very complicated formula.

The problem is if that what we have summarily recalled may be the only way to understand and configure rights; if on the contrary the intergenerational issue can above all be a key to rethinking and redefining the language of rights how to represent them and build around them a system of legal protections.

First of all, I think that a right can be fundamental even if the tools of legal defence and mechanisms of enforceability are not yet perfect and consolidated. The history of many rights we now consider fundamental and inviolable teaches us that the availability of legal remedies is not an original fact of the existence of a right, but rather an arrival point of a process that needs phases of adjustment and rooting both legal and cultural. Rights are not just individual resources but "goods", objective interests reflecting principles of justice supported by law.

They have an expansive and pervasive force in terms of objective law: they express a dynamic identity, a way of being of the law, indicating its essential purposes; in the same way, it seems simplistic not to allow the basic character but also the real existence of a right, depending on the degree and the intensity of protection instruments operated by the holders of that right.

The theme of the rights of future generations appears so at least "acceptable": we can talk about it, it is possible to use the moral force which undoubtedly belongs to the arguments "rights-based", in order to strengthen the other perspective of approach to the intergenerational issue, beginning with the duties, the responsibilities of the present generations towards future generations.

The other traditional objection to the configurability of future generations' rights is that of the unknowability, of "our ignorance" and of the unpredictability of the needs and the interests of future generations and at the same time of the ability and willingness of the future governments to satisfy them using the resources that we will be able to leave them (Golding 1972: 96).

Undoubtedly, it's a difficult argument which also may incur a criticism, that of risking ethical paternalism: who are we to shift onto the next and future generations our conception of life, our idea of what needs and interests should take priority? However, the extreme and intangible character of the interests that are part of the intergenerational issue, can in some ways alleviate the weight of this objection; these are goods that can be considered intrinsic to a necessary vision of humanity (Visser 'T Hooft 1999: 46).

Surely the intergenerational issue comes with an imprint that is one of the duties of the responsibility towards future generations. But it is a responsibility that has many points of contact with its inverted mirror, indeed the rights. First of all, human rights are inviolable and so earlier and later to the policy time. Secondly, on the background of the issue of future generations, there are goods certainly corresponding to rights, such as healthy environment, ecological balance, peace, retaining a cultural heritage. We can also talk about group rights, collective rights, that is, which could further overcome the "non-identity" paradox of Derek Parfit (1984). Collective rights don't represent an unknown figure to the law and its latest developments. I could do many examples indeed on the environment, of the protection of cultural and natural goods.

The relationship between generations and rights is to be envisaged even in another way. It can be a way of perceiving the rights, of understanding, that they should be exercised with responsibility, with awareness of the rights of others, even in the time.

According to this point of view, it changes also the concept of responsibility; it is subjected to a metamorphosis. It assumes meanings and implications not only retrospective and "curative" (e.g. the responsibility as a penalty, which in common law is called "liability"), but also a "perspective meaning", in the sense that it poses the problem of the preventive exercise of responsibility, of the orientation upstream of the actions and choices, of the subjective attitude. Being responsible means taking care, taking responsibility, the ability to target one's own behaviour towards the implementation of values that are common today and tomorrow (Gorgoni 2009: 257). A responsibility that is defined as a task; it is the role-responsibility (Hart 1965).

Seeking institutionalization of intergenerational justice

However, it is important that there has been this passage from the theoretical narrative to a first arrangement legal and constitutional of the future generations' issue. Indeed we begin to have not only constitutional rules, but laws, regional statutes, Constitutional Court's judgments, several decisions of the common judiciary.

In short the constitutional requirements mentioned above already at the current state appear capable of legitimizing the experimentation on the legislative and administrative plane, of (some) procedural rules, organizations, practices able to counter the "bias" towards the present decision-making structures of constitutional democracy.

There are many and various proposals that have been advanced in recent years and not only in the theoretical debate to define a line of translation of the intergenerational principle:

From the establishment of independent authorities and technical legitimacy to protect the interests of future generations (such as in the Maltese's proposal of the guardian of the future generations' interests or such as the "tribunes of posterity" imagined by Edith Brown Weiss (1989: 119)); to strengthening the role of NGOs in cooperation policies for the benefit of developing countries; to the conditioning of these policies to the adoption of policies aimed at protecting the environment and the cultural and natural heritage; to the insertion in the legislative and administrative procedures which may have a strong impact on the cultural and environmental heritage and on the natural resources, of subjects with intergenerational interests, through a conversion of information and participatory mechanisms that characterize these proceedings; to the provision of class actions in which the protection of the interests of "those who are not yet here" steps through the initiative of "substitute" individuals (such as the association of children in the Minors Oposa case decided by the Supreme Court of the Philippines in 1993 about the theme of conservation of rain forests), in which the applicants (they were children represented by their own parents) claimed to "represent their generation and also those who were yet to be born"

(Gatmaytan 2003: 457); until the proposal – this perhaps beyond the limits of tolerance allowed by the constitutional principles of the political representation and equal voting – to graduate the ownership of the voting rights on the basis of the major and minor household size in the belief that those who have children are naturally brought to reason in the intergenerational sense (Van Parijs 1998: 292).

Final remarks: intergenerational justice in action?

In conclusion, between constitutional law (and theory) and future generations (solidarity, responsibility, intergenerational justice) runs a complex and "promising" relationship in terms of developments and implications.

The constitutional language is "naturally" open to the future not only in the sense that the constitutional rules have the vocation to transform the society expressing a change project that constantly seeks its relationship with the problems and issues of the new times, but in the sense that the constitutional principles and constitutional structures (such as the state) are concerned about the man and his environment over successive generations in the continuity of time.

Around the Constitution as a legacy there is a mutual and bilateral bond. Who receives the inheritance has to respect it, keep it alive and pass it under conditions of effective applicability (at least in its fundamental principles) to those who come after. We do not yet possess perfect individual rights (of future generations) or duties legally sanctioned (of those present as well as States and the international community), but certainly the issue moves towards juridical consolidation that now needs always more bodies, procedures, verification tools. In a word, it needs actions, not only discussions.

Note

1 The first two "principles" of the Stockholm Declaration are: (1) "Man ... bears a solemn responsibility to protect and improve the environment for present and future generations"; and (2) "natural resources of the Earth, including the air, water, land, flora and fauna ... Must be safeguarded for the benefit of present and future generations, through careful planning and management".

References

Barry, Brian (1999) "Sustainability and Intergenerational Justice". In Andrew Dobson (ed.), *Fairness and Futurity*, Oxford University Press, New York, pp. 93–117.

Beckerman, Wilfred (2006) "The Impossibility of a Theory of Intergenerational Justice". In Joerg Chet Tremmel (ed.), *Handbook of Intergenerational Justice*, Edward Elgar, Northampton, pp. 53–71.

Bifulco, Raffaele (2008) *Diritto e generazioni future*, Franco Angeli, Milan.

Casavola, Franco (2004) "Il tempo del diritto". *Studium* 100(4–5): 687–692.

Gatmaytan, Dante B. (2003) "The Illusion of Intergenerational Equity: *Oposa v. Factoran* as a Pyrrhic Victory". *Georgetown International Environmental Law Review* 15(3): 457–485.

Golding, Martin Philip (1972) "Obligations to Future Generations". *The Monist* 56(1): 85–99.

Gore, Al (2013) *Il mondo che viene: Sei sfide per il nostro futuro*, Rizzoli, Milan.

Gorgoni, Giudo (2009) "La responsabilità come progetto. Primi elementi per un'analisi giuridica di responsabilità prospettica". *Diritto e Società* 2: 243–292.

Häberle, Peter (2006) "A Constitutional Law for Future Generations – the 'Other' Form of the Social Contract: The Generation Contract". In Joerg Chet Tremmel (ed.), *Handbook of Intergenerational Justice*, Edward Elgar, Northampton, 215–229.

Hart, Herbert (1965) *Il concetto di diritto*, Einaudi, Turin.

Haverkate, Görg (1992) *Verfassungslehre: Verfassung als Gegenseitigkeitsordnung*, Beck, Munich.

Jonas, Hans (1979) *Das Prinzip Verantwortung: Versuch einer Ethik für die technologische Zivilisation*, Insel, Frankfurt.

Kumar, Rahul (2007) "Torti inflitti a chi vivrà nel futuro: un'analisi contrattualista". *Filosofia e questioni pubbliche* 12(1): 51–68.

Meadows, Donella, Dennis Meadows and Jorgen Randers (2006) *I nuovi limiti dello sviluppo: La salute del pianeta nel terzo millennio*, Mondadori, Milan.

Moltmann, Jürgen (1971) *Teologia della speranza*, Queriniana, Brescia.

Mulgan, Tim (2006) *Future People*, Oxford University Press, Oxford.

Parfit, Derek (1984) *Reasons and Persons*, Clarendon Press, Oxford.

Passmore, John (1974) *Man's Responsibility for Nature*, Scribner's Sons, New York.

Pope Francis (2015) "Encyclical Letter *Laudato Si'* of the Holy Father Francis on Care for Our Common Home". 24 May 2015, available at http://w2.vatican.va/content/francesco/en/encyclicals/documents/papa-francesco_20150524_enciclica-laudato-si.html.

Posner, Eric A. and David Weisbach (2010) *Climate Change Justice*, Princeton University Press, Princeton, NJ.

Rawls, John (1971) *A Theory of Justice*, Belknap, Cambridge, MA.

Steiner, Sarah K. and Leslie Madden (2008) *The Desk and Beyond: Next Generation Reference Services*, Association of College and Research Libraries, Chicago, IL.

Stone, Christopher D. (1972) "Should Trees Have Standing? Toward Legal Rights for Natural Objects". *Southern California Law Review* 45(2): 450–501.

Tremmel, Joerg Chet (2006) "Establishing Intergenerational Justice in National Constitutions". In Joerg Chet Tremmel (ed.), *Handbook of Intergenerational Justice*, Edward Elgar, Northampton, ch. 10.

Van Parijs, Philippe (1998) "The Disfranchisement of the Elderly and Other Attempts to Secure Intergenerational Justice". *Philosophy and Public Affairs* 27(4): 292–333.

Visser 'T Hooft, Hendrik Philip (1999) *Justice to Future Generations and the Environment*, Kluwer Academic, The Hague.

Weiss, Edith Brown (1989) *In Fairness to Future Generations: International Law, Common Patrimony, and Intergenerational Equity*, United Nations University, New York.

Weiss, Edith Brown (1990) "Our Rights and Obligations to Future Generations for the Environment." *American Journal of International Law* 84(1): 198–207.

Zagrebelsky, Gustavo (2011) "Nel nome dei figli se il diritto ha il dovere di pensare al futuro". *La Repubblica*, 2 December.

18 The need for a new regulatory approach for the promotion of ecological sustainability

Massimiliano Montini and Francesca Volpe

Introduction

The present chapter explores the role of regulation for the promotion of ecological sustainability, within the context of a reappraisal of the sustainable development concept. Our analysis starts from the consideration that sustainable development has nowadays become a flagship word of our society. At the same time, with regard to the current use of the term sustainable development, two major risks, strictly interrelated, may be detected.

On the one side, there is "too much ado about nothing" while, on the other side, the "sustainable development building" is not built on a solid ground. In fact, there is presently a race to elaborate new solutions, to develop new technologies, to find new slogans concerning sustainable development, which often result in mere "green washing" practices, having very little in common with truly sustainable development. As noted, for instance, by Pope Francis with regard in particular to sustainable growth:

> in this context, talk of sustainable growth usually becomes a way of distracting attention and offering excuses. It absorbs the language and values of ecology into the categories of finance and technocracy, and the social and environmental responsibility of businesses often gets reduced to a series of marketing and image-enhancing measures.[1]

This distracting green washing trend, masked under the sustainable development "mania", is, in our opinion, closely connected to the second risk, namely the risk of a new kind of consumerism: a "sustainable development consumerism". This could be represented by a house where a first, second and third floor have been built, but do not lay on a solid, shared, ground. The risk is to undermine the entire building. Therefore, first of all we should agree on a solid ground, a theoretical basis upon which truly sustainable floors may be built. To this end, it is necessary to restart from the very roots, through a reappraisal of the sustainable development concept.

In the Rio+20 Outcome Document "The Future We Want", the signatory states agreed that sustainable development goals must address and incorporate in a balanced way all three dimensions of sustainable development and their interlinkages. But one substantive, preliminary, doubt emerges: after so many years since its inception, is sustainable development a concept adequately understood? Is it a concept which entails a shared vision for the future or is it just an omni-comprehensive slogan, an umbrella term encompassing a plethora of different, even contrasting, interpretations and claims? In this respect, it must be highlighted that different interpretations of the same concept may lead, and frequently lead, to different approaches and actions. This remark, although self-evident, entails serious consequences. In fact, it is exactly the ambiguity of the sustainable development concept that, in our opinion, has opened the way to the dichotomous behaviour which may be detected at international level: from one side, in theory, there is a general widespread plea for a more sustainable development model, the 2015 replacement of the Millennium Development Goals with the Sustainable Development Goals being a signal of this; from the other side, in reality, international governance promotes a development model fundamentally centred on the business as usual scenario, based on the main-stream "growth mantra".[2] What is even worse is that such a growth model is strengthened exactly by the fact that it is hidden under the sustainable develop-ment plea, although, as we will see in greater detail in the next paragraphs, it is far from being sustainable.

In such a context, two major problems arise. On the one side, a *common* understanding of sustainable development is missing, leading to the paradox that everything may be allegedly presented as sustainable, depending on the different interests at stake; from the other side, a *correct* understanding of sustainable development is missing, as there is no clear benchmark for assessing sustainability. Therefore, it seems that the time to seriously address three basic questions concerning sustainable development has come. They can be expressed as follows:

- What does sustainable development stand for?
- What should sustainable development properly mean?
- How can a truly sustainable development model be achieved?

Sustainable development: the need for a reappraisal

Grounded on the aforementioned premises, we start by addressing the first research question, namely "what does sustainable development stand for?" Originally, "sustainability" was the reference term. Its origins trace back to the beginning of the eighteenth century, when Von Carlowitz, the author of *Sylvicoltura Oeconomica*, who is considered to be the father of sustainable yield forestry, used the term sustainability to refer to the necessity of keeping the renewable resources levels of extraction within their regeneration capabilities through natural reproduction.[3] However, progressively, such strong original eco-logical connotation of sustainability has been replaced by a lighter environmen-tal characterization, which eventually turned "sustainability" into "sustainable development". Within such a new concept, the environmental dimension has been coupled, and overwhelmed, by the economic and social ones. According to Bosselmann, this has probably been the greatest misconception of sustainable development.[4]

In this way, if sustainability was still the reference concept in the 1972 *Limits to Growth*,[5] which introduced the term sustainability into the political arena, only fifteen years later sustainable development was unquestionably established as the leading concept by the 1987 Brundtland Report. Despite the fact that in the original mandate of the Brundtland Commission there was a certain awareness of the need to contribute to build a future based on "policies and practices that serve to expand and sustain the ecological basis of development"[6] and that in the Report there is a clear understanding that the global economy and the global ecology have become interconnected and locked together,[7] the centrality of the ecological dimension seems to be absent in the well-known definition of sustainable development contained in the Report. In the Brundtland definition, where sustainable development is the kind of development that meets the needs of the present without compromising the ability of future generations to meet their own needs,[8] the real compromised element seems to be sustainability in its original meaning.[9] As one can see, in the Brundtland definition there are no references to proper sustainability concepts such as ecological integrity, ecosystem health or biodiversity protection.[10] The situation has not changed meanwhile. Therefore, nowadays, sustainably developed is still often understood as merely promoting a sustained economic growth, somehow just made more environmentally friendly.[11]

Having briefly clarified what sustainable development stands for, we can now move to our second research question, namely "what should sustainable development properly mean?" This question could be expressed also as "what kind of balance should be found among the well-known three pillars of sustainable development?", namely the environmental, the social and the economic ones. Such inquiry has been called in the literature the "sustainability dilemma".[12] However, in our opinion the dilemma is only a political one, because actually the answer should be self-evident, being deeply grounded on the biophysical reality. The existence of biophysical limits is not a bias, rather it is clearly recognized by science, especially by the branch of physics known as thermodynamics, in particular with the so-called second law of thermodynamics, better known as "entropy law". This demonstrates that the availability of energy to perform useful work for human purposes is progressively reduced every time it passes through transformations such as the ones brought about by economic activities.[13] However, the mainstream economic model calls for a potentially limitless economic growth, an expansive vision where the goal is the maximization of profits, alongside the increase of gross domestic product. Such a model, however, which is no longer a purely economic model, having become the mainstream development model, does not take into proper account its inherent contrast with the biosphere, the "safe-operating space", characterized by limited natural resources and limited sinks for waste and pollution.[14]

On the same line of reasoning, Pope Francis in his encyclical letter *Laudato Si'* affirms that:

> the idea of infinite or unlimited growth, which proves so attractive to economists, financiers and experts in technology . . . is based on the lie that there is an infinite supply of the earth's goods, and this leads to the planet being squeezed dry beyond every limit. It is the false notion that "an infinite quantity of energy and resources are available, that it is possible to renew them quickly, and that the negative effects of the exploitation of the natural order can be easily absorbed".[15]

In the mainstream model, the economy is conceived as an isolated system, without dependency from the biosphere. Actually, the reality is exactly the opposite: the economy is a sub-system of the biosphere and "no subsystem can expand beyond the capacity of the total system of which it is a part".[16] The economy depends on the biosphere both as a source of inputs and as a sink for waste. However, both sources and sinks, although abundant, are physically limited. In such a context, technology, invoked by many as the solution, can only postpone the depletion or reduce the dependence, but cannot change the biophysical reality. As a consequence, the biological imperative "to learn to live sustainably on this planet . . . is an absolute imperative in that it is determined by the laws of nature and, hence, is non-negotiable".[17] These assertions may appear self-evident; however, the mainstream development model completely and dramatically ignores them.

In our opinion, starting from the Brundtland definition, in the mainstream approach Nature is neither, or no more, a value per se, nor yet a commodity. In fact, even disregarding the intrinsic significance of Nature, which is far from being considered a value per se in the mainstream approach, the economic system should at least take into account that without the ecosystem services there would be no economic, nor social or political system. Actually, it does not and the eco-system services are usually assumed to be free in the mainstream economic model.[18] This is what Sukhdev calls the "economic invisibility of Nature", meaning that "most of what nature provides is not transacted in markets whether it's clean air or fresh water or whether it's the pollination of bees for fruit trees".[19] In this regard, he usually asks in his talks: "When did a bee ever send you an invoice?" The pollination service of bees, for instance, is economically invisible but its total value was found to be 200 billion dollars, which is almost 8 per cent of the total agricultural output on Earth.[20] Dramatically, however, bees are progressively disappearing due to environmental pollution and pollination is nowadays hand-made by farmers in some regions of the world. In this way, economic growth is becoming uneconomic growth.[21]

Therefore, as very clearly stated by Bosselmann, who well conveys the proper meaning sustainable development should be given, "there is only ecological sustainable development or no sustainable development at all".[22] In more detail, "no economic prosperity without social justice and no social justice without economic prosperity, and both within the limits of ecological sustainability".[23] In our opinion, this is the solid ground, the theoretical basis sustainable development should be based upon, as "not only is the pursuit of biophysical sustainability non-negotiable; it's preconditional".[24] In brief, a society based on the mainstream development model which does not take into account ecological sustainability is like the house built on the sand, whereas a revised development model grounded on ecological sustainability is like the house built on the rock which does not fall down, as in the Holy Gospel.[25]

Therefore, we may conclude that the expression sustainable development should be interpreted in the light of ecological sustainability, in order to properly highlight and give the right value to its ecological core.[26]

From theory to practice: the need for a new regulatory approach

Having clarified in the previous paragraph how sustainable development should be correctly interpreted and understood, it is now time to move to our third research question, namely: "how can a truly sustainable development model be achieved?" This is necessary in order to translate theory into practice. In fact, as Pope Francis clearly stated in his encyclical letter, "the establishment of a legal framework which can set clear boundaries and ensure the protection of ecosystems has become indispensable".[27] In this sense, it seems interesting to explore and re-appraise the role that regulation can play to promote a shift towards a truly sustainable development, as correctly interpreted in the light of its ecological core.

The subject matter of regulation has been analysed mostly from the legal and the economic point of view, often with different patterns and results. A pivotal reference on this topic are the writings of Ogus, who has tried to combine the legal and economic theory on the subject within a single framework of analysis.[28] In such a context, Ogus starts from the definition of the term regulation provided by a social scientist, namely Selznick, according to which regulation can be defined as "a sustained and focused control exercised by a public agency over activities that are valued by a community".[29] As one can see, in such a definition the focus is on the one side on the actors holding the regulatory powers, that consist in a public authority or public agency, while on the other side it is on the activities which are being regulated, corresponding to those which have a certain value and importance for a given community. In Ogus's opinion, such a definition is a good starting point for the analysis. However, it needs to be narrowed further, in order to be applied to the legal and economic context. On such a basis, the author argues that "regulation is fundamentally a politico-economic concept and, as such, can be best understood by reference to different systems of economic organization and the legal forms which maintain them". In such a definition, as one can see, the focus shifts to the kind of economic organization in place in a certain society, it being more or less oriented towards a market-based system or a collectivist system, and on legal forms which are adopted to make it work.[30]

Building on Ogus's work, in our opinion, the centrality of regulation lies in its steering and framework-creating role and it should be understood as a comprehensive system of norms / provisions aimed at a common goal.

Quite surprisingly, however, if one observes the current regulatory system in the environmental field, it seems that such a legal regime rather than promoting a common goal, namely the sustainability of the ecological basis of social and economic development, merely deals with externalities associated with environmental pollution or other negative consequences on the environment, through a piecemeal approach.[31] This piecemeal approach theoretically aims at promoting environmental protection in a more tailor-made way but, in practice, often simply generates a lot of bureaucracy and is not suitable for promoting any concrete results in terms of promoting sustainability.

Environmental law historically started in the seventies of the last century as a response to the environmental emergencies created by the unregulated human activities on the environment. In fact, following the outcomes of the 1972 Stockholm Conference, in most jurisdictions the first environmental statutes were passed, dealing with the management and control of externalities with regard to specific environmental media, such as air, water and waste. Later on, the same piecemeal approach was extended also to the management of land planning activities, in order to protect the environment from the negative consequences which may derive from the implementation of projects as well as plans and programmes; this was made through the introduction of regulatory instruments such as EIA and SEA.

The piecemeal approach to environmental regulation in most countries has been characterized by a lack of a unifying and comprehensive vision. This is demonstrated by the fact that it is hard to find any clear and convincing definition of the term "environment" in most jurisdictions. For instance, within the European Union legal framework, an official definition of the term is missing in the Treaties. However, according to Kramer, it may be argued that, on the basis of art. 191 and 192(2) of the Treaty on the Functioning of the European Union, "the environment includes human beings, natural resources, land use, town and country planning, waste and water".[32] In such a context, the environment seems to be conceived as the sum of all the possible negative externalities that may affect the different environmental media. In other words, it may be said that environmental regulation is lacking an overall aim. This may be considered as one of the main inherent problems of environmental regulation, stemming from the present limited ecological culture, which, as Pope Francis argues, should not be "reduced to a series of urgent and partial responses to the immediate problems of pollution, environmental decay and the depletion of natural resources".[33]

Moreover, it must be acknowledged that in the last 20 years, in parallel with an increased globalization, there has been a global and widespread trend towards deregulation. Such a trend has been driven by the advent of the liberal agenda promoted *in primis* by Margaret Thatcher in the United Kingdom and Ronald Reagan in the United States. The underpinning value of such a deregulatory trend is the perception that an excess of regulation may hinder the competitiveness of the economic actors on the market as well the recognition that a command and control approach cannot solve all the environmental problems by itself. Therefore, the supporters of deregulation call for a reduction and streamlining of existing legislation in many sectors, including environmental protection.[34]

The combination of the lack of a unifying vision for environmental protection on the one side and the deregulatory trend on the other side has led to a paradoxical situation where environmental regulation is designed and implemented within the framework of a clear deregulatory trend. As the deregulatory agenda has become the reference scenario of the global economy, this has negatively affected the effectiveness of environmental law in most jurisdictions.

The effectiveness of environmental law, as it has been correctly proposed by Bodansky drawing on Young's theoretical analysis, should be analysed distinguishing among three dimensions: legal effectiveness, behavioural effectiveness and problem-solving effectiveness.[35] As advocated by Bodansky, the effectiveness of environmental law is often measured and analysed from a formal point of view. In fact, there is a certain tendency to concentrate merely on the legal effectiveness of environmental law, which focuses on whether the obligations written in a certain treaty are formally respected by the Parties, without dealing with the other two forms of effectiveness, namely the behavioural effectiveness and the problem-solving effectiveness. As a consequence, environmental law, even if formally respected, may end up contributing very little to environmental protection in a substantive way.[36]

As a consequence, we nowadays have a regulatory system for the protection of the environment that may end up protecting unsustainable conducts. This trend was well highlighted by Westerlund, according to whom "unless law is made sustainable, it protects unsustainable conducts".[37] The analysis conducted above fits within the statement provided by Westerlund. In fact, even a well-designed and correctly implemented environmental legislation does not necessarily lead to positive results in terms of achieving (ecological) sustainability in practice. This is due to the combination of the two factors highlighted above, namely the lack of a unifying vision within environmental regulation and the presence of a deregulatory agenda which limits the effectiveness of environmental law and policy. Therefore, as Westerlund correctly points out, sustainability may be achieved only if regulation is reframed in order to clearly protect and promote sustainable conducts, rather than tolerating and possibly even indirectly facilitating unsustainable conducts, by failing to place the ecological sustainability objective at the core of environmental regulation.[38] On the basis of this reasoning, in the next paragraph we will deal with the main instruments that may be relied upon in order to promote sustainability.

The regulation for the promotion of ecological sustainability: an outline

The correct regulation for the promotion of ecological sustainability should be premised on a change of perspective from at least three points of view:

the methodological perspective, the temporal perspective and the substantial perspective.

The methodological perspective

Ecological problems can hardly be tackled within the traditional legal structures and disciplinary perspectives. Our society has been deeply influenced by and still is deeply embedded within the Cartesian scientific paradigm, which focuses on a mechanistic approach. According to such an approach, nature can be analysed by partitioning reality into small parts and pieces.[39] This reductionist approach has led to a subdivision of science in several small disciplinary clusters. On the contrary, sustainability challenges go beyond any disciplinary framework, being extremely complex and multifaceted. Therefore, a correct approach to sustainability needs to overcome the disciplinary barriers towards a holistic and comprehensive vision. In this sense, a transdisciplinary approach is the proper methodology to analyse, frame and regulate the several multipolar and interconnected issues posed by sustainability.[40] Moreover, it could play a crucial role in addressing and solving the traditional fragmentation of environmental law caused by the already mentioned piecemeal approach, which characterizes legislation, policies and measures.

The temporal perspective

A new regulatory approach for the promotion of ecological sustainability needs a shift from a short-term to a medium–long-term perspective. The current environmental regulatory system is strongly characterized by short-termism as a consequence of its traditional "emergency approach", whereby environmental legislation is designed and implemented with a view to tackle the most relevant negative environmental externalities, not necessarily within a long-term reference theoretical framework. Such a short-termism, coupled with the absence of an overall vision and aim for environmental protection, necessarily leads to a widespread lack of effectiveness of environmental legislation and policies at a general level. As already mentioned above, in order to achieve sustainability it is instead absolutely necessary to conceive and implement environmental regulation in the framework of a holistic and comprehensive sustainability vision. This necessarily entails the need to shift from a short-term to a medium–long-term perspective.

Moreover, such a shift also represents a fundamental evolution of the environmental regulation from an economically oriented to an ecologically based perspective. In fact, the traditional emergency based environmental regulation was conceived as a tool to control and manage the environmental externalities caused by human activities in order to protect the human environment, following the anthropocentric approach of the 1972 Stockholm Conference and the 1992 Rio Conference.[41] Conversely, the new ecologically based environmental regulation should primarily aim at the protection of the ecosystems and the coexistence of human activities with all other living beings on an equal footing, being based on ecological sustainability as a foundational principle,[42] as aptly affirmed by the 2000 Earth Charter.[43]

The substantial perspective

On the basis and along the lines of the methodological and temporal perspectives seen above, it is necessary to proceed to a brief evaluation and reassessment of the existing regulatory instruments used so far for environmental regulation. In this sense, the starting point has to be the recognition that the current environmental regulation scenario is characterized by the use of two main types of instruments: command and control instruments and economic instruments.[44] In the last decades, the evolution of environmental regulation in most jurisdictions has been characterized by a gradual evolution from a simple regulatory system dominated by command and control instruments to a more complex regime where those instruments are intertwined with economic instruments. This evolution was mainly due to the fact that the original environmental regulatory system, purely based on a command and control approach, has shown through the time severe shortcomings in terms of excessive burdens on economic activities coupled with a not fully satisfactory record in the attainment of environmental objectives.[45] This has paved the way for the gradual introduction of economic instruments to deal with the most relevant environmental externalities.[46]

There are three main forms of economic instruments: taxes and charges, subsidies and tradable emission rights.[47] Among these types of instruments, in recent years the third one seems to be the most widely supported by both the business community and public authorities. In fact, taxes and charges, which in the nineties were widely advocated by economists and endorsed by some countries around the world,[48] have gradually lost their appeal due to the strong opposition of economic actors that fear their possible negative impacts on competitiveness. On the contrary, subsidies are welcomed by economic actors and have been sometimes used in several jurisdictions, for instance with regard to the promotion of renewable energies on the way towards a gradual decarbonization of the economy. Unfortunately, however, their widespread use has proven to be too costly and its generalized application seems not to be feasible. This might explain the relative success of tradable emission rights in recent years, which may be ascribed to their capability to please both the business community and public authorities. It is not possible to analyse in detail the several examples of tradable emission rights and permits devised and implemented around the world in recent times. However, it should be mentioned that the existing experiences have shown contradictory results.[49] For instance, the EU emission trading scheme (EU-ETS) designed for controlling and reducing CO_2 emissions in Europe, has been characterized by twofold results: on the one side, strictly speaking, it may be said to be characterized by a high level of compliance, as argued by the European Commission;[50] on the other side, it may be questionable whether it has truly achieved positive results in terms of its contribution to environmental protection and (ecological) sustainability, as noted by several scholars.[51] This shows that economic instruments should not be perceived as a panacea for all the possible shortcomings experienced with the use of command and control instruments.[52]

For the above-mentioned reasons, the search for a new regulatory approach based on ecological sustainability should in our opinion be accompanied by a reassessment of the existing environmental regulatory instruments and a new and better balancing between market-based instruments and command and control tools. The question to be answered therefore is the following: how to proceed to such a reassessment and new and better balancing? First of all, it is necessary that environmental regulation is conceived and implemented within a single holistic and comprehensive (ecological) sustainability framework. This does not necessarily mean that the currently existing environmental legislation must be reframed or repealed; what is most relevant is a correct application with a unifying vision in mind.[53] Therefore, the focus for the implementation of environmental regulation should shift from the formal compliance aimed at the protection of the single environmental medium to the substantial compliance related to the contribution of each single piece of legislation to the overall aim of ecological sustainability.

Within this new mind-set, the application of the existing regulatory instruments should be re-evaluated and re-assessed in the light of ecological sustainability. In such a context, for instance, the use of economic instruments should not be conceived as an alternative to command and control tools, as it has been often perceived so far. Quite on the contrary, the complementarity between the two types of instruments should be highlighted and enhanced. For instance, the use of a cap and trade system for the control of CO_2 emissions could still play a pivotal role in such a new scenario, but it should be placed in the framework of a properly designed command and control scenario. Within such a system, the regulatory authorities and agencies should have continuous monitoring on the correct functioning of the economic instrument in the light of the achievement of the ecological sustainability goal and should have the power to complement and substitute the application of the economic instrument, in case there is evidence of a certain lack of environmental effectiveness of a given economic instrument in place.

Moreover, in the light of ecological sustainability, some of the existing environmental regulatory instruments might be also reframed and partially redesigned. For instance, as it has been argued in greater detail elsewhere,[54] the two traditional instruments used for preventive evaluation purposes, such as the environmental impact assessment (EIA) procedure for the preventive evaluation of the possible negative effects of certain projects as well as the strategic environmental assessment (SEA), (sometimes also named strategic impact assessment, or SIA), for the preventive evaluation of the possible negative effects of certain plans and programmes, could be revised in the light of ecological sustainability. This revision could be made by merging the existing EIA and SEA procedures into a new comprehensive instrument based upon a holistic sustainability approach which could be named holistic impact assessment (HIA). Within such a context, the two types of existing evaluations will continue to exist and be conducted separately, still dealing respectively with the upstream (SEA) and downstream (EIA) assessment, but will be placed under a single framework, inspired by a common approach, governed by substantially the same rules and managed in a coordinated way. In such a way, the HIA, as the new comprehensive instrument for the preventive evaluation of projects as well as of plans and programmes, may become the common reference framework for the prior assessment of all the activities likely to have significant adverse effects on a certain land.[55]

Finally, the re-organization of environmental regulation along the lines of the ecological sustainability might sometimes imply the necessity to come up with the introduction of new regulatory measures especially designed to further the ecological sustainability objective. However, it is not the aim of the present chapter to further explore this issue that might be the object of further research in the future.

Conclusions

The new regulatory approach for the promotion of ecological sustainability should be premised on the recovery of the proper meaning of sustainable development as grounded on its ecological sustainability core. On this basis, firstly, the main focus of the process for the establishment of such a new regulatory framework should be centred on the definition of the correct methodological, temporal and substantial dimensions to further the ecological sustainability concept. Subsequently, the existing regulatory instruments should be re-evaluated and may be partially re-framed in the light of ecological sustainability. This should include the reassessment of the existing regulatory instruments and the promotion of a new and better balancing between economic instruments, including most notably market-based instruments, and command and control techniques, within a clearly defined regulatory scenario. Moreover, the revision of environmental regulation through ecological sustainability might sometimes imply the necessity to come up with the introduction of new regulatory measures that may be designed so as to concretely promote and further the ecological sustainability objective.

Notes

1 Pope Francis, "Encyclical Letter *Laudato Si'* of the Holy Father Francis on Care for Our Common Home", 24 May 2015, available at http://w2.vatican.va/content/francesco/en/encyclicals/documents/papa-francesco_20150524_enciclica-laudato-si.html, para. 194.
2 On the mantra of economic growth see, for instance, T. Jackson, *Prosperity Without Growth: Economics for a Finite Planet*, London: Earthscan, 2009.
3 U. Grober, *Sustainability: A Cultural History*, Totnes: Green Books, 2012, pp. 80–85.
4 K. Bosselmann, *The Principle of Sustainability: Transforming Law and Governance*, Aldershot: Ashgate Publishing, 2008, p. 23.

5 D. H. Meadows, D. L. Meadows, J. Randers and W. W. Behrens III, *The Limits to Growth*, Dartmouth: Potomac Associates, 1972.

6 Brundtland Commission's mandate, in World Commission on Environment and Development (WCED), *Our Common Future*, Oxford: Oxford University Press, 1987, p. 356. See also H. C. Bugge, "1987–2007: Our Common Future Revisited", in H. C. Bugge and C. Voigt (eds), *Sustainable Development in International and National Law*, Groningen: Europa Law Publishing, 2008, pp. 3–20, in particular pp. 5–10.

7 WCED, *Our Common Future*, para. I.2.15.

8 Ibid., para. I.3.27: "Humanity has the ability to make development sustainable to ensure that it meets the needs of the present without compromising the ability of future generations to meet their own needs".

9 Bosselmann, *Principle of Sustainability*, pp. 1–2.

10 B. Callicott and K. Mumford, "Ecological Sustainability as a Conservation Concept", *Conservation Biology* 1997, vol. 11, no. 1, pp. 32–40, in particular p. 35.

11 R. Douthwaite, *The Growth Illusion: How Economic Growth has Enriched the Few, Impoverished the Many, and Endangered the Planet*, Tulsa, OK: Council Oak Books, 1993, p. 286.

12 P. Martens, "Sustainability: Science or Fiction?", *Sustainability: Science, Practice, and Policy*, 2006, vol. 2, issue 1, pp. 36–41, in particular p. 37.

13 On this issue see, for instance, N. Georgescu-Roegen, *The Entropy Law and the Economic Process*, Cambridge, MA: Harvard University Press, 1971; N. Georgescu-Roegen, "The Entropy Law and the Economic Process in Retrospect", *Eastern Economic Journal*, 1986, vol. 12, issue 1, pp. 3–25.

14 J. Rockström, W. Steffen, K. Noone, Å. Persson, F. Stuart III Chapin, E. Lambin, T. M. Lenton, M. Scheffer, C. Folke, H. J. Schellnhuber, B. Nykvist, C. A. de Wit, T. Hughes, S. van der Leeuw, H. Rodhe, S. Sörlin, P. K. Snyder, R. Costanza, U. Svedin, M. Falkenmark, L. Karlberg, R. W. Corell, V. J. Fabry, J. Hansen, B. Walker, D. Liverman, K. Richardson, P. Crutzen and J. Foley, "Planetary Boundaries: Exploring the Safe Operating Space for Humanity", *Ecology and Society*, 2009, vol. 14, issue 2, no. 32, at www.ecologyandsociety.org/vol14/iss2/art32/; J. Rockstrom, W. Steffen, K. Noone, Å. Persson, F. Stuart III Chapin, E. Lambin, T. M. Lenton, M. Scheffer, C. Folke, H. J. Schellnhuber, B. Nykvist, C. A. de Wit, T. Hughes, S. van der Leeuw, H. Rodhe, S. Sörlin, P. K. Snyder, R. Costanza, U. Svedin, M. Falkenmark, L. Karlberg, R. W. Corell, V. J. Fabry, J. Hansen, B. Walker, D. Liverman, K. Richardson, P. Crutzen and J. Foley, "A Safe Operating Space for Humanity", *Nature*, 2009, vol. 461, no. 7263, pp. 472–475. On this issue see also E. Tiezzi, *Tempi storici, tempi biologici*, Pisa: Donzelli editore, 2005; H. E. Daly, *Beyond Growth: The Economics of Sustainable Development*, Boston, MA: Beacon Press, 1996; R. Costanza and H. E. Daly, "Natural Capital and Sustainable Development", *Conservation Biology*, 1992, vol. 6, no. 1, pp. 37–46.

15 Pope Francis, "Encyclical Letter *Laudato Si'*", para. 106.

16 J. Porritt, *Capitalism As If The World Matters*, London: Earthscan, 2007, p. 56. On this issue see also Daly, *Beyond Growth*.

17 Porritt, *Capitalism As If The World Matters*, p. 3.

18 On this issue see, for instance, R. Costanza, "Assuring Sustainability of Ecological Economic Systems", in R. Costanza (ed.), *Ecological Economics: the Science and Management of Sustainability*, New York: Columbia University Press, 1991, pp. 331–343.

19 P. Sukhdev, "Banking Nature", at www.alliancesud.ch/fr/infodoc/downloads/nature-eldorado-de-la-finance/Conducteur%20VA%20Banking%20Nature.pdf.

20 Ibid.

21 H. E. Daly, "Che cos'è lo sviluppo sostenibile?", in *Lettera internazionale*, 2007, pp. 20–24; New Economics Foundation, *Growth Isn't Possible: Why We Need a New Economic Direction*, Schumacher College, 2010; H. E. Daly, "Toward Some Operational

Principles of Sustainable Development", *Ecological Economics*, 1990, vol. 2, issue 1, pp. 1–6.

22 Bosselmann, *Principle of Sustainability*, p. 23.

23 Ibid., p. 53.

24 Porritt, *Capitalism As If The World Matters*, p. 8.

25 See Matthew 7:21–27.

26 Bosselmann, *Principle of Sustainability*, p. 53.

27 Pope Francis, "Encyclical Letter *Laudato Si*'", para. 53

28 A. I. Ogus, *Regulation: Legal Form and Economic Theory*, Oxford: Clarendon, 1994.

29 P. Selznick, "Focusing Organizational Research on Regulation", in R. Noll (ed.), *Regulatory Policy and the Social Sciences*, Berkeley, CA: University of California Press, 1985, p. 363, cited in Ogus, *Regulation*, p. 1.

30 Ogus, *Regulation*, p. 1.

31 Ibid., pp. 4–5.

32 L. Kramer, *EC Environmental Law*, London: Sweet and Maxwell, 2007, p. 1.

33 Pope Francis, "Encyclical Letter *Laudato Si*'", para. 111.

34 On the deregulation of environmental law and its risks see K. Bosselmann and B. J. Richardson, "Introduction: New Challenges for Environmental Law and Policy", in K. Bosselmann and B. J. Richardson (eds), *Environmental Justice and Market Mechanisms*, London: Kluwer Law International, 1999, pp. 3–18, in particular pp. 3–4; E. Rehbinder, "States Between Economic Deregulation and Environmental Responsibility", in Bosselmann and Richardson, *Environmental Justice and Market Mechanisms*, pp. 93–109.

35 See D. Bodansky, *The Art and Craft of Environmental Law*, Cambridge, MA: Harvard University Press, 2010, pp. 253–257. On this issue, see also M. Montini, "Revising International Environmental Law through the Paradigm of Ecological Sustainability", in F. Lenzerini and A. Vrdoljak (eds), *International Law for Common Goods: Normative Perspectives in Human Rights, Culture and Nature*, Oxford: Hart Publishing, 2014, pp. 271–287, in particular pp. 271–275.

36 On this issue see in greater detail Montini, "Revising International Environmental Law through the Paradigm of Ecological Sustainability", in particular pp. 272–275.

37 S. Westerlund, "Theory for Sustainable Development", in Bugge and Voigt, *Sustainable Development in International and National Law*, pp. 49–66, in particular p. 54.

38 Ibid., p. 52.

39 See F. Capra and P. L. Luisi, *The Systems View of Life: A Unifying Vision*, Cambridge: Cambridge University Press, 2014, pp. 22–26.

40 On transdisciplinarity see, for instance, T. Jahn, M. Bergmann and F. Keil, "Transdisciplinarity: Between Mainstreaming and Marginalization", *Ecological Economics*, 2012, vol. 79, issue C, pp. 1–10; D. J. Lang, A. Wiek, M. Bergmann, M. Stauffacher, P. Martens, P. Moll, M. Swilling, J. Christopher and C. J. Thomas, "Transdisciplinary Research in Sustainability Science: Practice, Principles, and Challenges", *Sustainability Science*, 2012, vol. 7, issue 1, pp. 25–43; P. Stock and R. J. F. Burton, "Defining Terms for Integrated (Multi-Inter-Trans-Disciplinary) Sustainability Research", *Sustainability*, 2011, vol. 3, issue 8, pp. 1090–1113; M. A. Max-Neef, "Foundations of Transdisciplinarity", *Ecological Economics*, 2005, vol. 53, issue 1, pp. 5–16; B. Nicolescu, "Gödelian Aspects of Nature and Knowledge", in G. Altmann and W. Koch (eds), *Systems: New Paradigms for the Human Sciences*, Berlin: de Gruyter Verlag, 1998, pp. 385–403.

41 See art. 1 and 2 of the 1972 Stockholm Declaration and article 1 of the 1992 Rio Declaration. The latter provision is particularly relevant in this sense, since it reads as follows: "Human beings are at the centre of concerns for sustainable development. They are entitled to a healthy and productive life in harmony with nature."

42 See K. Bosselmann, "Grounding the Rule of Law", in C. Voigt (ed.), *The Rule of Law for Nature: New Dimensions and Ideas in Environmental Law*, Cambridge: Cambridge

University Press, 2013, pp. 75–93, in particular pp. 87–90; Westerlund, "Theory for Sustainable Development", p. 60.

43 See art. 1 of the 2000 Earth Charter, which reads as follows: "1. Respect Earth and life in all its diversity. a. Recognize that all beings are interdependent and every form of life has value regardless of its worth to human beings. b. Affirm faith in the inherent dignity of all human beings and in the intellectual, artistic, ethical, and spiritual potential of humanity."

44 It should be noted that according to the well-known tripartition proposed by Stewart, there are three main types of regulatory instruments currently used for environmental regulation: command and control instruments, economic instruments and information based approaches. See R. B. Stewart, "Instrument Choice", in D. Bodansky, J. Brunnée and E. Hey (eds), *Oxford Handbook of International Environmental Law*, Oxford: Oxford University Press, 2007, pp. 147–181, in particular p. 149.

45 See Bosselmann and Richardson, "Introduction"; Rehbinder, "States Between Economic Deregulation and Environmental Responsibility".

46 See Ogus, *Regulation*, pp. 245–256.

47 Ibid., pp. 246–250.

48 See the example of the controversial introduction of carbon taxation in several European countries and the failed attempt to introduce an EU-wide carbon tax by the European Union in the nineties. On this issue see M. Grubb, J. C. Hourcade and K. Neuhoff, *Planetary Economics: Energy, Climate Change and Three Domains of Sustainable Development*, Abingdon: Routledge, 2014, pp. 207–231.

49 Ibid., pp. 237–274.

50 European Commission, at http://ec.europa.eu/clima/policies/ets/index_en.htm.

51 See S. Borghesi and M. Montini, "The European Emission Trading System: Flashing Lights, Dark Shadows and Future Prospects for Global ETS Cooperation", in F. Francioni and C. Bakker (eds), *The EU, the US and Climate Governance*, Farnham: Ashgate, 2014, pp. 115–126, in particular pp. 117–119 and literature cited there.

52 See Bosselmann and Richardson, "Introduction", pp. 4–9; Rehbinder, "States Between Economic Deregulation and Environmental Responsibility", pp. 97–100.

53 See M. Montini, "Revising International Environmental Law through the Paradigm of Ecological Sustainability", p. 282.

54 See M. Montini, "Towards a New Instrument for Promoting Sustainability beyond the EIA and the SEA: The Holistic Impact Assessment (HIA)", in Voigt, *The Rule of Law for Nature*, pp. 243–258, in particular pp. 253–257.

55 See Montini, "Towards a New Instrument for Promoting Sustainability beyond the EIA and the SEA", pp. 253–257. See also M. Montini, "Promoting the Ecological Sustainability of Climate Change Related Investments through the Holistic Impact Assessment (HIA)", in L. Westra, J. Gray and V. Karageorgou (eds), *Ecological Systems Integrity: Governance, Law and Human Rights*, Abingdon: Routledge, 2015, pp. 96–105, in particular pp. 103–104.

19 Migration with dignity for climate justice

The situation of small island developing states

Susana Borràs

There has been a tragic rise in the number of migrants seeking to flee from the growing poverty caused by environmental degradation. They are not recognized by international conventions as refugees; they bear the loss of the lives they have left behind, without enjoying any legal protection whatsoever. Sadly, there is widespread indifference to such suffering, which is even now taking place throughout our World.

(Pope Francis, "Encyclical Letter *Laudato Si'*", 24 May 2015)[1]

Introduction: the tragedy of small island developing states

Environmental degradation has been a significant factor in forced population movements, as people flee to survive natural disasters or move due to difficult and deteriorating environmental conditions, in search of opportunities elsewhere. In this sense, it is possible that climate change aggravates sudden and underlying disasters as well as gradual environmental degradation and affects migration all over the world.[2]

The already known effects of climate change will particularly affect small island developing states,[3] through the loss of territory and the forced movement of people. It is estimated that there are currently about 634 million people in the world who live in low-lying territories and, therefore, they are highly vulnerable to the effects of climate change, particularly rising sea levels.[4]

This chapter considers this forced movement of people from small island states who, as a consequence of the effects of climate change, are forced to survive in another place. This analysis seeks to highlight how responses to the problem of forced migration are frequently not based on attempts to address causes (of migration) but rather on the negative consequences of migration for public order or national security. Thus, solutions are developed to eradicate the consequences and not the causes, making public order the primary objective and not protection of human rights. Two concerns arise here:

- the obligation to protect vulnerable people and cause the states which have contributed most to that vulnerability to take responsibility for their actions; and

- recognition that the protection of dignity and the fundamental rights of such vulnerable people is caused, at least in part, by climate change.

Following these initial observations, we discuss: the vulnerability of the populations of small islands, at risk of losing their habitat and their state; the movement of population as a necessary adaptation strategy to the adverse effects of climate change; adequate legal protection of fundamental human rights, considering the application of refugee status, protection through International Human Rights Law, and the application of the 1954 Convention relating to the Status of Stateless Persons.[5]

We also include a further section on the concept of the need to be able to "migrate with dignity" assuring the safety and wellbeing of the population of these small island developing states.

The vulnerability of the small island developing states

According to the Intergovernmental Panel on Climate Change (IPCC) there are "sharp differences across regions"[6] in regard to the threats of climate change. Developing states, which are already facing multiple stress factors, are most likely to be affected. This is particularly so in Africa, Asia and Oceania where there are mega-deltas, small island developing states, low-lying coastal areas and drylands; all providing conditions which contribute to migration trends in the climate change context.

Elevation data in island states is revealing those states' vulnerability. The average elevation is much lower, about one metre for the Maldives and Tuvalu and about two metres for the rest of the states which are low-lying atolls. The IPCC concludes that sea-level rise "impacts on the low-lying Pacific Island atoll states of Kiribati, Tuvalu, Tokelau and the Marshall Islands may, at some threshold, pose risks to their sovereignty or existence".[7]

The first time that the small island developing states were acknowledged as a well-defined group of developing countries was in the United Nations Conference on Environment and Development held in Rio de Janeiro in 1992. A total of 179 countries adopted Agenda 21, which recognizes that "Some island developing states . . . are a special case both for environment and development . . . [and] they are considered extremely vulnerable to global warming and sea-level rise".

In April 1994, the Global Conference on the Sustainable Development of Small Island Developing States adopted the Barbados Programme of Action.[8] This Programme established specific actions and measures at national, regional, and international levels to support the small island developing states. Some areas which require action are: climate change, sea-level rise, natural and environmental disasters, energy resources, tourism, biodiversity, marine resources, transport and communication, and science and technology.

Later on, in September 1999, the General Assembly evaluated progress after the Barbados Programme of Action had been implemented for five years. The Assembly recognized the progress was "uneven" and identified the main tendencies, particularly increasing globalization, rising inequalities in terms of income, and the constant deterioration of the global environment. Likewise, it also warned that rising sea-levels could submerge low-lying islands and pointed out the importance of preparation against natural disasters.

In 2005, the Mauritius Strategy for Further Implementation of the Programme of Action for the Sustainable Development of Small Island Developing States was approved. The strategy establishes measures in 19 priority areas, including the original subjects from the Barbados Programme of Action.[9]

These international initiatives are due to the fact that many small island states or states with low-lying coastal areas are particularly vulnerable to the impacts of climate change, even eventually causing the displacement of the entire nation's population. Other effects are the loss of coastal land and infrastructure due to erosion, floods, rising sea-level and storms; increased frequency and severity of cyclones, with the consequent threat to life, health and housing; the detriment of coral reefs with the resulting impact on food safety and ecosystems on which many islanders' livelihoods depend; changing rainfall patterns, causing floods in some areas, droughts in others and threats to drinking water supplies; intrusion of salt water into the agricultural fields; and extreme temperatures.[10]

Over time, the cumulative effects, when aggravated by pre-existing pressures such as overcrowding, unemployment, poor infrastructure, pollution, fragility of environment, etc., can make these territories uninhabitable. In this sense, climate change can be a "turning point". Therefore, it is likely that the vast majority of the population will have to leave long before the land is actually submerged by the rising sea-level. Just as the process of climate change, so the movement of people out of the small island states and low-lying coastal states will be slow and gradual, whereas some events such as cyclones or high tides, can cause more abrupt movements, though probably temporary (and internal).

Thus, climate change could affect population movements in various ways. The increasing frequency and intensity of sudden and latent natural disasters, through climate change causes, results in emergency humanitarian action, and often consequently the movement of populations. Adverse consequences of global warming, climate variability and other effects of climate change on the means of sustenance, public health, food security and water availability, can aggravate pre-existing vulnerabilities and bring about migration. Rising sea-level can make many coastal areas and low-lying islands become uninhabitable. Competition for scarce natural resources could lead to tensions and conflicts, which in turn, may lead to a forced migration.

So, a consequence of this climatic vulnerability is the need to adapt, through relocation and/or planned migration of the affected populations. In 1990, the Intergovernmental Panel on Climate Change warned that the forced movement of people could be one of the most serious consequences of global warming, estimating that between 150 and 250 million people could be affected by the year 2050.[11]

One of the key cases is Kiribati, an archipelago consisting of 33 atolls and a volcanic island located 2,152 km south of Hawaii, in the Pacific Ocean.

The highest point is 4 metres above sea level. Not more than 103,000 people live on the different islands within 3.5 million square metres, making the archipelago a place of high population density. Kiribati is facing the consequences of climate change: it is estimated that every year water level rises at least 3.7 millimetres. Due to the topography of the islands, several metres of beach are lost per year.[12] This is serious for the people of Kiribati, as its population is concentrated entirely one kilometre from the sea. In this sense, in the case of Kiribati, as many other small island states, the movement of population is like a forced solution, as a survival and adaptation strategy.

Migrating as a strategy to adapt to climate change

Population movements have been a natural strategy for human adaptation to environmental variability. But the legal (and sometimes physical) barriers currently imposed by the states limit considerably many people's migration options. Anticipation and planning for population movement in the face of climate change can avoid disorders, loss of property and loss of lives, as well as the abrupt exodus of people moving spontaneously.

The United Nations Framework Convention on Climate Change (UNFCCC) of 1992 and its Kyoto Protocol of 1997 does not explicitly refer to displacement or migration.[13] These instruments focus on adaptation and mitigation of climate change and the relevant financing and support mechanisms. However, the Cancun Agreements of 2010 encourage all the parties to undertake adaptation plans of action, including "Measures to enhance understanding, coordination and cooperation with regard to climate change induced displacement, migration and planned relocation, where appropriate, at the national, regional and international levels."[14]

Given the vulnerability of the situation, some countries, Tuvalu among them, are already negotiating agreements with neighbouring countries to relocate their populations.[15] Other states, such as Maldives, have started saving to buy land for their populations to use in the future.[16] Kiribati, for instance, is looking for nearby territories where it could transfer its population. After negotiating with Fiji, Kiribati has bought 3000 hectares of land.

In this regard, in the Fifth Assessment Report of the Intergovernmental Panel on Climate Change (IPCC) it was recognized that migration can be an effective strategy of adaptation to environmental and climate change.[17] In the case of Vanuatu, for example, the relocation of a complete settlement in one of the islands was described as a climate change adaptation project.[18] Migration can undoubtedly help people to manage the risks, diversify their means of livelihood and face the environmental changes that affect their lives.

In this sense, preventative movement, while still being forced, is a rational answer of adaptation.[19] In this regard, it is important to remember that some population movements are temporary, circular or possibly seasonal; and many people do not wish to leave or cannot. In the case under analysis, it is a definite movement, with no possibility of return, and it is also forced. Further most people do not want to leave their communities but those who cannot leave or choose not to leave may be particularly vulnerable to climate change effects. Support needs to be given to those who leave and those who remain.

Protection through the status of refugees

The International Organization for Migration (IOM) defines migration as "a process of moving, either across an international border, or within a State. Encompassing any kind of movement of people, whatever its length, composition and causes; it includes refugees, displaced persons, uprooted people, and economic migrants."[20]

Most of the definitions coincide in interpreting environmental migration, regardless of the specific legal status, as any person who leaves their territory of usual residence mainly or very significantly due to environmental impact, whether it is gradual or sudden and within the same state or across international borders.[21]

Although there are many ways of referring to forced population movements depending on their particular circumstances, the desperate situation of small island developing states raises questions about the adequacy of protection of their citizens in applying the Status of Refugees.[22] The legal definition of "refugee" and the rights of the refugees appear in the Convention relating to the Status of Refugees of 1951 and its Protocol of 1967. According to this legal acquis a "refugee" is defined as someone who: "owing to well-founded fear of being perse-cuted for reasons of race, religion, nationality, membership of a particular social group or political opinion, is outside the country of his nationality and is unable or, owing to such fear, is unwilling to avail himself of the protection of that country; or who, not having a nationality and being outside the country of his former habitual residence as a result of such events, is unable or, owing to such fear, is unwilling to return to it."[23]

There are difficulties in characterizing "climate change" as a type of "persecu-tion". "Persecution" implies a serious violation of human rights "by its nature or repetition".[24] Part of the problem in the context of climate change is to identify a "persecutor". However, if one considers that the governments of the small island developing states such as Kiribati and Tuvalu are not responsible for climate change, and do not greatly develop policies which increase its negative effects on certain sectors of the population, it could be argued that the "persecutor" in this case would be the "international community". Industrialized countries in particular with their history of greenhouse gas emissions and their failure to reduce them, have contributed to the difficult situation these states are facing.[25]

According to the Convention, refugees run away from their own government (or private agents the government cannot or does not want to protect them from), but a person who runs away from the effects of climate change is not running away from the government but is instead often seeking protection in the countries which have actually contributed to climate change. This presents another problem in terms of the legal definition of "refugee" in the cases where the government is still capable and willing to protect its citizens.

The Refugee Convention requires persecution to be for reasons of race, religion, nationality, political opinion or membership of a specific social group. This is also a problematic issue as the impact of climate change is mostly indiscriminate, not linked to particular characteristics such as people's backgrounds or beliefs. Climate change has a negative effect on some countries by virtue of their geography or resources and not nationality and race. Consequently, it will be difficult to establish the argument that people affected by climate change could make up a "certain social group", because the law requires the group to be connected by a fundamental and immutable characteristic, other than the risk of persecution itself.

As a result, while the 1951 Convention, as amended by its Protocol of 1967, is considered the main instrument for protection of refugees, as a rule of customary international law,[26] it is recognized that the terms "climate refugees" and "environmental refugees" are recognized not to be, legally speaking, accurate or useful and therefore should be avoided. However it should be noted that the Convention of 1951 can apply to specific situations such as when "the victims of natural disasters flee because their government has consciously withheld or obstructed assistance in order to punish or marginalize them on one of the five grounds [of the Convention]".[27] These can take place during armed conflicts, situations of generalized violence, public disorder or political instability and even in peacetime. In relation to this concept it is necessary to note, firstly, that the definition of refugee only applies to people who have crossed an international border. This requirement is true in the case of many the small island developing states, as the disappearance of their national territories would inevitably lead to movement of their populations to territories under the jurisdiction of other states.

The inadequacy of the concept "refugee" has been confirmed by some cases in Australia and New Zealand, where some residents of Tuvalu and Kiribati applied for the recognition of refugee status because of the impact of climate change.[28] However, in none of these was the recognition of such a condition as "refugee" admitted.[29] For instance, in one of the cases which took place in New Zealand, the Refugee Status Appeals Authority explained that "this is not a case where the appellants can be said to be differentially at risk of harm amounting to persecution due to any one of these five grounds. All Tuvalu citizens face the same environmental problems and economic difficulties associated with living in Tuvalu. Rather, the appellants are unfortunate victims, like all other Tuvaluan citizens, of the forces of nature leading to the erosion of coastland and the family property being partially submerged at high tide."[30]

In another case coming before the Australian Refugee Review Tribunal it was declared: "In this case, the Tribunal does not believe that the element of an attitude or motivation can be identified, such that the conduct feared can be properly considered persecution for reasons of a Convention characteristic as required. It has been submitted that the continued production of carbon emissions from Australia, or indeed other high emitting countries, in the face of evidence of the harm that it brings about, is sufficient to meet this requirement. In the Tribunal's view, however, this is not the case. There is simply no basis for concluding that countries which can be said to have been historically high emitters of carbon dioxide or other greenhouse gases, have any element of motivation to have any impact on residents of low lying countries such as Kiribati, either for their race, religion, nationality, membership of any particular social group or political opinion".[31]

The matter is, nonetheless, disturbing: the uncertain legal situation of the people of the small island developing states only increases their vulnerability and insecurity in the face of the consequences of climate change. Given this reality, the question is how to proceed in protecting these people who are in real danger and avoid the disappearance of their state which increases their helplessness and vulnerability, trying to safeguard their dignity as individuals.

Protection through international human rights law

Climate change can have an impact on a range of human rights: on the right to life, the right to adequate food and the right not to be hungry, the right to drinkable water, the right to the highest level of health and the right to adequate housing.

Despite the limitations of obligation to protect by a host country, the law related to human rights has extended the protection obligations of countries beyond the "refugee" category. International human rights laws also create a basis for a "complementary protection" (i.e. a protection based on human rights additional to those intended by the Refugee Convention of 1951).

Under the International Covenant on Civil and Political Rights,[32] it is necessary that each state party recognizes, first of all and as a rule, civil and political rights to "all individuals within its territory and subject to its jurisdiction . . . without distinction" (Article 2).[33] Moreover, the International Covenant on Economic, Social and Cultural rights guarantees social, economic and cultural rights of all people without discrimination. In this regard, these agreements ensure migrants the application of fundamental rights including the right to life,[34] among other basic rights, such as the right to an adequate standard of living and health.[35]

One of the most important consequences of this additional protection is the prohibition of a return to a situation of real risk, of arbitrary deprivation of life or of inhuman or degrading treatment. Again, the question is to determine whether fleeing the impact of climate change can satisfy the remit of protection established in existing case law on human rights. However, national practice in many countries in granting some kind of permission to stay for people fleeing from natural disasters is based on the idea that these people are in need of international protection, even if only temporarily.

Although in theory, any violation of human rights could lead to an obligation of *non-refoulement*,[36] in most cases it is virtually impossible for an applicant to establish that the control over her/his migration was disproportionate in relation to any breach of a human right.[37] This is because, unlike the absolute prohibition to return someone to a place where it would be subject to inhuman or degrading treatment, most provisions on human rights allow a test of balance between the interests of the individual and the state, and this way they place the protection from being forced to return out of reach of everybody except the most exceptional cases.

Violation of a human right, such as the right to an adequate standard of living, may be regarded as a form of inhumane treatment leading to international protection. However, the question is to determine whether it is considered that violations not inflicted by the State from which people flee can lead to protection, or they are considered mistreatments leading to an obligation of protection by a third state. Case law has determined that the meaning of "inhuman or degrading treatment" cannot be used as a remedy for widespread poverty, unemployment or the lack of resources or medical care, except in the most exceptional circumstances.[38] And even though this existing law does not deny the possibility that climate impact is recognized as a source for inhuman treatment,[39] this seems grossly inadequate for situations of climate-induced displacement. In these, the responsibility for displacement is very diffuse, attributable to a large number of polluting states during many years, instead of direct mistreatment by any particular government towards a particular person. The large number of displaced people may require group-determined solutions rather than individual ones.

Moreover, unlike traditional protection, which responds to flight from damage inflicted or ordered by the state of origin, the protection sought for climate-induced displacement is the reverse: people can demand protection in industrialized countries precisely because they have the responsibility to help those who have suffered as a result of those countries' emissions over time.[40]

Protection through the status of stateless persons

The potential disappearance of island states through the effects of climate change increases the risk of generating the phenomenon of "large-scale *de facto* statelessness, which could turn into *de jure* statelessness should the affected state be considered to have ceased existence"[41] and in the case where the person had not acquired another nationality (for example moving to another country and becoming citizens). It is clear that the number of states in danger would affect the amount of eventually displaced people; however, the total population of the already mentioned states of Kiribati, Tokelau, the Maldives and the Marshall Islands adds up to less than 600,000 people. This number could be considered small in relation to the total number of people who could be temporarily or permanently displaced due to flooding, which the United Nations Development Programme (UNDP) estimated at 330 million if global temperatures rise by three to four degrees Celsius.[42] However, the outcome could be considerable and lead to statelessness for some affected populations.

The possible disappearance of small island developing states, through the loss of territory by sinking, would determine, according to international law, the disappearance of the country as a legal entity. It could also occur that long before the disappearance of the physical territory, the state becomes uninhabitable because of the precarious situation over subsistence of its population. Consequently, the lack of population or even the loss of effective government, before the physical disappearance of the state, would be determining factors for the legal disappearance of its international subject status.

The legal definition for a "stateless" person is established in Article 1 of the Convention relating to the Status of Stateless Persons of 1954, which is deliberately restricted to the person who "is not considered as a national by any state under the operation of its law".[43] This refers to a country which has actually "denied or deprived" a person of nationality. In this sense, in the case of the possible disappearance of small island states, the 1954 Convention would not protect its inhabitants unless the country in question had formally withdrawn the nationality from people, which seems unlikely due to the obligations imposed by the law on human rights.

However, if a country is recognized as non-existent, then its former population would fit in the "stateless" definition, as long as people had not acquired a new nationality. This would oblige signatory states to provide these people with the rights contained in the mentioned treaty within their territory, including that "they shall as far as possible facilitate the assimilation and naturalization".[44] The application of this provision requires that the population of the extinct state leave their home country and reach a signatory country before being able to claim its benefits.[45] This possibility seems remote because the situation is recent and not clear if states would be prepared to consider that a pre-existing country has "disappeared". The Convention only binds the few states that have ratified it, and thus only few states recognize this statute.

Human dignity and climate justice as a basis for the protection of "climate migrants"

Given the legal uncertainty and lack of protection of climate migrants, the value for inherent "dignity" for human beings, in its various expressions, provides a basis for the protection of these people against major emitters of greenhouse gases responsible for climate change. In this sense, climate justice is one of the forms of environmental justice, and seeks nothing less than fair treatment of all people and countries to prevent discrimination arising out of certain decisions and projects.

One of the first signs of dignity regarding the small island developing states is to avoid the use of terms like "disappearance" or "sinking" of the islands, although, as it has been suggested, some small island states and low coastal areas, can cease to exist due to sea levels rising and the impact thereof on the state and its people.[46] In this regard, despite the situation of loss of territory as a recent event, there is a general presumption of continuity of statehood and international legal personality under international law. Thus, statehood is not automatically lost with the loss of habitable land, and it is not necessarily affected by the movement of population.

Moreover, in keeping with the value of "dignity", the recognition of a person's rights as having nationality or holding a visa should not be the exclusive basis for respecting human rights, as all human beings have an intrinsic value, dignity, which should never be deprived to anyone. It is unlikely that many countries disappear completely due to rising sea levels; but there remains a very real concern that some of these countries may become uninhabitable, probably due, for example, to insufficient freshwater resources.

The second sign of "dignity" in the migration process is the necessary assumption that basic human rights correspond equally to each person, whether national or foreign. The rationale for protection of these rights universally recognized is human dignity itself beyond any accidental circumstances. Thus, it must be assured that migrants, as human beings, have "dignity" and value and cannot be limited to or reduced to mere commodities, interchangeable objects of trade in the global business market, where migration is seen as a strategy for development.[47]

Based on this generalization and considering that the planned resettlement of entire populations may be necessary in the case of the small island developing states, the process of ensuring a dignified movement for populations should guarantee the enjoyment of safety and applicable rights for those who are resettled. These include the rights to enjoy and practise their own culture and traditions and the right to continue to exercise the economic matters in their own areas or countries of origin. In particular, individuals should have access to information about reasons and procedures for their relocation and, where applicable, compensation and resettlement.[48]

Climate change certainly poses particular issues around shared state responsibilities and international cooperation, especially because of the distinct responsibility in causing climate alteration. The impact of climate change on small island developing states is clear and can be caused by the misuse of natural resources by developed nations, who are responsible for providing compensation and respecting the rights of the population affected, forced to leave their land. In response to this responsibility, the main but not exclusive duty of states is to alert people and protect them from displacement, mitigating its consequences, providing humanitarian aid and finding long term solutions.

According to the UN, migration is a right when it is specified that the migrant seeks decent living conditions and of a reasonable moral quality.[49] Article 13.1 of the Universal Declaration of Human Rights recognizes that "Everyone has the right to freedom of movement and residence within the borders of each state". If migration is a right, and in addition in this case a necessity for survival, it should not criminalize those who exercise it. Thus, in order to guarantee this right, it is necessary that immigration policies safeguard the dignity of those who migrate. In this way, they should focus on the organization of environmental migration to minimize its effects on the human rights of those affected and ensure that their vulnerability is not exacerbated due to the migration process. In this regard, states which are the destination for environmental migrants have the obligation not to discriminate and to adopt policies to ensure effective equality among migrants and protection of their most fundamental rights.

Likewise, the dignity of a climate migrant should be translated, under the concept of climate justice, as an obligation of providing humanitarian aid to people affected by climate change. Such aid should support migrants who move as a result of environmental changes at the time they migrate. It can have different forms, either as an emergency response to a sudden disaster or as actions planned in advance to accompany sustained movements of migrants, or to assist resettlement. Humanitarian aid should ensure that the most fundamental rights of migrants moving for environment causes are respected, in accordance with the principles of human rights, and that appropriate attention is paid to fundamental principles of non-discrimination, participation, empowerment and accountability.[50]

However, the lack of international guarantees for protection and the need to provide a dignified process of relocation for those populations moving to other foreign territories leaving their ancestral lands, have caused several small island developing states to propose migration policies to ensure future respect for their population as victims of the activities of industrialized states. Thus, the President of Kiribati, Anote Tong, besides asking the international community for aid and support needed to help his country, has also proposed a migration policy called specifically "migration with dignity" or "migrate with dignity".[51] In order to minimize the impact that the permanent relocation would have on their nation, this strategy proposes the creation of programmes to provide the necessary education and training for the people of Kiribati in order to take more advantage of the economic opportunities abroad in developed nations such as Australia or New Zealand.

Within the scope of this policy, the government of Kiribati refuses to consider its inhabitant as refugees, considering that the "refugee" concept is a response to an unexpected event, whereas climate change is not only expected, but its reality is evident. The objective, regardless of names, is to achieve the necessary conditions to enable its people to migrate with dignity, and not at a disadvantage.[52] This policy is based on the Dhaka Principles of 2013,[53] adopted by the Institute for Human Rights and Business which, in turn, are based on the Guiding Principles on Business and Human Rights and International Human Rights standards, as well as the Labour Standards of 2011. Its aim is to protect human and labour rights within the promotion of business, especially the recruitment industry, in order to respect the rights of migrant workers.

The initiative of Kiribati is in response to the indifference of the community of states responsible for having contributed to some aspects of climate change and the inadequacy of international instruments for the protection of people forced to migrate. And while this policy is a very important step to give some dignity to the population of Kiribati, it is not enough: on the one hand, it only helps to smooth the way for those who are ready and willing to emigrate, but doesn't reach everyone, especially those with very limited literacy skills or those with very poor livelihoods; and on the other hand, responsibility reverts to the victims of climate change.

This is a first step in building the concept of climate justice, fundamental for assuming the responsibilities in the face of causes and effects of climate change and a preventative protection for the most vulnerable populations.

Conclusions

Nowadays, one cannot deny the existence of a causal link between past emissions of greenhouse gases in many countries of the global North, which have generated changes in the global environment and the consequences of those emissions which generally affect the countries of the global South.

Dignity, as a value within climate justice, is a concept particularly relevant, especially when those most affected by anthropic climate change are the least responsible for the greenhouse emissions that caused the problem. Instead, living standards for people in developed countries (countries which are largely responsible for generating global warming) are the least likely to be affected negatively by the direct consequences of such warming.

Forced population movements from small island developing states is a reality. It is necessary to engage the major emitters of greenhouse gases in the defence and guarantee of the rights of these vulnerable groups and establish policies to ensure that such rights are protected. In this sense, and from the human rights perspective, welfare, safety and the sustainability of migrants' lives are urgent in their fight for justice.

The seriousness and injustice of the situation generated requires a proactive response to protect the dignity of the victims of climate change, particularly in the populations of small island developing states. In this regard, it is also necessary that international law provides a framework for the protection of these vulnerable people without relying only on the political will of states.

Consequently, the challenge for migration in the context of climate change requires the development of a new strategic approach in policies; a rethinking of the categories of human rights relevant to migrants; and the eventual establishment of effective protection mechanisms for people who are forced to move.

Acknowledgements

This chapter has been prepared under the DER2013-44009-P Project, entitled "Del desarrollo sostenible a la justicia ambiental: hacia una matriz conceptual para la gobernanza global" (2014–2016), whose main researcher is Dr. Antoni Pigrau Solé.

Notes

1 Pope Francis, "Encyclical Letter *Laudato Si'* of the Holy Father Francis On Care for Our Common Home", 24 May 2015, available at http://w2.vatican.va/content/francesco/en/encyclicals/documents/papa-francesco_20150524_enciclica-laudato-si.html.

2 The ex-Special Reporter on the human rights of migrants, Jorge Bustamante, in his final report of 2011 to the Human Rights Council pointed out the growing relevance of climate change and its effects on people's movement and recommended the studies on the effects of climate change and environment be extended to the mobility of human beings (A/HRC/17/33, paras 47–62).

3 These small island states are in three main regions: in Africa and parts of the Indian Ocean and the China Sea (Cape Verde, Comoros, Guinea Bissau, Maldives, Mauritius, São Tomé and Principe, Seychelles, Singapore), in the Caribbean Region (Antigua and Barbuda, Bahamas, Barbados, Belize, Cuba, Dominica, Grenada, Guyana, Haiti, Jamaica, Dominican Republic, Saint Kitts and Nevis, Saint Lucia, Saint Vincent and the Grenadines, Suriname, Trinidad and Tobago) and in the Pacific region (Fiji, Cook Islands, Marshall Islands, Solomon Islands, Kiribati, Micronesia, Nauru, Niue Palau, Papua New Guinea, Samoa, Timor-Leste, Tonga, Tuvalu, Vanuatu).

4 The exact amounts regarding environmental migrants varies considerably, and the estimated number of people likely to be displaced due to climate change is between 50 and 250 million by 2050. N. Stern, *Stern Review on the Economics of Climate Change* (Cambridge: Cambridge University Press, 2006), p. 77, shows that the number 250 million reflects a "cautious" hypothesis as a base for quantifying the movements induced by climate change.

5 Adopted on 28 September 1954 by a Conference of Plenipotentiaries convened by Economic and Social Council Resolution 526 A (XVII) of 26 April 1954, entry into force: 6 June 1960, in accordance with article 39.

6 See Intergovernmental Panel on Climate Change (IPCC), *Climate Change 2007: Synthesis Report* (compilers: R. K. Pachauri and A. Reisinger), Working Groups I, II and III, Contribution to the Fourth Assessment Report of the Intergovernmental Panel on Climate Change (Geneva: IPCC, 2007), p. 65.

7 See IPCC, "Assessment of International Working Group II: Impact, Adaptation, and Vulnerability", in *Climate Change 2007* (Geneva: IPCC, 2007), p. 736.

8 Established by the UN General Assembly Resolution 47/189, the UN Global Conference on the sustainable development of small island developing States took place in Barbados from 25 April to 6 May 1994.

9 Strategy adopted at the International Meeting to review the implementation of the Programme of Action for the Sustainable Development of Small Island Developing States, held in Port Louis from 10 to 14 January 2005, A/CONF.207/11 United Nations.

10 Otin Taai Declaration: Declaration and Recommendation of the Pacific Conference of Churches on Climate Change, Tarawa, Kiribati, March 2004.

11 IPCC, First Assessment Report (Geneva: IPCC, 1990).

12 See J. McAdam, "'Disappearing States', Statelessness and the Boundaries of International Law", in J. McAdam (ed.), *Climate Change and Displacement: Multidisciplinary Perspectives* (Oxford: Hart Publishing, 2010), pp. 105ff.

13 See *ILM* 31 (1992): 851.

14 Decision 1/CP.16, The Cancun Agreements: Outcome of the work of the Ad Hoc Working Group on Long-term Cooperative Action under the Convention, in the Conference of the Parties Report on its sixteenth session, Supplement, part II: Measures adopted by the Conference of the Parties FCCC/CP/2010/7/Add.1, 15 March 2011, par. 14 (f).

15 B. Crouch, "Tiny Tuvalu in 'Save Us Plea' Over Rising Seas", *Sunday Mail* (Adelaide, Australia; 5 October 2008).

16 A. Revkin, "Maldives Considers Buying Dry Land if Sea Level Rises", *The New York Times* (November 2008). See also Report of the Rapporteur on the Human Rights of internally displaced persons, Mission to Maldives (A/HRC/19/54/Add.1).

17 IPCC, *Climate Change 2014: Impacts, Adaptation, and Vulnerability – Summary for Policymakers*, Working Group II Contribution to the Fifth Assessment Report of the IPCC, (Geneva: IPCC, 2014).

18 "National Adaptation Programme of Action (NAPA) for Vanuatu", conducted under the sponsorship of the Framework Convention, available at http://unfccc.int/national_reports/napa/items/2719.php (accessed 25 January 2011).

19 For more details, see International Organization for Migration (IOM), *Migration, Climate Change and the Environment* (Geneva: IOM, 2009).

20 See IOM, *Glossary on Migration* (Geneva: IOM, 2004); IOM, *Climate Change, Environmental Degradation and Migration*, no. 18 (Geneva: IOM, 2012).

21 E. El-Hinnawi, *Environmental Refugees* (Nairobi: United Nations Environment Programme, 1985), p. 4; A. Suhrke and Visentin, "The Environmental Refugee: A New Approach", *Ecodecision* (September 1991): 73–74; A. Suhrke, "Environmental Degradation and Population Flows", *Journal of International Affairs* 47(2) (1994): 473–496, among others.

22 The usual definition of "environmental refugee" comes from El-Hinnawi's United Nations Environmental Programme "Environmental refugees are defined as those people who have been forced to leave their traditional habitat, temporarily or permanently, because of a marked environmental disruption (natural and/or triggered by people) that jeopardized their existence and/or seriously affected the quality of their life. By 'environmental disruption' in this definition is meant any physical, chemical and/or biological changes in the ecosystem (or the resource base) that render it, temporarily or permanently, unsuitable to support human life" (El-Hinnawi, *Environmental Refugees*, p. 4).

23 Convention relating to the Status of Refugees (taken on 28 July 1951) 189 UNTS 137, Article 1 A (2), and its Protocol relating to the Status of Refugees of 1967 (taken on 31 January 1967) 606 UNTS 267.

24 See Council Directive 2011/95/EU of the European Parliament and of the Council of 13 December 2011 on standards for the qualification of third-country nationals or stateless persons as beneficiaries of international protection, for a uniform status for refugees or for persons eligible for subsidiary protection, and for the content of the protection granted [2011], *Official Journal of the European Union* 337, 20.12.2011, pp. 9–26, Article 9.1.a.

25 IPCC, *Climate Change 2007: Synthesis Report*, 5, 6, 12, 13.

26 Declaration of the State Parties of the 1951 Convention and/or 1967 Protocol Relating to the Status of Refugees, par. 4, Doc. ONU HCR/MMSP/2001/09, 16 January 2002.

27 UNHCR "Forced Displacement in the Context of Climate Change: Challenges for States under International Law", 20 May 2009, pp. 9–10.

28 In New Zealand the cases were: Refugee Appeal No. 72189/2000, RSAA (17 August 2000); Refugee Appeal No. 72179/2000, RSAA (31 August 2000); Refugee Appeal No. 72185/2000, RSAA (10 August 2000); Refugee Appeal No. 72186/2000, RSAA (10 August 2000); Refugee Appeal No. 72313/2000, RSAA (19 October 2000); Refugee Appeal No. 72314/2000, RSAA (19 October 2000); Refugee Appeal No. 72315/2000, RSAA (19 October 2000); Refugee Appeal No. 72316/2000, RSAA (19 October 2000); Refugee Appeal No. 72719/2001, RSAA (17 September 2001). Cases in Australia: 0907346 [2009] RRTA 1168 (10 December 2010); N00/34089 [2000] RRTA 1052 (17 November 2000); N95/09386 [1996] RRTA 3191 (7 November 1996); N96/10806 [1996] RRTA 3195 (7 November 1996); N99/30231 [2000] RRTA 17 (10 January 2000); V94/02840 [1995] RRTA 2383 (23 October 1995).

29 See J. McAdam, "Our Obligations Still Apply Despite High Court Win", *Sydney Morning Herald* (30 January 2015).

30 Refugee Appeal No. 72189/2000, Refugee Status Appeals Authority of New Zealand, 17 August 2000, par. 13.

31 See 0907346 [2009] RRTA 1168 (10 December 2009) par. 51 (Refugee Review Tribunal of Australia).

32 United Nations, *Treaty Series*, vol. 2296, No. 40906.

33 See Resolution 2200A (XXI), annex.

34 United Nations, *Treaty Series*, vol. 2296, No. 40906, Article 6.

35 See Resolution 2200 A (XXI), annex, Articles 11 and 12.

36 *Refoulement* means the expulsion of persons who have the right to be recognized as refugees. The principle of *non-refoulement* was first laid out in 1954 in the UN Convention relating to the Status of Refugees, which, in Article 33(1) provides that: "No Contracting State shall expel or return ('refouler') a refugee in any manner whatsoever to the frontiers of territories where his life or freedom would be threatened on account of his race, religion, nationality, membership of a particular social group or political opinion." R c. Special Adjudicator, ex parte Ullah [2004] UKHL 26, paragraphs 24–25 (Lord Bingham), 49–50 (Lord Steyn), 67 (Lord Carswell) in English.

37 Kacaj c. Secretary of State for the Home Department of the UK [2002] EWCA Civ 314, paragraph 26, in English.

38 D c. the United Kingdom (1997) 24 EHRR 423; consult also the opinions of the Committee Against Torture, as in AD c. The Netherlands, Communication No. 96/1997 (January 24, 2000), UN Doc. CAT/C/23/D/96/1997, paragraph 7.2.

39 R c. Special Adjudicator ex parte Ullah [2004] UKHL 26, in English; Human Rights Committee, General Comment 15: The position of aliens under the Covenant (11 April 1986), paragraph 5; see also Human Rights Committee, General Comment 18: Non-discrimination (10 November 1989).

40 This is a variation on the argument contained in the request of the Inuit petition to the Inter-American Commission on Human Rights seeking relief from violations resulting from global warming caused by acts or omissions by the United States, December 7, 2005, available in English at: www.earthjustice.org/library/legal_docs/petition-to-the-inter-american-commission-on-humanrights-on-behalf-of-the-inuit-circumpolar-conference.pdf.

41 S. Park, *Climate Change and the Risk of Statelessness: The Situation of Low-lying Island States*, Legal and Protection Policy Research Series, (Geneva: Division of International Protection, ACNUR, 2011), p. 33.

42 UNDP, *Human Development Report 2007/2008: Fighting Climate Change: Human Solidarity in a Divided World* (New York: UNDP).

43 Convention Relating to the Status of Stateless Persons (28 September 1954) 360 UNTS 117, Article 1.

44 See Article 32 of the Convention of 1954.

45 In this regard, it is important to note that, although little ratified and implemented, the 1961 Convention to reduce statelessness requires states to ensure that no person shall become stateless as a result of the transfer of territory (Article 10). Also refer to the draft articles of International Law Commission on nationality, note 91, Article 1, which contains "the right to nationality"; Article 4 requires states to take measures to prevent statelessness resulting from succession.

46 J. McAdam, "'Disappearing States': Statelessness and the Boundaries of International Law", in J. McAdam (ed.), *Climate Change and Displacement: Multidisciplinary Perspectives* (Oxford: Hart Publishing, 2010), pp. 105–130.

47 The ILC has already confirmed, for instance, the importance of the basic principles of international law for climate-related displacement. These principles include humanity and human dignity, whereas the principle of international cooperation merits further examination. Any documents related to the work of the ILC on the protection of people on disasters is available at: http://untreaty.un.org/ilc/guide/6_3.htm.

48 Guiding Principles on Internal Displacement, Principles 7–9.

49 McAdam, "'Disappearing States'", pp. 105ff.

50 See the Report of the Special Rapporteur on Human Rights of migrants, A/67/299, 13 August 2012.
51 For more information on this policy see www.climate.gov.ki/2013/02/12/i-kiribati-want-to-migrate-with-dignity (accessed 3 March 2015).
52 See J. McAdam and M. Limon, *Policy Report: Human Rights, Climate Change and Cross-Border Displacement: The Role of the International Human Rights Community in Contributing to Effective and Just Solutions* (Geneva: Universal Rights Group, August 2015).
53 For more information see www.dhaka-principles.org (accessed 3 March 2015).

20 Weapons of mass distraction

Rose A. Dyson

Introduction

Unbridled consumerism presents humankind with multiple challenges as the global population grows. Advanced capitalist economies generate not only an ever increasing array of products but growing emphasis on creating new demands (Whitly 2010). Clearly, a delicate balance must be struck as more people inhabit the planet. Fewer needs, desires and wants are a must as our collective ecological footprints grow. Changing behaviour and common practices to ensure a sustainable future is now the primary role for educators, policy makers and researchers. It means bridging the gap between theory and practice. Nowhere does the disconnect between the two persist more than in the field of media content, communications technologies and cultural policy – where they are leading us, how to harness their enormous potential and where the balance between regulation and cherished freedoms ought to lie?

Time is running out. As Naomi Klein points out in her latest book, *This Changes Everything: Capitalism versus the Climate*, in which she amplifies the latest warnings from the International Panel on Climate Change, we no longer have decades left to turn things around – only years (Klein 2014). Building safe, liveable communities now includes the added overlay of unfolding complications due to climate change, shrinking resources and financial instability. Adaptability is becoming the new normal. The need for a new economic order that encompasses modes of social and economic organization based on co-operation and social responsibility designed to serve life rather than accumulation requires examination of how profit driven, corporate media block such objectives with emphasis on materialistic definitions of desirability and success. Countless studies have been done and books written on the subject. The task ahead is to translate this accumulated knowledge into effective policy before it is too late.

Peter Nicholson, President of the Council of Canadian Academies, tells us that as we become information-rich, we are becoming attention-poor, an inevitable side effect of the digital revolution. "Economics teaches us", he says, "that the counterpart of every new abundance is a new scarcity – in this case, the scarcity of human time and attention" (Nicholson 2009). Additional observations and reports warn us of radiation overload from too much cell phone use and screen time, ways in which ever intensifying stimulants first initiated by televi-sion diminish our capacity for imagination and creativity, and how these accumulating side effects still receive little attention in policy making circles (Robbins 2010).

Pope Francis, in his *Laudato Si'* encyclical letter, challenges us to look beyond our prevailing economic practices and modes of thought that rely on markets to do their magic. We are asked to re-examine the creed of "individualism, unlimited progress, competition, consumerism, and the unregulated market" (Pope Francis 2015). Appropriate boundaries are needed to redefine our notion of progress. In this context, laws and regulation should not only provide barriers to wrong doing but amplify best practices (ibid.).

Distractions from reality

Chris Hedges, warns us of the diminishing distinction between reality and illusion (Hedges 2009). Such trends lead to death and we are at a crossroads. Either we will wake up from our state of induced childishness, where trivia, gossip and celebrity worship pass for news and information in our search for an elusive and unattainable happiness and confront the stark limitations before us or we will continue our headlong retreat into fantasy and ultimate demise. In his book, *Consumed: How Markets Corrupt Children, Infantilize Adults and Swallow Citizens Whole*, Benjamin Barber adds to the message of urgency. He offers examples of a global economy that overproduces goods and increasingly targets children as consumers because there are never enough shoppers. The primary goal is no longer to manufacture required goods but to create needs and desires (Barber 2007).

Meanwhile, the propaganda machines grind on with too many reporters functioning as stenographers for corporate elites rather than hard working investigative journalists (McChesney and Nichols 2010). Yet, the extent to which our cultural environment is controlled by large conglomerates, dominated by marketing and advertising interests, is poorly understood by the public at large. Two examples of how this occurs in the Canadian media emerged in early 2015. Bell Media's president, Kevin Crull, found himself having to apologize for intervening in CTV's news coverage of a landmark regulatory decision on the television industry involving pick and pay rather than bundles of programs as options for consumers, brought down by the Canadian Radio-Television and Telecommunications Commission (CRTC). Miffed because of the profits his corporation stood to lose, he directed senior news staff to exclude chairman Pierre Blais from coverage of the story on Bell-owned networks (Bradshaw 2015). Another example involved a "cautionary note" from Advertising Standards Canada, on 25 March 2015, about ads that disguise themselves as articles in the media – a technique associated with what is known as branded or sponsored content, or "native" advertising (Krashinsky 2015).

Educators, activists and media scholars have for years chronicled from the margins the systemic deterioration of journalism as well as entertainment media, increasingly dominated by violent content because it sells well in a global market and translates easily into any language. The prevailing norm in mainstream news coverage means if it bleeds, it leads. A deep seated and long term crisis has resulted because media owners have made the commercial and entertainment values of the market dramatically higher priorities than the civic and democratic values essential to good journalism and, in turn, a good society. An erosion of standards has also led to a rise in stories about sex scandals and celebrities, giving the illusion of controversy. This has led to an increasing displacement of good journalism with sophisticated propaganda which tells people what they need to know to consume products and support spurious wars or tougher law and order measures but little of what they need to know to be voters and responsible citizens.

According to David Suzuki, over US$500 billion is spent annually by the advertising industry to get any of us on the planet to buy things (The11thHour. com). Record profits are frequently announced by Microsoft and other gaming manufacturers who socialize our youth to amuse themselves with endless sedentary, interactive screen time with video and computer games such as Killzone Liberation and World of Warcraft. Many, such as Manhunt 2, banned in the UK are produced in Canada and subsidized by our own tax dollars. In December 2011 *The Economist* predicted that video games would be the fastest growing form of mass media, estimated at around $82 billion by 2015. The action packed video game Call of Duty: Black Ops, for example, had in one month taken in more than $1 billion in sales as fans in countries around the world queued for blocks on the first day of its release in 2010, to purchase a coveted early copy – an example of what Barber (2007) calls *infantilized consumerism*.

Demonizing government regulation

One of the biggest challenges for educators in this digital age is to dissect the myth that has sprung up in the last century and a half about the evils of government regulation of media. The notion that only the corporate sector is the true guardian against censorship has developed into a woeful under-examination of the privileges media moguls enjoy and perpetuated the myth of "self-regulation" as the only viable option. Public interest obligations in the news media, originally promised in exchange for these privileges, have virtually disappeared in recent decades. Freedom of the press rests with those who own one, or control the broadcast microphones or own the Internet servers.

Responsible journalism ought to be regarded as a public good like health care, national parks and defence. Historians have tended to avoid attempts to reconcile the fact that the same enlightened sages of the past who laid the foundations for democratic governance, also created a partisan press system, subsidized by political parties and government contracts. Furthermore, these have never been "free" markets. News, advertising and entertainment, either in print or electronic form, have been recipients of huge direct and indirect subsidies, tax breaks and exemptions. Few people understand this.

On the eve of China's 60th anniversary celebrations in 2008, a front page article ran in the *Toronto Star* on Beijing's vast and effective propaganda apparatus (Schiller 2009). But the reporter's description of how inhibited Chinese journalists are compared to those in the West was overstated. Granted, the firing of Chinese journalist Liu Yuan for breaking a story on the death of a 15-year-old boy at a camp to cure Internet addiction was inexcusable. Comparatively speaking, such a story in Canada might well have resulted in an award for the journalist. But that is as far as it would be allowed to go. Corporate media interests are far more likely to block any progress toward serious policy development than to address the problem of Internet addiction, itself.

This is precisely what occurred in 2008 when a tsunami of opposition orchestrated by corporate media interests broke out in the mainstream media over a bill proposed in the House of Commons which would have eliminated tax credits for audio-visual productions deemed to be harmful to the public interest. Industry lobbyists rejected the argument that such discretionary funding is expected of any democratically elected government entrusted to set policy on how public money is spent. In 2008, it was pointed out to the Canadian Senate Committee on Banking, Trade and Commerce that, in the previous 12 years, taxpayers contributed over $22 billion to the audio-visual industry regardless of the nature of the content. In its 2015 budget, the Ontario Government underscored this backward trend. On 30 April, the *Globe and Mail* reported that the province's video game companies were "declaring a victory after a much sought-after reform to a digital media tax credit was unveiled" (Dingman 2015). It was explained that this credit will only apply to "entertainment products and educational products for children under the age of 12 years" (ibid.).

The upside-down world of journalism is hardly unique to China. The difference in Canada is that it is the corporate media rather than party officials who call the shots. Indeed, in China, the problem of Internet addiction, among youth in particular, is at least acknowledged. Within the mental health community in Canada, growing concerns from parents, educators and family therapists about increasing evidence of Internet addiction among youth tend to be ignored. That is, with the possible exception of ways in which the Internet is now acknowledged as a primary source of sex education for children. In Ontario the introduction of new sex education curricula in schools, in part to address the problem, remains fraught with controversy (Brown 2015). Remarkably, parents, teachers and policy makers bent on protecting children, all continue to overlook the degree to which the unregulated Internet robbed youth of its innocence long ago. Too often, the seduction of children by new media technologies is a problem considered too messy and difficult to address despite news reports that Canadian children are engaged in 3 times the medically recommended amount of any kind of screen time (Ogilvie 2008). In Japan the government has warned parents and teachers of new legislation restricting screen time for youth as a result of evidence of widespread addiction to computer use. A growing number of countries in the developed world, including South Korea and Switzerland, have added to the list of those who have banned violent video and computer games (Grossman 2010).

So far, Quebec is the only jurisdiction in North America to have at least banned advertising to children 13 years and under on the basis of research showing harmful effects. How effectively these laws are enforced is another matter.

Parents are now up against enormous odds in rearing their children. Ways in which the advertising industry knowingly undermines family cohesion through the use of marketing tactics such as "the nag factor" and "pester power" need to be better understood and resisted. Essentially children are targeted to wear down the resistance of their parents on billions of products involving toys, junk food, digital entertainment and much more. Aggressive marketing yields US$1 billion in annual sales for Brainy Baby and Baby Einstein videos despite protests from the Campaign for a Commercial Free Childhood, based at the Harvard University Medical School. Both the Canadian and American Pediatrics Academies recommend no screen time at all for children under the age of two, yet only 6 per cent of parents are aware of this. Evidence demonstrates that children who watch television or video games before the age of two actually show slower vocabulary development than those who experience no screen time (CCFC 2007a).

The problems are mushrooming. It turns out that teenagers with large social media followings are easily bought and enormously effective as the new power brokers of digital advertising. Much of it is highly personalized and disguised as advertising entirely as young people move away from the "in your face" marketing on Facebook and gravitate toward emerging platforms like Twitter, Instagram and Snapshot. The ads become a part of the content itself. This is also known as "advocacy advertising" or "product placement". Concerns are growing that by enlisting young people as stealth messaging mules, companies are crossing an ethical line. For a few dollars or a t-shirt, a company can avoid having to pay out millions to supermodels and star athletes. University of Toronto's editor-in-chief of *The Varsity*, Danielle Klein, says she is deluged with online solicitations from students who have struck promotional deals with marketers of various products, local clubs or other advertisers. In the process, so called codes of conduct promoted by advertising standards organizations are being further eroded (Lorinc 2015).

Digital technologies and carbon emissions

Harmful media content also fuels climate change and social upheaval in ways that go far beyond obvious concerns about fair and accurate reporting or the effects of violent entertainment. Despite the popularity of Al Gore's celebrated documentary, *An Inconvenient Truth*, vast amounts of fossil fuel continue to light up billboards, and power vehicles for a single film, television or commercial shoot. In 2007, Canada's national newspaper, *The Globe and Mail*, ran an article on a television recycling plant near Toronto's Pearson International Airport where overworked technicians scramble to keep up with the demand for removal of lead and other harmful metals from old fashioned tube style sets as viewers switched to flat screens. But for every old set shredded, it was reported, 600 new ones are manufactured for distribution in China. Similar stories abound about the life cycles of computers, cell phones and other forms of communications technologies.

Ironically, at the 2008 Conference of the Canadian International Council, held in Toronto, Wenran Jiang from the University of Alberta spoke of how China was starting to wake up to the ills of environmental devastation and climate change and is especially concerned about becoming the dumping ground for 90 per cent of the world's electronic garbage.

In California's Silicon Valley, it has been reported that many data centres appear on the state government's Toxic Air Containment Inventory, a roster of the area's top diesel polluters. In Virginia and Illinois, according to state records, Amazon was cited with more than 24 violations over a three-year period. These included running some of its generators without an environmental permit. These realities comprise an industry dirty secret that several senior executive insiders have acknowledged as "just not sustainable" (Glanz 2012).

Current demands for net neutrality and the right to equal, unlimited access to the Internet, regardless of energy costs or the purposes of its use are short-sighted. What we need, instead, are new lifestyles and changes in basic modes of production and consumption with restriction of advertisements to essential information and compulsory rationing of products. Pornography, now estimated to involve 40 per cent of all Internet use, fuels violence against women and children. The worldwide sex slave trade is identified by the United Nations as the largest illegitimate form of business in the world today. Meanwhile, alternative sources of livelihood, in poverty stricken developing countries, in particular, continue to shrink in the aftermath of globalization, climate change and growing food and fuel shortages. The pornography industry has actually won the culture war fought with feminists to free women from sexual tyranny. This was amply demonstrated in the reprimand given by the CRTC in 2014 to three Toronto based erotica TV channels for failing to broadcast sufficient levels of Canadian-made pornography to meet established Canadian content quotas (Hopper 2014). As Hedges points out, "Stripping, promiscuity, S&M, exhibitionism, and porn are now mainstream chic" (Hedges 2009: 86). Pornography is the same disease involving violence and domination that glorifies the cruelty of war and celebrates it in other forms of "action filled" popular culture commodities. These, in turn, are metaphors for the disease of corporate and imperial power.

On the whole, the dots remain stubbornly unconnected. In both the 1000-page Falconer Report on Violence in Toronto Schools released in January 2008, and the yards of ink which followed, the focus was on the code of silence and fear among teachers and superintendents, with none on the culpability of the media industries themselves. The result has been a gradual erosion of our right to feel safe and our responsibility to make others feel safe as the boundaries between entertainment and victimization fade. Other ominous findings involving the use of MRI techniques demonstrate that brain cells which normally counsel empathy are shut down in teens who play violent video games (Linn 2004). It is clear that better coordinated regulatory policy for digital technologies could not only drastically reduce carbon emissions but in the process provide safer, healthier communities as well.

Conflict tourism a growing threat

The threat to our survival goes well beyond the issue of climate change. Violent video and computer games are also helping to fuel terrorism. It is estimated there are now over 8,000 websites associated with extremist groups, many used for recruiting young converts (Dyson 2007). Almost all terrorist activities in recent years have been executed by members of diaspora communities, whether in Spain, England, Holland or Canada. Often they are well educated, technologically savvy and look up to al-Qaeda. The war on terror is now virtual. The threats are no longer contained in any one country.

Some see involvement in terrorism as a graduation from gangsta rap for young testosterone propelled male teens to a kind of status symbol. Many bring with them skills and lifestyles associated with urban youth gangs where guns, violence and extremism are the norm. At the 6th Annual Summit on Emergency and Disaster Planning for colleges, universities and K-12 schools held in Toronto in 2008, Bill Byrd, Safe Schools Inclusion Administrator for the Toronto District School Board, reported on trends toward more gangs and youth violence while family control is deteriorating (Strategy Institute 2009). Speaking at the same summit, Craig Peddle, who studies and investigates youth gangs, said he has no doubt that popular culture is a causal factor and reported that for every single website that addresses the problem, there are 100 promoting it, not only on the Internet itself, but through periodicals, rap music and other forms of popular culture. The gangs are growing in number, many are now inducting girls and the average age level is being pushed up as high as 22 years of age. In 2013, the Simon Wiesenthal Centre reported a 50 per cent growth in online forums for hate and terror at 20,000, up from 15,000 a year earlier (Macmillan 2013).

One of the central themes in popular culture for young people for decades has been a tendency to undermine all authority figures, parents and teachers included. The subsequent rise in government promises for tougher law and order measures and the increasing focus on national security issues was predictable. It resonates with the late George Gerbner's definition of "the mean world syndrome" as one of the harmful effects from entertainment violence to the community at large (Morgan 2002). Other manifestations of the syndrome were discussed in response to a number of on-going new events that unfolded, both in mainstream and social media, in November 2014 (Nowak 2014). These included the allegations of sexual violence by former CBC radio talk show host, Jian Ghomeshi, his subsequent firing, the killing of a soldier in Quebec and the fatal shootings on Parliament Hill in Ottawa, all in a space of three weeks. But Nowak had no solutions to offer, other than quotes from experts recommending use of the "off" switch and "seeking out content that can counter some of the crushing weight of digesting traumatic events, and Mean World Syndrome" (ibid.: L4).

We now have a renewed emphasis on security and the need to address rising evidence of terrorism among youth radicalized by Muslim extremists not only in the war torn Middle East, but throughout the Western world. In his interview with John Stackhouse at the Munk Centre at the University of Toronto for a series titled "The State of the World", on 20 March 2015, Gelber prize winner Paul Collier called youth radicals in the developed world dashing off to join the Islamic State in the Middle East as engaging in "conflict tourism". He sees this as an inevitable outgrowth of countless hours spent playing violent video games. An example of the growing synergy between the military industries and the entertainment industries emerged at the 2015 Hot Docs film festival in Toronto. Norwegian director Tonje Hessen Schei's film *Drone* showed how the impersonal industrialized killing program, employed by the US military in Pakistan, has long range consequences throughout the world. Military expert Lawrence Wilkerson, chief of staff to former Secretary of State Colin Powell, claims that for every 4 terrorists eliminated, 10 more are created. It was the video game connection itself that led to production of the film which includes scenes of young people attending video gaming conventions while army personnel in fatigues hover about assessing their potential as recruits for drone operation (Lederman 2015).

In the aftermath of the two fatal attacks in October 2014 in Saint-Jean-sur-Richelieu, Quebec, and on Parliament Hill in Ottawa, Bill C-51 was introduced by the Harper Government in February 2015 to extend policing and surveillance powers. Predictably, it has been fraught with controversy over how it will diminish our cherished freedoms and over whether or not it is even required. Yet, remarkably, although there is an occasional call for a new counter narrative to offset youth radicalization by the Islamic State and other similar terrorist groups, no reference is made in the endless commentary from media pundits to the fertile ground for such radicalization provided by violence in popular culture. The code of silence has given way to wilful blindness. One notable exception comes from Marc Hecker, an expert in Paris who does not believe we can allow illegal content to proliferate over Twitter feeds. Massive surveillance is not justified, he said, "just surveillance of the tens of thousands of people that propagate jihadism" (Solyom 2015). According to one French anthropologist, who spoke to 160 families of affected youths, 90 per cent were considered to have been radicalized by content online. France is now regarded as the largest source of foreign fighters in the West.

What about the research?

Over the years, whenever the impact of violence in popular culture has been publically discussed at all, findings on the harmful effects of violent entertainment from thousands of studies have been quickly neutralized by industry sponsored studies to ensure that the debate never gets beyond proof of harmful effects and onto policy (Dyson 1995). In 2001, a leading Japanese brain specialist found that playing Nintendo video games renders parts of the brain inert. The corporate giant approached him and quickly became his number one research donor. Soon after, Dr Rutya Kawashima reciprocated by calling for more research, and for the gaming industry it was back to business as usual (Cameron 2006). A documentary released in March 2015 titled *Merchants of Doubt*, based on the book of the same title authored by scientists Naomi Oreskes and

Erik M. Conway, details the conundrum of "science versus spin" with emphasis on how public relations experts help to discredit the science on harmful, cancer and addiction causing effects of tobacco and, these days, on climate change. The same tactics have been employed by corporate interests on numerous products over the years, including popular culture commodities such as violent video games (Scott 2015).

What about censorship?

Demands for the First Amendment to be updated are growing. This is a central theme in McChesney and Nichols's book *The Death and Life of American Journalism* (2010). Scholars from various disciplines, particularly in the humanities, are calling for a reinterpretation of both the Amendment and the definition of healthy economic activity in response to mental and physical health issues, looming environmental disasters and the crisis in American journalism itself. Global warming and the coming energy crisis are not our main problems. They are symptoms of deeper, entrenched cultural and societal problems. Massive changes to our conceptions of truth, freedom, individual and human rights, and how we see ourselves in relation to the Earth as a whole must change dramatically in the next few years if we are to survive as a species. Corporate freedom of enterprise is not the same thing as individual freedom of speech. That needs to be better understood by the public at large.

The deadly attacks on the satirical magazine *Charlie Hebdo* early in 2015 briefly revived discussion on the issue of free expression, spawning examples of hypocrisy in the reactions of the French Government. Broad support for the right to satirical portrayal of the Prophet Mohammed or any other manner of expression appeared to have dissolved within days when a notorious performer for a Facebook post who identified with one of the attackers was arrested. University of London-based law professor Eric Heinze stressed the need to recognize limits to free expression throughout the developed world. Using speech to place people in imminent danger is not protected, nor is revealing information deemed vital to national security (Slater 2015).

Finally

Having tolerated the deterioration of our cultural environment despite countless studies warning us of harmful effects, how do we turn things around? First, we must break the international code of silence and fear on the subject. At the 2007 UNESCO annual meeting in Ottawa, Ahmed Djoghlaf, Executive Secretary of the Convention on Biodiversity and Culture, emphasized the importance of transformative change in large urban centres where reverence for nature and biodiversity must be cultivated. But he somehow managed to avoid entirely the issue of consumer driven lifestyles fuelled by billions of advertising dollars and government subsidies for harmful content with universities collaborating through skills training.

Plans for safe schools in large urban centres such as Toronto admittedly hinge on funding, but surely it is a no brainer that they must involve more than the millions earmarked to hire more social workers, policemen and psychologists to deal with cyberbullying, sexual predators and expelled students. One wonders if officials from Ministries of Culture ever talk to those in Ministries of Education. All applications for government funding – anywhere – should be carefully monitored for their impact, on both the cultural and natural environment. A lesson can be learned from Leonardo DiCaprio, who concludes his film *The 11th Hour* with the statement that it was produced with the smallest possible impact on the environment.

Health care providers, on the whole, must become better advocates for change. Given the evidence of the potential harm of commercialized culture, it is essential for them to take on more responsibility for the education of parents about the negative effects of media and marketing on children and to work with parents to limit their access to screen media. But that is not enough. Limits need to be set on the access marketers have to children as well. As pointed out in April 2010 by educational psychologist Susan Linn, Director of the Campaign for a Commercial-Free Childhood (www.commercialfreechildhood.org), "Toward that end, health care providers need to move beyond their offices and become public advocates for policies that restrict and/or prohibit advertising and marketing to children."

Canadian organizations such as Edupax, the Canadian Centre for Policy Alternatives, and Canadians Concerned About Violence in Entertainment have taken public stands against media violence marketed to children, in particular, to in-school commercialism and to advertising directed to children. The Canadian Paediatric Society and its individual members should also become more proactive and join with municipal bodies such as the Toronto Board of Health, organizations such as the Ontario Public Health Association, the Toronto Elementary Teachers Association, and the Ottawa based Centre for Science in the Public Interest and become part of a growing movement to extend Quebec's ban on advertising directed to children 13 years and under to the rest of the country.

Policy making everywhere in the developed world must measure up to standards in the province of Quebec and other parts of the world where advertising to children is banned. Such rules apply to violence as entertainment, the marketing of junk food and the sexual exploitation of children. The Scandinavian countries, Malta, Greece, Turkey and New Zealand adopted such standards years ago, with the UK, Switzerland, Italy and France joining the list more recently. The Canada-wide ban called for by the Toronto Board of Health which led to the introduction of private members bills in the Ontario Legislature and at the federal level in Canada must be revived, quickly passed and implemented (Goar 2010; Jeffery 2007).

The CRTC and their regulatory counterparts around the world should exhibit greater boldness with regulations that better protect the public interest not only with pick and pay television purchasing packages but in restriction of content known to fuel youth violence, terrorism, growing problems such as obesity from the sedentary nature of computer use and the relentless marketing of junk food.

In 2004, researchers at Laval University found that, in 10 years, acts of violence on Canadian TV rose 286 per cent with 81 per cent of it before 9:00 p.m., the watershed hour established for the protection of children. This happened despite the Canadian Association of Broadcaster's creation of a Broadcast Standards Council in 1993. Clearly, industry self-regulation does not work without a little help from government and the rule of law. Cultural policy must become more topical during all election campaigns and better connected with global warming and other looming disasters. Only then will we begin to move toward real change.

Unfettered capitalism, regardless of the consequences, will not result in a sustainable future. Technological and economic development that does not leave in its wake a better world and an integrally higher quality of life cannot be considered progress. As Pope Francis points out, too often references to sustainable growth absorb the language and values of ecology into the categories of finance and technocracy, but the social and environmental responsibility of businesses gets reduced to a series of marketing and image-enhancing measures (Pope Francis 2015). It is in this context that the media industries bear exceptional responsibilities. More than any others, they control the tools of image-enhancing measures. We are confronted, at this juncture in our human history, with a great cultural, spiritual and educational challenge. It demands that we set out on a path of renewal.

References

Barber, B. (2007) *Consumed: How Markets Corrupt Children, Infantilize Adults and Swallow Citizens Whole*, W. W. Norton, New York.

Bradshaw, J. (2015) "Bell Chief Apologizes for Intervening in News Coverage". *The Globe and Mail* (26 March): A1.

Brown, L. (2015) "Experts Bust Myths around New Sex-Ed". *The Toronto Star* (6 May): GT1.

Cameron, D. (2006) "Moving to the Dark Side of the Screen". *Sydney Morning Herald* (13 May).

Dingman, S. (2015) "Game Developers Laud Changes to Media Tax Credit". *The Globe and Mail* (30 April): B3.

Dyson, R. (1995) "The Treatment of Media Violence in Canada Since Publication of the Lamarsh Commission Report in 1977". Doctoral Thesis, OISE/UT.

Dyson, R. (2007) "Hate on the Internet". *The Learning Edge*, January, available at www.oise.utoronto.ca/CASAE.

Glanz, J. (2012) "The 'Dirty Secret' of the Digital Revolution". *The Toronto Star* (23 September): A1.

Goar, C. (2010) "Shielding Youth from Pressure to Buy". *The Toronto Star* (26 February).

Grossman, D. (2010) "Trained to Kill". Available at www.killology.com/print/print_trainedtokill.htm.

Hedges, C. (2009) *Empire of Illusion: The End of Literacy and the Triumph of Spectacle*, Knopf, Canada.

Hopper, T. (2014) "Porn Not Canadian Enough, CRTC Warns". *The National Post* (6 March): A1.

Jeffery, B. (2007) "Hitting the Easy Mark: The Law on Marketing to Children is Ripe for Reform". In *Our Schools/Our Selves: Media Education and Educating the Media*, Canadian Centre for Policy Alternatives, Ottawa.

Klein, N. (2014) *This Changes Everything: Capitalism versus the Climate*, Simon & Schuster, New York.

Krashinsky, S. (2015) "Ad Watchdog Warns against 'Disguised' Marketing Attempts". *The Globe and Mail* (26 March): B4.

Lederman, M. (2015) "A Searing Indictment of US Drone Strikes". *The Globe and Mail* (6 May): L2.

Linn, S. (2004) *Consuming Kids: The Hostile Takeover of Childhood*, New York: The New Press.

Linn, S. (2010) "The Commercialization of Childhood and Children's Well-Being: What is the Role of Health Care Providers?". *Paediatrics and Child Health* 15(4) (April).

Lorinc, J. (2015) "Your Kids, The Influencers". *Corporate Knight* (Spring): 50–53.

Macmillan, D. (2013) "Twitter Fuels Rise in Web-Based Hate Forums, Report Says". *The Toronto Star* B.1.

McChesney R. W. and J. Nichols (2010) *The Death and Life of American Journalism*, Nation Books, Philadelphia, PA.

Morgan, M. (2002) *Against the Mainstream*, New York: Peter Lang Publishing.

Nicholson, P. (2009) "Information-Rich and Attention-Poor". *The Globe and Mail* (12 September).

Nowak, P. (2014) "The Rise of Mean World Syndrome". *The Globe and Mail* (7 November): L4.

Ogilvie, M. (2008) "Kid's Couch-Surfing Hits New High". *The Toronto Star* (28 May).

Pope Francis (2015) "Encyclical Letter *Laudato Si'* of the Holy Father Francis on Care for Our Common Home", 24 May 2015, available at http://w2.vatican.va/content/francesco/en/encyclicals/documents/papa-francesco_20150524_enciclica-laudato-si.html.

Robbins, J. (2010) "Missing the Big Picture: Studies of TV's Effects Should Consider How HDTV is Different". *The New Atlantis* 27 (Spring).

Schiller, B. (2009) "The Aim is to Make People Docile". *The Toronto Star* (27 September).

Scott, A. O. (2015) "Tobacco to Climate Change: Science vs Spin". *The New York Times* (6 March).

Slater, J. (2015) "Nothing's Clear-Cut on issues of Free Speech". *The Globe and Mail* (17 January): A4.

Solyom, C. (2015) "Extremism Runs Wild on the Web: Censor it: Expert". *The National Post* (27 June): A4.

Strategy Institute (2009) 6th Annual Summit on Emergency & Disaster Planning, 6–7 October.

Whitly, J. (2010) "Population: The Last Taboo". *Mother Jones* (May), available at motherjones.com/environment/2010/05population-growth-india-vatican.

Conclusion

Governing the commons –
can states be trustees?

Klaus Bosselmann

Introduction

This chapter summarizes some ideas of my new book, *Earth Governance*.[1] In many ways, the Global Ecological Integrity Group was instrumental for these and, in fact, for much of what I have worked on over the last 12 years since I first joined the Group. So a special thank you to Laura Westra for her inspirations to all of us.

The contents of the book circle around the central question of how ecological integrity can be implemented in law and governance. If it is true that long-term survival depends on our ability to live within planetary boundaries and to maintain and restore the integrity of planetary ecological systems, then how we control and govern ourselves becomes – literally – vital.

Earth governance proclaims a shift from state-centred governance to multi-actor governance involving states, but not exclusively. The new approach emphasizes the role of the citizen rather than nation-state as the source of legitimacy. As democratic governance is rooted in citizenship, consensus building must start there, not at the governmental level. In the Anthropocene that we are in now, citizenship has ecological and global dimensions. This calls for transnational, truly global processes of forming the collective will. In this way we can experience earth as the common reference point and develop a strong sense of stewardship or guardianship for the earth.

I will begin my thoughts with asking a simple question.

Who owns the earth?

Who owns the earth? In a broad sentimental way we can say all of us living today and who ever will come after us own it. And not just humans. All inhabitants of the planet "own" the earth in a sense that they need spaces to live in. But such an idea of ownership refers to a biological condition and does not tell us anything about power and control. Once we talk about legal ownership, power and control come into the picture and the question arises what it means to legally own earth.

For a start, only land can be owned in a legal sense, not water (including the oceans) or air (including the atmosphere). The earth overall is 123 billion acres in size, of which 37 billion acres are land. These 37 billion acres are currently shared by 7.3 billion people, so each of us theoretically owns 5 acres. This is plenty of space per capita and should theoretically allow humanity to utilize available resources without overshooting the earth's life-supporting capacity. In reality, there is a single person who legally owns about 6.6 billion acres (i.e. one-sixth of the earth's land surface). This person is Queen Elizabeth II, the queen of 32 countries and head of a Commonwealth of 54 countries.[2] She owns, for example, the second-largest country on earth, Australia, and also the third largest country, Canada.

Legal ownership means control and power, but a lot depends on whether land is owned individually or collectively and whether ownership involves obligations of care and stewardship. In the case of Queen Elizabeth, she doesn't control the land herself, of course, but her countries do. Thanks to state sovereignty Australia and Canada can do with their land whatever they like, and they have done that extensively, and in recent times in the most exploitative way: Australia's coal mines and Canada's oil sands are responsible for a considerable chunk of carbons emitted into the global atmosphere, which incidentally is not owned by anyone. The atmosphere – like the oceans – is *ius nullis* and does not have any legal status that could be used to protect against interferences such us greenhouse gas emissions or – in the case of oceans – acidification, pollution, overfishing and biodiversity loss.

Remember, legal ownership means power and control. And as each of the 196 countries is the owner of its territory, they not only can do within their own territories whatever they like, they can also externalize any waste and pollution originating from their respective territories. Apart from a few global treaties and the legal doctrine of state responsibility – both rather weak instruments – there is nothing that could legally prevent states from completely destroying the earth.

Countries do not intentionally destroy the earth, of course, but they allow it to happen. The reason is that governments continue to produce laws – domestically and internationally – that are essentially geared to secure "their" property at the exclusion of all others. "The other" comes in many forms: other states, other people (non-citizens, foreigners), other beings (animals and plants), other areas (global commons) and other times (future generations). Fundamentally, state sovereignty is about excluding "the other" and cooperation between states is hampered by a counterproductive me-over-you attitude called national interest.

In this way, not just national laws, but the world's entire legal system was developed on the basis of protecting the individual ownership of states, corporations and people. In other words, national and international laws are largely about competing property rights. In today's culture of competition and rights, success is determined by ownership. You either own something in which case you are somebody or you own nothing in which case you are nobody.

What at a personal level may hardly be noticeable – most of us own, at least, "something" – at a collective and global level appears as a massive imbalance: the richest 10 per cent of the world population own more than half of the global assets and the three richest people – Bill Gates, Carlos Slim and Warren Buffett[3] – own more than the combined annual GDP of 47 countries.[4]

At nation-state level, the combined GDP of the world's 10 richest countries – US, China, Japan, Germany, UK, France, India, Brazil, Italy and Canada [5] – is nearly the same as the combined GDP of the remaining 186 countries. Small wonder that these 10 countries, with only a further 15–20 rich countries, are firmly in charge of everything that affects the lives of the world's entire population: what is being done about climate change, nuclear weapons, poverty, food security or the internet including our personal data. The richest countries shape the international agenda and nothing would ever be accepted that could jeopardize their specific economic and strategic interests. What is morally right doesn't seem to matter. There are no signs that states follow a universal morality of care for the well-being of today's and future generations and of the entire planet. Such morality, even if it exists in a rudimentary way, is completely outdone by economic and financial power structures backed up by property rights.

Essentially, a few rich people and a few rich countries dictate what is possible in this world and what isn't. This has been said many times before and is no longer news to anyone. What is so strange about the concentration of wealth and power in the hands of a few (1%) is that the rest of us (99%) seem to stay passive. Even stranger, some of the 1 per cent, including Warren Buffett and Bill Gates, have been at the forefront of an initiative called "We are the 1 percent; we stand with the 99 percent".[6] They call for drastic increases of taxes for the rich and a global taxation system to capture multinational companies. So if almost 100 per cent of humanity feel that the massive gap between rich and poor is no good, then why are we not doing something about it? Again, power and influence of the financial world may have something to do with it, but most people seem to be in a state of paralysis – unable to move even though we know that we have to make a move. Perhaps it is human nature to do wait until someone else makes the first move. And may be there needs to be a critical mass before actually change can occur. But surely this shouldn't stop us from trying.

In my own field – environmental law and legal theory – many of my colleagues seem to be content with the world as it is. Or why is it that most books and articles are still being written as if "the law" is a given that cannot be fundamentally changed, only gradually improved?[7] A revolution in legal thought is not really happening.

This is not to say that environmental lawyers can't be revolutionaries. There is, in fact, an ever-growing ecological approach to environmental law as visible, for example, in earth jurisprudence, earth law and rights of nature.[8] This approach – or movement – has its main focus on attacking the dominant growth and property paradigm.

Property rights are, as mentioned, at the heart of the legal system globally and in almost every country. Domestically expressed in the form of private property[9] and internationally in the form of state sovereignty, property rights define most legal relationships. In fact, the Western concepts of property and sovereignty emerged in tandem. One couldn't have appeared without the other. They both express exclusivity and both are invoked with respect to the natural environment whether locally (land, water, forests) or globally (sea, air space).[10] What cannot be expressed through these legal terms and its underpinning you-or-me paradigm, does not really count. This is why property owners have a strong case for having "their" environment protected, while the general public has a weak case for having the environment protected as a commons: the public has no property rights, only individuals have.

But what happens if the public would, in fact, have property rights over the environment? What if the public at large were to own the natural environment – domestically as citizens and globally as humanity? It would change the game altogether. Every member of the public would have a case against anyone – states, corporations or individuals – threatening the integrity of ecological systems. Or even better, what if Nature or Mother Earth herself had rights, so that she has a legal defence against humans attacking her?

These kinds of questions have bothered me since the beginnings of my legal career in the 1980s. There are many answers to the basic question how property rights could be reconciled with the need for an ecological self-constraint of humans. The spectrum of possible answers includes collective ownership rights, ecological limitations to private property, rights of nature or a total abolishment of property with respect to land, water and natural features. Each of these rights models have their merits, however, their effectiveness depends on the wider cultural context they suppose to be operating in. What works within indigenous cultures and traditional societies may not work in the United States of America or in Europe or within the United Nations. The trick is to find workable solutions in a given legal and cultural context.

But regardless of what legal solutions we may favour, fundamental to them all is a deeply felt ethics of care and stewardship for the earth. Without such an ethics, rights of nature or the Pachamama constitutions of Ecuador or Bolivia fall on parched soil and will remain ineffective. This ethics is of course instrumental for our group and advocacy of ecological integrity. It should also be instrumental for law-makers and government officials. There are many forms in which stewardship or guardianship ethics can be articulated in relation to property as mentioned.

Reclaiming earth: the global commons

A particularly powerful form of applied stewardship ethics is to claim legal ownership over the commons, in particular the global commons. The atmosphere belongs to all of us, as most would agree, but the problem is that it is currently being treated as an open access resource without legal status. The atmosphere is widely regarded as *ius nullius*. This legal nullity has worked to the advantage of property owners who filled the vacuum by exercising their property rights. Property rights may not include a right to pollute, but the absence of someone who could claim violation of own rights means that actual pollution goes without any sanction. In fact, it is free. You and I or the fossil fuel industry can freely emit greenhouse gases into the atmosphere as nobody's property rights are affected. This is the status quo. It will only be qualified if, and in so far as, the law sets rights-limiting emissions standards. To-date, this has been an uphill battle.

But we can win the battle, and it is quite simple. By asserting that we all own the atmosphere, we begin to use the institutions of law working in our favour. As legal owners we can charge for damage of our common property, provide rewards to those who protect it (e.g. producers and users of renewable energy) and in this way eliminate greenhouse gases. All we have to do is to insist on public common goods – owned by us all – to be protected from the fossil fuel industry and their supporters (states, banks, corporations). Positively speaking, private property or state sovereignty continue to exist, but end where common property begins.

This is a very simple mechanism and could, for example, be supported by the well-established public trust doctrine. The public trust doctrine says that natural commons should be held in trust as assets serve the public good. It is the responsibility of the government, as trustee, to protect these assets from harm and ensure their use for the public and future generations. So nationally, the government would act as an environmental trustee, internationally states would jointly act as trustees for the global commons such as the atmosphere. Considering that only about 90 companies are responsible for two-thirds of carbons emitted into the atmosphere, a global trusteeship institution could quickly fix the problem of climate change.[11]

The idea of global nature's trusts has been promoted by environmental lawyers Mary Wood[12] and Peter Sand[13] or economist Peter Barnes.[14] Recently, the global petition "Claim the Sky" was started by Robert Costanza[15] with support from the Club of Rome and many other institutions and individuals.

Trusteeship governance is also advocated by the rich literature on the commons.[16] The "Reclaiming the Commons" movement has certainly found a momentum in recent times.

My new *Earth Governance* book tries to make the case that international law and the United Nations are not only in need, but ready to develop institutions of trusteeship governance. There is, for example, a tradition of UN institutions with a trusteeship mandate including the (now defunct) UN Trusteeship Council, the World Health Organization (WHO) with respect to public health and ironically also the World Trade Organization (WTO) with respect to free trade.[17] A number of other UN or UN-related institutions with weaker trusteeship functions exist also.[18] Quite obviously, states have been capable of, expressively or implicitly, creating international trusteeship institutions. These developments – and in particular the existence of supranational organizations such as the EU – demonstrate that sovereignty of states can be transferred to international levels. The underpinning motives are not so much of particular legal nature, but more driven by politics. And politics are driven by moralities that presently favour exploitation, but being moralities they can change. By insisting on the common good, civil society can reclaim lost ground and rebuild democracy.

The overall findings of my book suggest that a combination of ethically motivated activism and new political alliances, for example between particularly motivated progressive states, can make a crucial difference. Chances are that this combination will become very powerful as our global ecological, financial, political and democratic systems continue to disintegrate.

One major stumbling stone in the way of trusteeship for the global commons are the states themselves. Will they be ready to act as trustees? I do not expect trusteeship governance being initiated by the "top" (i.e. the UN and its member states themselves), but rather by forces outside the system, in particular global civil society. To this end, we can build on many years of activism and proposals for institutional change. Nor do I advocate states to be in charge of running and controlling global trusteeship institutions such as a World Environment Organization or a Global Atmospheric Trust. Rather I envisage their governance as jointly formed by representatives from global civil society, UN and states with an equal say in decision-making.

What I was particularly interested in when writing the book was how "resonances" between civil society and the UN system (including states) work. The chapters of the book explore the decline of democracy and rise of "corporatocracy", the commons movement, the history of commons governance, the public trust doctrine and the various trusteeship traditions within states and the UN system. The aim is to show that there are sufficient resonances between the world's legal cultures to make a compelling case for democratic and effective trusteeship governance (at global and local levels).

This chapter is focused on states. They are a major stumbling block as mentioned. Telling by their environmental behaviour so far, governments seem to be very slow learners and, most alarmingly, too close to corporate powers. The challenge for civil society is, therefore, to bring them back into a position that allows them to actually govern and help solve the crisis rather than just managing or even exacerbating it.

Sovereignty and trusteeship

At virtually every GEIG conference of the last 7–8 years, we have discussed how our democratic institutions have been hijacked by neoliberal economics. The unholy alliance between politics ("sovereignty") and private interests ("property") raises serious questions about the ability of the public to influence policy. Furthermore, as Barnes points out, "Not even seated at democracy's table – not organized, not propertied, and not enfranchised – are future generations, ecosystems, and nonhuman species."[19]

This reality of neoliberal governance has unsurprisingly affected how environmental policy and law is conceived *within* states also. Primarily, they are characterized by what Mary Wood calls a "discretionary frame".[20] This means that governments have positioned themselves as holding discretionary powers to permit resource exploitation.[21] Domestic environmental commons may be "government-owned" but this isn't to say that they are managed on behalf of future generations, nonhuman species, or ordinary citizens.[22] To the contrary, commons such as forests, water, energy etc. have been privatized and commercialized in most countries.

We can clearly see that "governance" today is about a *quid pro quo*, symbiotic relationship between politicians and corporations.[23] The rewards include property rights, friendly regulators, subsidies, tax breaks, and free or cheap use of the commons. What this ultimately means when issues such as environmental degradation arise, is that governments don't govern, rather create as little interruption to market forces as possible. In the words of Peter Barnes, "we face a disheartening quandary here. Profit-maximizing corporations dominate our economy. Their programming makes them enclose and diminish common wealth. The only obvious counterweight is government, yet government is dominated by these same corporations."[24] The notion that the state promotes "the common good" is sadly naive.[25]

To appreciate how we reached this point, we need to understand the ontological and epistemological separation of humans and nature that occurred within European history. Part of the Western *conditio humana* is the ego-cult (i.e. the separation and the exclusion of the social and ecological systems of which humans are part).[26] Conservation biology since the nineteenth-century largely developed along this dualistic tradition and remained separated from philosophy, science, democratic citizenship and politics.[27] Aldo Leopold called this separation of human and non-human communities (of plants and animals) one of the anomalies of modern day ecology and predicted that fusion of these two lines of thought would perhaps be one of the most outstanding advances of the century.[28] Alas, what we have is an enduring competition between two realities. In one reality, governments continue their modest environmental agenda of energy efficiency and climate measures, in another reality, they do everything to undermine even the most modest policies of environmental protection. Herein lies the problem, and the reason that states fail, time and time again, to reach agreement as to what must be done to save the global commons.

States' very purpose, their reason for existence and the driving force of all policy is, directly or indirectly, based on an empirical mistake or schizophrenia we might say. Faced with the modern environment "problematique" governments have never engaged in ecological thinking and instead settled for the superficial idea of "sustainable development". Understood as somehow mixing environmental, social and economic concerns this idea allows the fantasy that you can eat your cake and save it too, just eat with greater care. Denial is a strong characteristic of the human psyche, but it probably can't last forever.

What this certainly shows is a poverty of imagination, a crisis of ideas and a blaring lack of critical analysis into what really defines the scope of a government's mandate to govern. And this is the point, the failure of international environmental law is not simply a failure of cooperation, a display of distrust and competition, it stems from a lack of critical self-reflection. Essentially states are unable to look beyond the sanctimonious monopoly of economics, despite the fact that governance of society is far more complex than mere material accumulation. If institutions of governance believe there is no alternative to economic growth even in the face of ecological and human disaster, then we must doubt their ability to govern responsibly. It is not that states are incapable of governing at all. Recently we have seen just how willing states have been to help banks in times of crisis. The response of governments to the current environmental crisis reveals not just helplessness, short-sightedness and denial, but genuine signs of schizophrenia. No wonder that they rapidly lose any credibility, trust and legitimacy.

Fundamentally, the legitimacy of the state rests on its function to act for, and on behalf of, its citizens. This requires consent with the governed.[29] Governmental duties can therefore be understood as fiduciary obligations towards citizens.[30] Such fiduciary obligations are recognized typically in public law,[31] exist in common law and civil law (although in varying forms and degrees[32]) and are also known in international law.[33] The fiduciary function of the state can also be described as a trusteeship function.[34]

In the following I want to show how state sovereignty can be reconciled with trusteeship. *Prime facie* both seem to have different purposes, yet as we will see they are part of the same basic function of the state (i.e. to serve the citizens it depends on and is accountable to).

The environmental crisis and the state of the global commons gives rise to the need for revisiting the relationship between sovereignty and trusteeship.[35] Trusteeship must be pursued at both the international level and the domestic "internal" level. As Eyal Benvenisti notes, the private, self-contained concept of sovereignty is less compelling than it was in the past because of the "glaring misfit between the scope of the sovereign's authority and the sphere of the affected stakeholders."[36] This "glaring misfit" engenders inefficient, undemocratic and unjust outcomes for under- or unrepresented affected stakeholders.[37] Non-citizens, future generations and the natural environment all fall into such a category of "affected stakeholders".

Trusteeship for sustainability needs to be a ubiquitous principle at all levels of governance down to even the family unit, in the same way that other fundamental ethical principles such as fairness are implicit in the our social lives. There are two challenges to advancing the idea of trusteeship and both boil down to sovereignty. On the one hand, to propose a system of international trusteeship is to directly challenge the principle of non-interference in states' domestic affairs. To propose that states become trustees themselves, in addition to an international system of trusteeship, is again an intrusion into states' sovereign right to determine their approach to the environment. However, without the latter we will not achieve the former. Regardless of what one thinks about the legitimacy of sovereignty and the entire makeup of international relations, the reality is that states call the shots. Absent a radical reorganization of global politics we need to work within the state-centric context.

Trusteeship is an idea which softens the blow of what would otherwise be seen as an unprecedented intrusion into sovereign state affairs. Consider what an international trustee would be. Trustees are not states; a trust council might not even be an intergovernmental institution if it were composed of individuals rather than drawn exclusively from "states". Arguably this represents a less threatening intrusion into sovereignty. After all, this is a type of intervention that was not envisaged in the UN Charter, but is desirable and legitimate.[38] As Catherine Redgwell explains, "trust arrangements do not challenge sovereignty directly, for one of the advantages of trusteeship arrangements is the absence of sovereignty in the exercise of trusteeship functions – there is no transfer of sovereignty to the trust authority".[39]

But what if trust arrangements were perceived as a significant intrusion into sovereignty? The many proposals of trusteeship arrangements at the level of the United Nations have been, more often than not, greeted with hostility. States seem too attached to the principle of non-interference to appreciate cooperation of this kind. Yet, the very origins of the concept of state sovereignty are closely linked with humanitarian concerns. The Peace of Westphalia, as the foundation of state sovereignty, was a key instrument for upholding humanitarian precepts relating to freedom of conscience and religion.[40] To the extent that it resolved a crisis of freedom of conscience and equality before the law and many pre-existing institutions had lost their legitimacy and ultimately collapsed, sovereignty has been and can be justified. But it should also be remembered that humanitarian concerns were at the root of the crisis that the new order resolved. Where new crises emerge, can the principle of non-interference really be justified?

Likewise, with regards to the state itself as environmental trustee for those over whom it governs, it could hardly be refuted that a democratically elected government does not owe its citizens a duty to govern their natural wealth and resources in a sustainable way.[41] The first step, then, is reminding ourselves as citizens and society that these rights and responsibilities rest with us, despite the state acting as our representative. The second step is convincing the consumer society of what these rights and responsibilities entail. This is no small feat.

Although there is a dedicated green movement, and ever more "lite" green sentiment, convincing people to alter their engrained, even unconscious neoliberal proclivities in favour of an ethic of stewardship and trusteeship is a very trying task, especially with the economic wind blowing in their faces. But whoever's interests are involved, they must pass through the individual countries like a camel through the eye of a needle. Without a mobilized civil society – a *demos* – that is willing to hold governments to account, and demand that they represent their ecological interests internationally, states will continue to behave in the way they always have – reacting to the global environmental crisis according to the conflict model of international law that they are so used to.

Fiduciary duties of the state

The only way to turn things around and move international law from the Westphalian conflict model to a twenty-first-century cooperation model is to re-define states as trusteeship organizations. Sovereignty and trusteeship must be seen as complementary, not mutually exclusive. The argument in favour of states as trustees goes as follows.

The state gains its legitimacy exclusively from the people who created it. While the legality of a state depends on recognition by other states, once in existence a state can only ever legitimize its continued existence through ongoing trust by its people. The core idea of the modern democratic state is that it acts through its people, by its people and for its people. This implies a fiduciary relationship between people and state and is arguably the only legitimate basis for political authority in the English civil war, American Revolution, and then again confirmed in the French Revolution.[42] It is echoed in constitutional documents such as the 1776 Pennsylvania Declaration of Rights: "[A]ll power being . . . derived from the people; therefore all officers of government, whether legislative or executive, are their trustees and servants, and at all times accountable to them."[43] John Locke had famously asserted that legislative power is "only a fiduciary power to act for certain ends" and that "there remains still in the people a supreme power to remove or alter the legislative, when they find the legislative act contrary to the trust reposed in them".

Likewise, Immanuel Kant drew the moral basis of fiduciary obligations from the duty-bound relationship between parents and children.[44] Kant claimed that children have an innate and legal right to their parents' care. In a similar sense, he believed that state legitimacy was the result of a contract that is necessarily created between people to form a Rousseauian "general will". Through this process, Kant claimed, we jointly authorize the state to announce and enforce law.

That state sovereignty is fundamentally a trust relationship cannot be dismissed as a Western ideal. As I have tried to show in my *Earth Governance* book, trusts and the implicit fiduciary relationship are traced back to Middle Eastern origins, Roman and Germanic law as well as being inherent in religious teachings. The idea is perhaps even more prevalent in non-Western societies than present day Western societies because the former emphasize collective identities (for example, family, clan, nation, religion) over individual freedom and dignity, imbuing implied fiduciary obligations into the structure of public and private legal institutions.[45]

Article 21(3) of the Universal Declaration of Human Rights states that, "the will of the people shall be the basis of the authority of government".[46] But as Ron Engel points out, "democracy" can have differing interpretations. There is the "thin" interpretation which includes procedural democracy, liberal democracy, representative democracy, or simply put "the democratic process".[47] He explains:

> In this view, the democratic ideal is a way of bringing free and equal but competitive individuals and groups (or "interests") into cooperative and stable relationships by such devices as constitutionally protected civil and political rights, limited government, basic fairness in the distribution of social goods, and opportunities for citizen participation in the formation of public policy through membership in the voluntary associations of civil society and voting in electoral politics.[48]

Engel contends that the problem with this approach is that "its principle and narrative, while essential components of the democratic inheritance, are not a complete account of the moral and spiritual requirements of human self-governance".[49] The role of government is to act as conscientious trustee of the citizens' resources.[50] We see that in our market-based societies governments fail in both the interpretation and application of what this means – this is a systematic failure of neoliberalism.[51] It has captured how we think about our political arrangements. The state has become, if not the "engine of market enclosure",[52] then most certainly the conductor.

So although we may have democracy (and many places do not) in its technical form we have lost sight of what duty the state owes those it governs. At its most simplistic, the state's legitimacy to govern is based on its ability to serve the common interest. Aristotle saw the purpose of the state as for the "common good". John Locke also hinted at such a purpose. But of course who defines common good and what does it include? According to Locke's definition the "common good" was what arose from there being surplus produce that could be sold in the marketplace.

As Sheila Collins explains, Locke's "definition of the common good is a quantifiable one, not a moral one. From this concept of quantity would flow the modern measure of the common good – the Gross Domestic Product – a poor measure of any society's real quality of life."[53] But because "common interests" are socially conceived they are not static and can be contested – we can argue that new functions and responsibilities ought to become a part of state's mandate to govern.

Putting aside for a moment the question of what the common good is, let us consider what the relationship between government and the governed ought to be. We have seen that government perceives its role largely as a facilitator of economic growth, seen as analogous with "prosperity", and thus the protector of private property;[54] that is, the belief that allowing individuals to pursue their own interests will result in the best possible social organization. Few governments could argue that they do not owe a fiduciary duty to their constituents. Indeed now more than ever governments are scrambling to reduce deficits in order to fulfil their obligation to the public not to overspend. The problem is that states have neglected the ecological aspects of their fiduciary duty. And we, as the voting public, have let them.

Benvenisti conceives of three other normative bases according to which we should ascribe a trusteeship function to states' mandate to govern. The first two grounds lend themselves most easily to the development of rights and obligations under a conception of state trusteeship limited to intra-generational concerns. A normative approach which grounds itself in global resource distribution may be more conducive to the realization of state trusteeship according to principles of inter-generational equity.[55]

Firstly, sovereignty should be viewed as a vehicle for the exercise of personal and collective self-determination.[56] Collective self-determination embodies the freedom of a group to pursue its interests, further its political status, and "freely dispose of [its] natural wealth and resources".[57] An outdated conception of sovereignty which equates the voting constituency with the affected stakeholders can undermine communities' ability to exercise their right to self-determination. Urbinati and Warren note that this "geography-based constituency definition" introduces an arbitrary criterion of inclusion/exclusion right at the outset of the discussion of sovereignty.[58] A key challenge in the practical implementation of this principle is thus a clear definition of the "affected stakeholders". The Aarhus Convention provides an interesting empirical reference point: it defines the "public concerned" for the purposes of notification of environmental decision-making[59] as "the public affected or likely to be affected by, or having an interest in, the environmental decision-making".

Second, Benvenisti refers to a conception of sovereign states as agents of humanity as a whole.[60] He bases this conception largely on the equal moral worth of all human beings[61] and the corresponding foundation of international law in human rights.[62] He argues that it is humanity at large that assigns to certain groups of citizens the power to form national governments.[63] Accordingly, states can and should be viewed as agents of a global system that allocates competences and responsibilities for the promotion of the rights of all human beings and their interest in the sustainable utilization of global resources.[64] As such, the corollary of states' authority to manage public affairs within their domestic jurisdictions is an obligation to take account of external interests and balance internal against external interests.[65]

Similarly we could say that the privilege of territorial sovereignty – when given to a group of people – can be legitimized (and respected by other groups and individuals) only when used to promote universal interests of humanity as a whole and not those of the sovereign's citizenry or its (majority or any hegemonic) subgroup. This argument is based on the observation that boundaries of states no longer coincide with boundaries of nationalities, or more generally, with the boundaries of the groups whose members commonly share a distinct interest or conception of the good.[66]

Benvenisti also refers to a conception of sovereignty as the power to exclude portions of global resources.[67] He notes that both ownership and sovereignty are claims for the intervention in the state of nature by carving out valuable space for exclusive use.[68] Such a perception of states as power-wielding property owners provides a solid normative foundation for the imposition of a positive obligation on states to take other-regarding considerations into account when managing the resources assigned to them.[69] Property law theory can thus provide us a framework within which we can translate these moral grounds into legal obligations.[70] Thus, we can and should conceptualize ownership of global resources as originating from a collective regulatory decision at the global level, rather than as an entitlement of sovereign states.[71]

Although we might all agree a government has fiduciary duties, there is seemingly little to establish a precedent of state *environmental* trusteeship.[72] However, what many people do not know is that the Magna Carta of 1215, the armistice in civil war between commoners and King John, "contained largely unappreciated calls against the exploitation of the forests as ordained by the king at that time". In a scenario potently familiar to many current governments, "the king wanted to degrade the forests to a source of lumber, convert the lumber into money, and invest it in those who promised him their loyalty".[73] The accompanying "Forest Charter" constitutes landmark statements of commoners' rights. It included statement of common rights of the forest (chapters 47 and 48), and the common right of the piscary, or fishing rights (chapter 33).[74] Yet in the 1870s, the champions of Anglo-American capital recast the Magna Carta to justify their imperial ambitions and racist politics. Certain portions of the Magna Carta have been celebrated and enshrined while other portions – especially those dealing with commoners' rights to the fruits of the commons – have been portrayed as feudal relics and local particularities.[75] Furthermore, if environmental trusteeship was a seemingly fluent part of ways of life around the world, why did this aspect of social organization not get translated into state governance? The short answer is that emerging capitalism has dramatically turned things around. It replaced nature – as the basis of everything including economics – with money. In the course of unfolding, unfettered capitalism business and governments lost interest in safeguarding natural cycles and ecological integrity. Only a blunt move to environmental trusteeship can change that.

Conclusion: states as environmental trustees

Global commons governance reverses the traditional rule that international law and governance ends where national borders begin. Such a dichotomy defies ecological reality. States need to exempt transnational aspects of the domestic environment from the concept of territorial sovereignty, making way for global commons governance.[76] Through environmental trusteeship at the state level, territorial sovereignty is conceptually restricted at the global level, leading to a paradigm shift in international environmental law. Instead of state sovereignty setting limits to environmental protection, environmental protection would set the limits to state sovereignty. This is not an unprecedented idea; "limiting the self-interest of states by taking into account global concerns of humanity has become a fundamental aspect of international law".[77]

At present, states are in a paradoxical situation. They cannot shake off the capitalistic logic of profits at all costs, even at social and environmental costs of suicidal dimensions. States may well want to avoid collective suicide, but they cannot resist the forces of global markets. These forces have heavily eroded state sovereignty – the same state sovereignty that is needed to resist complete dominance of global markets. The paradox of surrendering sovereignty to free trade and market forces, on the one hand, and on the other hand insisting on sovereignty when expected to protect the commons, has been described as the "sovereignty paradox".[78]

The way out of the sovereignty paradox is differentiation. More sovereignty where possible, less sovereignty where necessary. In a globalized world this means protecting citizens and the environment from global economic forces ("more sovereignty") and protecting the global commons through international rules controlling financial and economic markets ("less sovereignty"). The perspective of differentiated sovereignty, also referred to as "responsible" or "smart" sovereignty,[79] inevitably calls for reforming and strengthening global institutions. Nothing could be more urgent than matching political institutions to the global challenges that we face.

The drivers for responsible and smart sovereignty are not states *per se*, of course, but real people (i.e. citizens, activists, advocates and decision-makers in a bottom-up approach called democracy). But this requires an idea of citizenship that operates at all levels (i.e. locally, nationally and globally).

This is also the only way to reclaim the global commons that we are so rapidly losing. Arguably, the concern for the global commons is a unifying feature of humanity. If we see ourselves as stewards of the earth and states as trustees of the common good, then this is a crucial step towards earth governance, perhaps then to be called earth democracy.[80]

Notes

1 K. Bosselmann, *Earth Governance: Trusteeship of the Global Commons* (Cheltenham: Edward Elgar, 2015).
2 See K. Cahill, *Who Owns the World? The Hidden Facts behind Land Ownership* (Edinburgh: Mainstream Publishers, 2006).
3 "The World's Billionaires", *Forbes Magazine*, www.forbes.com/billionaires/list.
4 See "International inequality" at http://en.wikipedia.org/wiki/International_inequality.
5 Measured in US dollars; see http://knoema.com/nwnfkne/world-gdp-ranking-2015-data-and-charts.
6 See http://westandwiththe99percent.tumblr.com.
7 For an excellent critique see S. Gaines, "Reimaging Environmental Law for the 21st Century", *Environmental Law Reporter* 44(3) (2014): 10,188–10,215.
8 See K. Bosselmann and P. Taylor (eds), *Ecological Approaches to Environmental Law* (Cheltenham: Edward Elgar, forthcoming 2016).
9 With North Korea probably being the only exception (apart from indigenous and traditional societies).
10 A. Fitzmaurice, *Sovereignty, Property and Empire, 1500–2000* (Cambridge: Cambridge University Press, 2014); M. Mueller and F. Badiei, "Sovereignty and Property Rights: Conceptualizing the Relationship between ICANN, ccTLDs and National Governments", March 2015, available at http://ssrn.com/abstract=2575450; K. Bosselmann, "Property Rights and Sustainability: Can They Be Reconciled?", in D. Grinlinton and P. Taylor (eds), *Property Rights and Sustainability: The Evolution of Property Rights to Meet Ecological Challenges* (Leiden: Martinus Nijhoff, 2011), pp. 23–42.
11 P. Costanza, "Claim the Sky!", (2015) *Solutions* 6(1): 18–21, available at www.thesolutionsjournal.com/node/237301.
12 M. C. Wood, "Nature's Trust: A Legal, Political and Moral Frame for Global Warming," *Environmental Affairs* 34 (2007): 577; M. C. Wood, *Nature's Trust: Environmental Law for a New Ecological Age* (Durham, NC: North Carolina University Press, 2013).
13 P. Sand, "Sovereignty Bounded: Public Trusteeship for Common Pool Resources", *Global Environmental Politics* 4 (2004): 47; P. Sand, "The Rise of Public Trusteeship in International Law" (2013) *Global Trust Working Paper Series* 04/2013, 21; P. Sand, "The Concept of Public Trusteeship in the Transboundary Governance of Biodiversity" in L. Kotzé and T. Marauhn (eds), *Transboundary Governance of Biodiversity* (Leiden: Brill, 2014).

14 P. Barnes, *Capitalism 2.0: Who Owns the Sky? Our Common Assets and the Future of Capitalism* (Washington, DC: Island Press, 2001); P. Barnes, *Capitalism 3.0: A Guide to Reclaiming the Commons* (San Francisco, CA: Berrett-Koehler Publishers, 2006).

15 Costanza, "Claim the Sky!".

16 E.g. D. Bollier, *Think Like a Commoner: A Short Introduction to the Life of the Commons* (Gabriola Island: New Society Publishers, 2014); D. Bollier and B. H. Weston, *Green Governance: Ecological Survival, Human Rights and the Law of the Commons* (Cambridge: Cambridge University Press, 2013); S. Helfrich and J. Haas (eds), *The Commons: A New Narrative for Our Time* (Berlin: Heinrich Böll Stiftung, 2009); E. Ostrom, *Governing the Commons: The Evolution of Institutions for Collective Action* (Cambridge: Cambridge University Press, 1990).

17 Bosselmann, *Earth Governance*, pp. 198–232.

18 Ibid., p. 206.

19 Barnes, *Capitalism 3.0*, p. 38.

20 Wood, *Nature's Trust*.

21 Ibid., p. 592.

22 Barnes, *Capitalism 3.0*, 43.

23 Ibid., p. 37.

24 Ibid., p. 45.

25 Ibid.

26 K. Bosselmann, *When Two Worlds Collide: Society and Ecology* (Auckland: RSVP, 1995), p. 71.

27 R. Engel, "Contesting Democracy", in Ron Engel, Laura Westra and Klaus Bosselmann (eds), *Democracy, Ecological Integrity and International Law* (Newcastle upon Tyne: Cambridge Scholars, 2010), p. 35.

28 See C. Meine, *Aldo Leopold: His Life and Work* (Madison, WI: University of Wisconsin Press, 1988).

29 J. Locke: "(G)overnment is Not Legitimate Unless it is Carried on with the Consent of the Governed", R. Ashcraft (ed.), *John Locke: Critical Assessments* (London: Routledge, 1991), p. 524.

30 E. Fox-Decent, *Sovereignty's Promise: The State as a Fiduciary* (Oxford: Oxford University Press, 2012); T. Frankel, "Fiduciary Law" *California Law Review* 71 (1983): 795.

31 Including constitutional law, administrative law, tax law, criminal law and environmental law.

32 For example, the United States, Canada, Australia and New Zealand recognize them with respect to indigenous peoples, ratepayers and (with the exception of New Zealand) in the form of public trusts, whereas continental European countries more fundamentally rely on public law to assume fiduciary relationships between individuals and governments.

33 M. Blumm and R. Guthrie, "Internationalizing the Public Trust Doctrine", *UC Davis Law Review* 45 (2012): 741; H. Perritt, "Structures and Standards for Political Trusteeships", *UCLA Journal of International Law and Foreign Affairs* 8 (2004): 391; E. Brown Weiss, "The Planetary Trust: Conservation and Intergenerational Equity", *Ecology Law Quarterly* 11 (1984): 495.

34 P. Finn, "The Forgotten 'Trust': The People and the State", in Malcolm Cope (ed.), *Equity: Issues and Trends* (Leichhardt: Federation Press, 1995), pp. 131–151.

35 S. Stec, "Humanitarian Limits to Sovereignty: Common Concern and Common Heritage Approaches to Natural Resources and Environment", *International Criminal Law Review* 12 (2010): 361, 384–385, 378–380.

36 E. Benvenisti, "Sovereigns as Trustees of Humanity: On the Accountability of States to Foreign Stakeholders", *AJIL* 107(2) (2013): 295, 301.

37 Ibid.

38 I. Bantekas, *Trust Funds under International Law: Trustee Obligations of the United Nations and International Development Banks* (The Hague: TMC Asser Press, 2009), p. 19.
39 C. Redgwell, "Reforming the UN Trusteeship Council", in W. Bradnee Chambers and Jessica F. Green (eds), *Reforming International Environmental Governance: From Institutional Limits to Innovative Reforms* (New York: United Nations University Press, 2005), p. 179.
40 Stec, "Humanitarian Limits to Sovereignty", pp. 378–380.
41 See e.g. Declaration on Permanent Sovereignty over Natural Resources, GA Res 1803 (XVII) (1962) [1].
42 W. Reisman, "Sovereignty and Human Rights in Contemporary International Law", *AJIL* 84 (1990): 886, 867.
43 E. Criddle and E. Fox-Decent, "A Fiduciary Theory of Jus Cogens", *Yale Journal of International Law* 34 (2009): 331; Pennsylvania Constitution of 1776, art IV.
44 Criddle and Fox-Decent, "A Fiduciary Theory of Jus Cogens", 352.
45 Ibid., 378–379.
46 Universal Declaration of Human Rights, GA Res 217 A(III) (adopted 10 December 1948) (UDHR)
47 Engel, "Contesting Democracy", p. 28.
48 Ibid.
49 Ibid., p. 31.
50 D. Bollier, "The Commons: A Neglected Sector of Wealth Creation", in S. Heinrich (ed.), *Genes, Bytes and Emissions: To Whom Does the World Belong?* (Berlin: Heinrich Böll Stiftung, 2008); P. Barnes, J. Rowe and D. Bollier, *The State of the Commons 2003/04: A Report to Owners* (Minneapolis, MN: Tomales Bay Institute, 2004).
51 Bollier, "The Commons".
52 Ibid.
53 S. Collins, "Interrogating and Reconceptualizing Natural Law to Protect the Integrity of the Earth", in L. Westra, K. Bosselmann and R. Westra (eds), *Reconciling Human Existence with Ecological Integrity* (London: Earthscan, 2008), p. 455.
54 Ibid.
55 As initially expounded by Edith Brown-White, "The Planetary Trust: Conservation and Intergenerational Equity", *Ecology Law Quarterly* 11 (1984): 495.
56 Benvenisti, "Sovereigns as Trustees of Humanity", p. 301.
57 International Covenant on Civil and Political Rights, 999 UNTS 171 (adopted 16 December 1966, entered into force 23 March 1976), art 1 (ICCPR).
58 N. Irbinati and M. Warren, "The Concept of Representation in Contemporary Democratic Theory", *Annual Review of Political Science* 387 (2008): 397, cited in Benvenisti, "Sovereigns as Trustees of Humanity", p. 304.
59 Convention on Access to Information, Public Participation in Decision-Making and Access to Justice in Environmental Matters, 2161 UNTS 447 (adopted 25 June 1998, entered into force 30 October 2001), art 6.2 (Aarhus Convention).
60 Benvenisti, "Sovereigns as Trustees of Humanity", p. 305.
61 Ibid., referring to John Stuart Mill, *Considerations on Representative Government* (first published 1861).
62 J. Raz, "Human Rights in the Emerging World Order", (2010) *TLT* 1(31): 42, cited in Benvenisti, "Sovereigns as Trustees of Humanity", p. 306.
63 Benvenisti, "Sovereigns as Trustees of Humanity", p. 306.
64 Ibid., p. 308, paraphrasing Huber in *Island of Palmas (Netherlands v United States)* (1928) 2 RIAA 829, 869.
65 Ibid., paraphrasing Huber in *British Claims in the Spanish Zone of Morocco (Spain v United Kingdom)* (1925) 2 RIAA 615, 641.
66 C. Gans, *The Limits of Nationalism* (Cambridge: Cambridge University Press, 2003).
67 Benvenisti, "Sovereigns as Trustees of Humanity", p. 308.

68 Ibid.
69 Ibid., pp. 309, 310. Also when making rival claims on transboundary and public resources.
70 "Property Rights and Sustainability: Can They Be Reconciled?".
71 Benvenisti, "Sovereigns as Trustees of Humanity", p. 309.
72 See generally K. Bosselmann, *The Principle of Sustainability: Transforming Law and Governance* (Farnham: Ashgate, 2008), 145–174.
73 S. Helfrich, "Commons: the Network of Life and Creativity", in Heinrich *Genes, Bytes and Emissions*, p. 1.
74 I. Kaul, "Meeting Global Challenges: Assessing Governance Readiness", in Hertie School of Governance (ed.), *Governance Report 2013* (Oxford: Oxford University Press, 2013), pp. 33–58; K. Bosselmann, "Earth Democracy: Institutionalizing Ecological Integrity and Sustainability", in Engel et al. (eds), *Democracy, Ecological Integrity and International Law*, pp. 319–330; see also W. Baber and R. Bartlett, *Global Democracy and Sustainable Jurisprudence* (Cambridge, MA: MIT Press, 2009).
75 D. Bollier, "The Future of the Commons: Notes from a Retreat Exploring the Potential of the Commons to Fight Enclosures and Build Commons-Based Alternatives", Retreat on the Future of the Commons, Crottorf Castle, Germany (25–27 June 2009), available at http://commonstrust.global-negotiations.org/resources/Bollier,%20 Crottorf%20retreat.pdf.
76 Bosselmann, *The Principle of Sustainability*.
77 Stec, "Humanitarian Limits to Sovereignty", p. 364.
78 D. Zaum, *The Sovereignty Paradox* (Oxford: Oxford University Press, 2007), pp. 226–231; Kaul, "Meeting Global Challenges", pp. 33–34.
79 Ibid., pp. 34–58.
80 Bosselmann, "Earth Democracy"; see also Baber and Bartlett, *Global Democracy and Sustainable Jurisprudence*.

Index